FAULT DETECTION & RELIABILITY

KNOWLEDGE BASED & OTHER APPROACHES

International Series on

SYSTEMS AND CONTROL, Volume 9

FAULT DETECTION & RELIABILITY

KNOWLEDGE BASED & OTHER APPROACHES

Proceedings of the Second European Workshop on Fault Diagnostics, Reliability and Related Knowledge Based Approaches, UMIST, Manchester, April 6-8, 1987

Organised within the framework of a joint research project sponsored by the Commission of the European Communities under the Stimulation Action Programme aimed at promoting collaborative research work in the community

M G SINGH
UMIST, Manchester, UK

K S HINDI
UMIST, Manchester, UK

G SCHMIDT
Technical University of Munich, FRG

and

S G TZAFESTAS
Technical University of Athens, Greece

PERGAMON PRESS

OXFORD · NEW YORK · BEIJING · FRANKFURT
SÃO PAULO · SYDNEY · TOKYO · TORONTO

U.K.	Pergamon Press, Headington Hill Hall, Oxford OX3 0BW, England
U.S.A.	Pergamon Press, Maxwell House, Fairview Park, Elmsford, New York 10523, U.S.A.
PEOPLE'S REPUBLIC OF CHINA	Pergamon Press, Room 4037, Qianmen Hotel, Beijing, People's Republic of China
FEDERAL REPUBLIC OF GERMANY	Pergamon Press, Hammerweg 6, D-6242 Kronberg, Federal Republic of Germany
BRAZIL	Pergamon Editora, Rua Eça de Queiros, 346, CEP 04011, Paraiso, São Paulo, Brazil
AUSTRALIA	Pergamon Press Australia, P.O. Box 544, Potts Point, N.S.W. 2011, Australia
JAPAN	Pergamon Press, 8th Floor, Matsuoka Central Building, 1-7-1 Nishishinjuku, Shinjuku-ku, Tokyo 160, Japan
CANADA	Pergamon Press Canada, Suite No. 271, 253 College Street, Toronto, Ontario, Canada M5T 1R5

First edition 1987

Library of Congress Cataloging in Publication Data

Fault detection and reliability.
(International series on systems and control; v. 9)
Proceedings of the Second European Workshop on Fault Diagnostics, Reliability, and Related Knowledge-Based Approaches, held at the University of Manchester Institute of Science and Technology, 4/6-8/87.
1. Fault location (Engineering) - Congresses.
2. Reliability (Engineering) - Congresses. 3. System analysis - Congresses. I. Singh, Madan G. II. European Workshop on Fault Diagnostics, Reliability, and Related Knowledge-Based Approaches (2nd: 1986: University of Manchester Institute of Science and Technology)
III. Series.
TA169.6.F38 1987 620'.00452 87-7261

British Library Cataloguing in Publication Data

Fault detection and reliability:
knowledge based and other approaches
— (International series on systems and control; V.9).
1. System failures (Engineering)
I. Singh, Madan G. II. Series
003 TA169.5

ISBN 0-08-034922-6

Printed in Great Britain by A. Wheaton & Co. Ltd., Exeter

Preface

The increasing complexity of man-made systems, such as computer and communication networks, manufacturing systems and electric power systems, faces system planners, dispatchers and controllers with difficult problems. The complexity arises not only from the high dimensionality of these systems and the large volumes of the necessary information flows, but also from the randomness of faults and failures. Carefully constructed optimal or sub-optimal operational strategies can be easily rendered null and void by an unexpected failure. Hence the crucial importance of reliable and fast fault detection, location and isolation.

The use of estimation and other techniques rooted in control theory for fault detection has by now become recognised and established. However, the advent of expert and knowledge—based systems opened up new possibilities. This is particularly so in view of the fact that the most successful applications of knowledge-based approaches have been in diagnosis.

The above considerations constituted the rationale of a research project carried out jointly by the Group of Professor M.G. Singh at the University of Manchester Institute of Science and Technology (UMIST) in the United Kingdom, the group of Professor G. Schmidt at the Technical University of Munich in the Federal Republic of Germany and the group of Professor S.G. Tzafestas at the Technical University of Athens in Greece. Within the framework of this joint research project, which was sponsored by the Commission of the European Communities under the Stimulation Action Programme aiming at promoting collaborative research work in the community, two European Workshops on Fault Diagnostics, Reliability and Related Knowledge-based Approaches were held. The first was held in Rhodes in August 1986, and the second at UMIST in Manchester from the 6th to the 8th of April 1987. The present volume provides the proceedings of this second European Workshop on Fault Diagnostics, Reliability and Related Knowledge-based Approaches.

These proceedings are split up into four major parts in order to reflect the most significant aspects of Reliability and Fault Detection as viewed at present. The first part deals with Knowledge-Based Systems and comprises eleven contributions from leading experts in the field. The emphasis here is primarily on the use of Artificial Intelligence, Expert Systems and other Knowledge-Based Systems for fault detection and reliability. The second part is devoted to Fault Detection of Technological Systems and comprises thirteen contributions dealing with applications of fault detection techniques to various technological systems such as gas networks, electric power systems, nuclear reactors and assembly cells. The third part of the proceedings, which consists of seven contributions, treats Robust, Fault Tolerant and Intelligent Controllers and covers methodological issues as well as several applications ranging from nuclear power plants to industrial robots to steel grinding. The fourth part treats Fault Tolerant Digital Techniques and comprises five contributions.

The support of the Commission of the European Communities is gratefully acknowledged; without it neither the joint research project nor the consequent two European Workshops on Fault Diagnostics, Reliability and Related Knowledge-Based Approaches would have been possible.

M. G. Singh

K.S. Hindi

G. Schmidt

S. G. Tzafestas

Manchester
June 1987

Contents

I. Knowledge Based Systems

A Survey of Expert Systems for Equipment Maintenance and Diagnostics

William T. Scherer and Chelsea C. White, III

University of Virginia
School of Engineering and Applied Science
Department of Systems Engineering, Thornton Hall
Charlottesville, VA 22901 USA

ABSTRACT

Considerable research has been conducted in the area of expert systems for diagnosis, with the early work concentrated in medical diagnostic systems and the more recent efforts expanding the field to equipment maintenance and diagnostics. We present an extensive survey of application efforts in this area and discuss issues critical to diagnostic expert systems.

KEY WORDS

Diagnostics, expert systems, equipment maintenance, knowledge-based systems, built-in test [BIT], automatic test equipment [ATE]

INTRODUCTION

Considerable research is being conducted in the area of expert systems for diagnosis. Early work was concentrated in medical diagnostic systems [Clancey and Shortliffe, 1984], with the now well-known MYCIN [Buchanan and Shortliffe, 1984] expert system. Current efforts are expanding to the area of equipment maintenance and diagnostics, with numerous systems having been built during the past several years. We concentrate our effort in this paper on expert systems for diagnosis and refer the reader to [Hayes-Roth, 1983], [Waterman, 1986], and [Charniak and McDermott, 1985] for an introduction to the area of expert systems. We remark that it is common for diagnostic systems to be comprised of elements from the fields of artificial intelligence [expert systems], decision theory and operations research.

EXPERT SYSTEMS: GENERAL

The basic components of an expert system [ES] include the knowledge base [KB], which consists of factual knowledge and preferential knowledge, and the inference engine, which directs [controls] the use of the KB. Numerous supporting components such as the user interface may also be incorporated into an ES. Various types of knowledge representation schemes can be employed: frames, production rules, lists of facts, semantic networks, logic predicates, etc.. The complexity of most diagnostic systems makes knowledge representation a critical issue - see [Pau, 1986] for an evaluation of knowledge representation schemes. We note, however, that a ES will often require a combination of the above types. With regard to the inference mechanism for ESs, the various options include: forward-backward chaining [still the most common], generate/test methods, heuristic search, and meta-rules [Pau, 1986]. As with the knowledge representation, a combination of the above inference mechanisms is common.

Three basic types of ES, in order of increasing generality, are production systems, structured production systems, and distributed systems [Pau, 1986]. Production systems have a single KB and a single inference engine, while structured production systems use meta-rules to move between several KBs. Distributed systems use a hierarchically structured network of cooperating specialists [Bylander, Mittal, and Cahdrasekaran, 1985]. Distributed systems being the most general of the three, they are the preferred type for most ES.

EXPERT SYSTEMS: DIAGNOSTIC

The first diagnostic expert systems [DESs] for fault diagnosis were developed in the early 1970's at MIT. These systems were unique in that they modelled the normal behavior and structure of the system rather than a tree of possible faults. A significant benefit of the early AI based systems was the explanation capabilities, usually generated via a backtrace of the analysis. The early systems such as EL, LOCAL, and DESI had the following features: the device was seen as a collection of components with given component interactions, where the components were organized in an hierarchical fashion [King, 1982]. These systems were often based on a circuit-level description of the system. The DART project [Genesereth, 1984], however, approached the diagnosis problem from a different perspective. Using the EMYCIN framework for medical diagnosis, DART is a system for diagnosing IBM-370 class computers that was not based on the circuit level but on a system level. The philosophical basis for this approach, which could be termed "shallow-reasoning", came from the medical diagnostic systems, where a complete description at the "circuit" level is not fully understood or is highly complex. Often in medical diagnosis, rules are used that relate sets of symptoms to possible diseases. These "pattern->action" rules are based on problem-solving heuristics generated by the expert, and often employ a simple control structure in using the knowledge-base [Chandrasekaran,et al.,1984]. This is in contrast to the early DESs, where the model was one of normal functioning. It is also in contrast to the diagnostic systems that employ the notion of "deep-reasoning" to determine faults [see Davis, 1984a; Pan, 1984]. Deep-reasoning involves using lower-level reasoning on causal, functional, and spatial information, while shallow reasoning refers to the use of rules-of-thumb that are often acquired from experience. In many instances, however, it is difficult and too complex to define the interactions at the system level.

Conventional fault diagnosis, used by reliability engineers, employs fault trees and their successors, cause-consequence trees [CCT]. The CCT methodology models faults that can arise in a system, along with associated logical, causal, and temporal relationships. CCTs require a great deal of effort in their construction, e.g., a CCT for a nuclear plant requires approximately ten man-years [King, 1982]. Major difficulties with CCT models is that they do not allow feedback models, and a priori information is difficult to take advantage of effectively. Weaknesses in AI based techniques, as compared to CCT, include: the requirement of a complete model of the device, the assumption of reliable measurements, the lack of guidance with respect to what course of action to take other than which component to replace, and the lack of time-critical processing [King, 1982]. We note that many AI based expert systems for diagnosis employ structures and techniques from conventional fault diagnosis.

Experienced diagnosticians use a combination of techniques for diagnosis, including a familiarity with the system documentation, a functional understanding of the system components, an understanding of the system interrelationships, knowledge of the failure history of the device, along with numerous heuristics [Pau,1986]. It is this complex mixture of knowledge that complicates the diagnostic problem. Medical diagnosis and treatment expert systems are often hindered by the lack of a complete understanding of the human body, while equipment diagnostic systems can have available a complete range of information. Such information includes a complete schematic diagram of all circuits involved, functional models of behavior, spatial models of behavior, and reasoning by second principles. This rich amount of information allows for expert diagnostic systems of high accuracy and considerable complexity.

In current AI-based diagnostic systems there are several common notions and concepts. The diagnostic problem is often viewed as a discrepancy between predicted and observed behavior. Systems such as DART, however, often relate observed behavior to faults from a system viewpoint. Also, diagnostic systems attempt to exploit hierarchical behavior whenever possible. This allows a modular system where the degree of detail associated with any module within any level is not required to be consistent. The interactions between modules and levels is kept as simple as possible.

Critical issues in expert system design for electronic equipment diagnosis, therefore, are 1) the level at which to approach the design [e.g., at the schematic diagram level or the functional model level], 2) the knowledge representation scheme to be used once the level has been selected [e.g., IF-THEN rules, frames, etc.], and 3) the method by which uncertainty is to be handled [e.g., probabilities, certainty factors, fuzzy sets, etc.]. Important in the knowledge representation scheme is the notion of reasoning from first principles, which relates to the level at which the system being diagnosed is described. Davis, [1984b] argues for "models of causal interaction" rather than the traditional models of fault diagnosis. Often, an expert diagnostician is required to revert to "deep" knowledge when confronted with a particularly difficult problem or when required to explain their diagnostic process [Chandrasekaran and Mittal, 1984], [Davis, 1984b]. Also with regard to knowledge representation, the concept of functional versus behavioral models is an important distinction. A diagnostic system can model a component of the system with only the notion of the behavior of the component without regard to function. This allows the behavior of each component to be understood in a context-free manner, which allows for relatively easy interchange of modules in a hierarchical system. Uncertainty, one of the most important issues in DESs, is discussed in detail in the following section.

UNCERTAINTY IN DES

Diagnostic expert systems can be viewed as a search for a single fault or multiple faults, where the search is usually heuristic in nature. Inherent in most diagnostic systems is the notion of uncertainty, i.e., the knowledge the system has concerning the cause of the fault is not known exactly. Early DESs employed IF-THEN rules and a simple chaining mechanism as the knowledge base and inference engine [control structure], respectively. Current expert diagnostic systems are employing sophisticated techniques to deal with the issue of uncertainty. The reader is referred to the Proceedings of Uncertainty in Artificial Intelligence from the 1986 AAAI conference for an extensive collection of current research papers dealing with uncertainty in AI. Other surveys of uncertainty in diagnostic systems include [Dutta, 1985], [Kanal and Lemmer, 1986], and [Goodman and Nguyen, 1985].

Five general [non-exhaustive] techniques exist for dealing with uncertainty: 1) Probabilities, 2) Fuzzy Sets, 3) Certainty Factors, 4) Dempster-Shafer Theory, and 5) Set-Covering Techniques. These are general categories and most systems use a modified version of one of the above. We refer the reader to [Kanal and Lemmer, 1986] for a description of the above uncertainty models.

Dempster-Shafer theory is a generalization of Bayesian theory. For an introduction to Dempster-Shafer theory see [Shafer, 1976] and [Dempster, 1968]. Cantone and colleagues [1983], for example, describe a diagnostic system that uses probabilistic search for electronics troubleshooting. Their technique is not a classical Bayesian system in that they do not require the probabilities to sum to one and updating is done via Dempster's rule. However, the "probabilities" are between 0 and 1 and can be interpreted in the relative frequency sense. Two kinds of rule updating are allowed: evidence that is independent and combined with earlier evidence, and non-independent evidence that revises belief about other components in the system. Recent work on Dempster-Shafer theory includes [Zadeh, 1986], [Rowe, 1986] and [Yen, 1986].

Fuzzy sets [Zadeh, 1965] are receiving considerable attention, much of which is in the medical diagnostic system literature [Bandler and Kohout, 1985; Bolc and colleagues, 1985; Gazdik, 1985; Moon and colleagues, 1979; Negotia, 1985; Ralescu, 1984; Turksen, 1983; Umeyama, 1986; Kacprzyk and Tager, 1985; Baldwin, 1985; Kitowski and Ksiazek, 1985]. Advantages of fuzzy sets for medical diagnosis [and general diagnosis] include the ability to define inexact entities as fuzzy sets, the idea that fuzzy sets provide a linguistic approach that closely approximates medical texts, and the ability of fuzzy sets to draw approximate inferences [Adlassnig, 1986].

Probably the most common uncertainty description is the use of certainty factors associated with rules in the rule base. Certainty factors systems update beliefs by using heuristic indices, often with questionable assumptions [Yadrick and colleagues, 1986]. First described in the MYCIN project [Buchanan and Shortliffe, 1984] and in the PROSPECTOR system [Yadrick, and colleagues, 1986; Duda, Gaschnig, and Hart, 1979], numerous systems employ a modified form of certainty factors [Odryna, 1985; Merry, 1983]. Recently there has been considerable interest regarding the relationship between belief updating and probabilities [Heckerman, 1986].

Set-covering models for diagnostic problem solving are also receiving some current attention [Reggia, 1984, 1985a, 1985b; Kalme, 1986]. Major features of the set-covering model include the adaptability of the model to more general problems, and the ability of the model to handle multiple faults. Set-covering models attempt to explain all the symptoms with a minimum of complexity, e.g., the minimum number of faults [Reggia, 1984]. INTERNIST [Masarie, Miller, and Myers, 1985; Miller, Pople, and Myers, 1982] is a system that is closest to a set-covering model.

One of the earliest mechanisms for dealing with uncertainty in diagnostic expert systems was Bayesian based probabilistic reasoning [Clancey and Shortliffe, 1984]. In the 1960s many medical diagnostic systems were developed that employed a Bayesian framework. Several difficulties with Bayesian based systems brought about the decline in their use [Shortliffe, 1979]. Major difficulties with Bayesian systems include 1) the lack of generality, 2) the extensive data requirements if independence and mutual exclusiveness are not assumed, 3) the requirement of precise probabilities, 4) lack of an explanation capability and 5) the problematic acceptability of the systems. Despite these difficulties, Bayesian systems that have been developed often have excellent performance [Clancey and Shortliffe, 1984]. Recent efforts to address the issues involved with Bayesian systems include [White, 1986], [Snow, 1986], [Pearl, 1986], [Norton,1986], [Barth and Norton, 1986] and [Schwartz, Baron, and Clark, 1986].

DEVELOPMENT ENVIRONMENTS

Many diverse environments are being used for the development and implementation of expert systems. We noted based on this survey that the language and environment selected is usually based on availability of the hardware, the software (language and environment), and the expertise. It is also common to develop the system in one environment and then to transfer it to another environment for actual implementation, e.g., PUFF and AEGIS.

Three basic approaches can be used when selecting software for expert diagnostic system development. The first is to design the system from a fundamentally low-level language, such as LISP, PROLOG, or C [or BASIC, PASCAL or FORTRAN]. The second approach is to use an expert system shell program or expert system language that usually has a built-in knowledge base form and a relatively fixed control structure. Examples of such systems include EXPERT, KEE, and OPS5 [Brownston and colleagues, 1985]. There are also expert system shells designed exclusively for building diagnostic systems [see "shells" in the sample system list below]. See [Waterman, 1986] for a description of numerous general purpose expert system shells. The third approach is a combination of the first two - modifying a shell system for a particular application. This approach is advantageous, if the shell system allows user modification, in that the designer saves considerable development time by using as much of the existing structure as possible.

In the systems described in the applications section below, one of the most common software environments used was the OPS5 ES language [IDT, ANGY, PROD, Toast, The Diagnostician, MELD], while the EXPERT shell system was also common [AEGIS, APRES]. Other shell systems used were S.1 [Mark 45 FDA], SAGE [AMF], and Inference ART [APRES]. Languages used include FORTH [DELTA], FORTRAN [EXPERT], LISP [IN-ATE, ACES, DORIS, FIS], PROLOG [DIGS, APEX 3], and PASCAL [CATA]. Several of the systems were based on or similar to MYCIN or EMYCIN [DART, PUFF, PDS]. There is a current debate as to whether LISP will become popular as an implementation language or tend to remain as a development language [e.g., Jacobson, 1986; Gabriel, 1986]. With respect to hardware, there was a wide variety of machines used, with the most common being the DEC VAX computers. Also used was the DEC PDP, dedicated LISP machines [e.g., Symbolics], and various microcomputers.

APPLICATION AREAS

Application areas include: computer malfunction diagnosis [DART, IDT; Agre and colleagues, 1985], power networks [Toast; Liu and Tomsovic, 1986; Zwingelstein and Monnier, 1985], faults in VLSI chips [PROD, MIND], chemical plant fault diagnosis [Palowitch, 1985], communications networks [Williams, 1983], ATE diagnosis [Ryan and Wilkinson, 1985], software diagnosis [Jackson, 1985], aircraft diagnostics [Montgomery, 1986], and manufacturing cells [Chiu and colleagues, 1985].

Following are some sample DESs, followed by sample DES shell systems. The list is by no means exhaustive, and the brief descriptions present information that we feel is of interest. We remark that it is often difficult to determine whether a system is in actual use, in the developmental stages, or a paper model only.

Sample systems:

ACES: [Attitude Control Expert System] Performs diagnosis of failures in the attitude control system of the DSCS-III satellite. Uses failure-driven learning to improve diagnostic capability. System implemented in a combination of LISP and PROLOG on a Symbolics 3600. Developed by the Aerospace Corporation. [Pazzani, 1986]

AEGIS EXPERT: Built using the EXPERT shell development system (developed at Rutgers University for medical diagnosis). System adds diagnostic capabilities to the shipboard AEGIS system. Uses IF-THEN rules. The final version will be translated into C to run on an IBM AT. Developed at Advanced Technology Laboratories. [Adams and colleagues, 1986]

AMF: [Advanced Maintenance Facility] A DES for TXE4A telephone exchanges [England]. The system has approximately 1350 rules, 2200 questions, 2400 assertions, and 2300 items of advice, arranged into a net with 660 areas relating to specific fault indications. Developed using the SAGE expert system shell on a 16-bit microcomputer running under UNIX. [Thandasseri, 1986]

APRES: Performs an analysis of computer crash dumps in order to determine computer hardware faults. Originally coded in the EXPERT shell system, it has been translated into Inference ART, which resides on a VAX in a LISP environment. The system has approximately 1000 rules using the EXPERT shell system. [Jackson, 1985]

CATA: [Computer-Aided Test Analysis System] Helps engineers <u>design</u> electronic systems for testability. Written in PASCAL on a VAX 11-780. [Robach, Malecha, and Michel, 1984]

CRIB [Computer Retrieval Incidence Bank]: Designed for computer fault diagnosis using a pattern-directed inference system. [Hartley, 1984]

DART: [Diagnostic Assistance Reference Tool] Assists a technician in finding faults in a computer system. The first version was EMYCIN based, while the second version used meta-level reasoning system [MRS]. [Genesereth, 1984; Bennet and Hollander, 1981]

DELTA: [Diesel Electric Locomotive Troubleshooting Aid, field prototype is CATS-1, for Computer-Aided Troubleshooting System] Implemented in FORTH on a DEC PDP 11/23. Has approximately 530 rules and uses a mixed configuration inference engine that has forward chaining and backward chaining. [Bonissone and Johnson, 1983]

The Diagnostician: [see "Toast"] Diagnoses faults in a power network [Allegheny Power System]. Written in OPS5, the system has approximately 350 rules. Designed for a Distributed Problem Solving [DSP] environment, where there are human problem solvers that interact with a number of programmed problem solvers. [Talukdar, Cardozo, and Perry, 1986]

DIEX [Diagnostic Expert System]: Diagnoses chemical plant operations. Divides problem into subtasks, where different strategies are used for each subtask, including pattern-matching and model-based algorithms. [Palowitch and Kramer, 1985]

DORIS [Diagnostic Oriented Rockwell Intelligent System]: A rule-based system for aiding in fault diagnosis. Allows forward and backward chaining, and has a rule maintenance system and an explanation system. Various versions are coded in InterLisp on a VAX 11/750, Lisp-88 on an IBM PC, and ZetaLisp on a Symbolics. [Davis, 1986]

ET: [Expert Technician, Experience Trapper] Probabilistic, heuristic based diagnostic system for PRODUCT, a complex electronic device manufactured by Texas Instruments. Uses weighted predictive functions to rank actions that will remove symptoms. Also employs machine learning and adoption techniques to modify and add knowledge concerning its domain. Employs an extended version of the Dempster-Shafer theory of evidential reasoning. [Rowe, 1986]

Fieldserve: [Field Service Advisor] Diagnoses electromechanical systems, specifically in the field. Runs on a Vax 11/780 in Franz LISP and the general purpose inference engine Genie

[Vanderbilt]. Frame based, employs IF-THEN rules. [Hofmann and colleagues, 1986]

FIS: [Fault Isolation System] Uses a rule-based causal model to update beliefs and recommend tests for electronic equipment. Information provided in set description and probabilities. Written in Franz LISP on VAX 11/780, developed at US Naval Research Lab. [Pipitone, 1986]

Hitest: Knowledge-based automatic test generation system for digital circuits. [Bending, 1984]

IDT: [Intelligent Diagnostic Tool]: Identifies faults in PDP 11/03 computers. Uses OPS5 and was developed at Carnegie-Mellon University. [Shubin and Ulich, 1982]

IN-ATE: Electronic troubleshooting aid designed to guide a novice technician. IN-ATE uses probabilistic search to determine faults. The search is guided by three parameters: test cost, conditional probabilities of test outcomes, and proximity to a solution. Written in Franz LISP on a VAX 11/780. [Cantone and colleagues, 1983]

LES [Lockheed Expert System]: A system for diagnosing a large signal-switching network. Uses production rules and frames to describe least repairable units [LRUs]. A goal-driven, backward chaining system, LES also allows for data-driven, forward chaining reasoning. Has approximately 70 rules and 1200 facts. [Laffey, 1984]

MARK 45 FDA [Fault Diagnosis Advisor]: An integrated expert system for the diagnosis and repair of the Mark 45 Naval Gunmount. Implemented in the S.1 expert system programming language system developed by Teknowledge, Inc., which is a descendent of EMYCIN. Integrates maintenance history database, spare parts inventory database, BIT, ATE, and other KBS. [Powell, Pickering, and Westcourt, 1986]

MDX: Decomposes diagnostic knowledge into a classification hierarchy of cooperating specialists. Uses an "establish-refine" mechanism to control movement through the hierarchy. [Sticklen and colleagues, 1984; Chandrasekaran and Mittal, 1984]

MELD [Meta-Level Diagnosis]: Diagnosis of electromechanical systems. Two levels of reasoning are used - object-level and meta-level. Written in OPS5. [Thompson and Wojcik, 1984]

MIND: [Machine for Intelligent Diagnosis] An expert system for VLSI test system diagnosis. Hierarchically structured rule-based system designed to reduce the MTTR. [Wilkinson, 1985] [Ryan and Wilkinson, 1986] [Grillmeyer and Wilkinson, 1985]

NDS: [Network Diagnostic System] An expert system for fault isolation in a nationwide communications network [COMNET]. Based on ARBY [McDermott, 1982], the system performs a heuristic search for multiple faults through a space of hypotheses. NDS has approximately 150 ARBY rules and the performance level is that of an intermediate level diagnostician. [Williams, 1983]

PROD: A complete diagnostic system for determining faults in VLSI chips. Matches measurements on chip parameters to a library of faults. A production rule system written in OPS5 then generates a set of certainty factors for the most likely faults. Main system implemented using a UNIX file system. [Odryna and Strojwas, 1985]

PUFF: Highly successful system for interpreting pulmonary function test results. Originally developed using EMYCIN on Stanford's SUMEX-AIM computer, it has been translated into BASIC on a PDP-11 for on-site use. [Aikins, Kunz, Shortliffe, 1983]

REACTOR: An expert system for the diagnosis of nuclear reactor accidents. [Nelson,1982]

Toast: [see "The Diagnostician"] An expert system for simulating events in power networks and diagnosing problems. Uses Cops [super-set of OPS5] on a network of VAX computers. [Talukdar, Carduzo, and Leao, 1986a]

Sample Diagnostic Shell Systems:

APEX 3: An expert system shell for fault diagnosis. Uses forward and backward chaining as its main control mechanism. Employs the uncertainty system developed for PROSPECTOR - a mixture of Bayesian inference techniques and fuzzy logic. Written in PROLOG and POP-2 in a highly modular fashion. [Merry, 1983]

CSRL [Conceptual Structures Representation Language]: A language for building hierarchical diagnostic expert systems. Based on the paradigm of "cooperating diagnostic specialists". [Bylander, Mittal, and Chandrasekaran, 1985]

DIGS: A domain-independent expert system for the diagnosis of complex electronics. Uses a graph-like model of a diagnostic net and performs breadth-first and depth-first search based on the current symptoms. Developed in PROLOG running under the control of USCD PASCAL. [Agre and colleagues, 1985]

EXPERT: [with SEEK: (System for Experimentation with Expert Knowledge)] A FORTRAN based general purpose system that uses findings, hypotheses, and rules relating the findings and hypotheses to obtain a conclusion that is assigned a level of certainty. Developed at Rutgers University.

IDM: [Integrated Diagnostic Model]: Integrates "deep" and "shallow" sources of knowledge. Uses a semantic network for shallow knowledge representation and a hierarchical, functional model for deep knowledge. [Fink, Lusth, and Duran, 1984]

PDS [Process Diagnostic System] Application to Turbine-Generators. PDS is an expert shell system developed by M.S. Fox at the Carnegie-Mellon University Robotics Institute. PDS uses a rule base similar to the inference net in PROSPECTOR and has many features similar to MYCIN and PROSPECTOR. Uses certainty factors. Appropriate for on-line, sensor-based diagnosis. [Gonzalez and Lowenfeld, 1986]

ROGET: A knowledge-based system for acquiring the structure of a DES. Uses concepts common to most DESs. Developed at Stanford. [Bennett, 1985]

FUTURE DES RESEARCH AREAS

The areas of future research and development efforts include:

1. DES systems that have learning capabilities, i.e., where the system can improve its performance from experience [see Rowe, 1986].

2. Integrating DES into the design process [see Robach, Malecha, and Michel, 1984].

3. Improved uncertainty mechanisms.

4. Multiple fault handling capabilities [see De Kleer and Williams, 1986; Edgar and Petty, 1984; Williams, Orgren, and Smith, 1983].

5. Languages for describing systems for diagnostic purposes [see Bylander, Mittal, and Chandrasekaran, 1985].

6. Improved BIT DES.

BIBLIOGRAPHY

Adams, J.A., Gale, M., Dempsey, J. W., Kaizar, G. W., Straguzzi, N., "Artificial Intelligence Applications", RCA Engineer, Jan.-Feb. 1986, pp. 68-75.

Adlassnig, K., "Fuzzy Set Theory in Medical Diagnosis," IEEE Transactions on Systems, Man, and Cybernetics, Vol. SMC-16, No. 2, 1986, pp. 260-265.

Agre, G., Sgurev, V., Dochev, D., Dichev, C., and Markon, Z., "An Implementation of the Expert System DIGS for Diagnostics", Computers and Artificial Intelligence, Vol. 4, No. 6, 1985, pp. 495-502.

Aikins, J. S., Kunz, J. C., and Shortliffe, E. H., "PUFF: An Expert System for Interpretation of Pulmonary Function Data", Computers in Biomedical Research, Vol. 16, 1983, pp. 199-208.

Baldwin, J. F., "Fuzzy Sets and Expert Systems", Information Sciences, Vol. 36., 1985, pp. 123-156.

Bandler, W., and Kohout, L., "Probabilistic Versus Fuzzy Production Rules in Expert Systems," Int. J. Man-Machine Studies, Vol. 22, 1985, pp. 347-353.

Barth, S.W., and Norton, S.W., "Knowledge Engineering Within A Generalized Bayesian Framework", Uncertainty in Artificial Intelligence, AAAI-86 Workshop, Philadelphia, PA, 1986, pp.7-16.

Bending, M. J., "Hitest: A Knowledge-Based Test Generation System", IEEE Design and Test, May, 1984, pp. 83-92.

Bennett, J. S., and Hollander, C.R., "DART: An Expert System for Computer Fault Diagnosis", IJCAI-81, 1981, pp. 843-845.

Bennett, J. S., "ROGET: A Knowledge-Based System for Acquiring the Conceptual Structure of a Diagnostic Expert System", Journal of Automated Reasoning, Vol., 1, 1985, pp. 49-74.

Bolc, L., Kowalski, A., Kozlowska, and M., Strzalkowski, T., "A Natural Language Information Retrieval System With Extentions Towards Fuzzy Reasoning", International Journal Man-Machine Studies, Vol. 23, 1985, pp. 335-367.

Bonissone, P.P., and Johnson, H.E., "Expert System for Diesel Electric Locomotive Repair", IJCAI-83, 1983.

Brownston, L., Farrel, R., Kant, E., and Martin, N., Programming Expert Systems in OPS5: An Introduction to Rule-Based Programming, Addison-Wesley, New York, 1985.

Buchanan, B., and Shortliffe, E., Rule-Based Expert Systems: The MYCIN Experiments of the Stanford Heuristic Programming Project, Addison-Wesley, New York, 1984.

Bylander, T., Mittal, S., and Chandrasekaran, B., "CSRL: A Language for Expert Systems for Diagnosis," Comp. & Maths. with Appls., Vol. 11, No. 5, 1985, pp. 449-456.

Cantone, R. R., Pipitone, F.J., Lander, W. B., and Varrone, M.P., "Model-Based Probabilistic Reasoning for Electronics Troubleshooting", IJCAI-83, 1983, pp. 207-211.

Chandrasekaran , B., and Mittal, S., "Deep versus Compiled Knowledge Approaches to Diagnostic Problem-Solving", Developments in Expert Systems, Academic Press, London, 1984, pp. 23-34.

Charniak, E., and McDermott, D., Introduction to Artificial Intelligence, Addison-Wesley, New York, 1985.

Chiu, M., and Niedermeyer, E., "Knowledge-Based Diagnosis of Manufacturing Cells", Siemens Research and Development Reports, Vol. 14, No. 5, 1985, pp. 230-237.

Clancey, W.J., and Shortliffe, E.H., eds.,Readings in Medical Artificial Intelligence: The First Decade, Addison-Wesley, Reading, Mass., 1984.

Davis, R., "Diagnostic Reasoning Based on Structure and Behavior",Artificial Intelligence, Vol. 24, 1984a, pp. 347-410.

Davis, R., "Reasoning from First Principles in Electronic Troubleshooting", Developments in Expert Systems, Academic Press, London, 1984b, pp. 1-21.

Davis, K., "DORIS: Diagnostic Oriented Rockwell Intelligent System", IEEE AES Magazine, July 1986, pp. 18-21.

De Kleer, J., and Williams, B., "Reasoning About Multiple Faults",AAAI-86: Proceedings of the Fifth National Conference on Artificial Intelligence, Philadelphia, PA, August 11-15, 1986, pp.132-139.

Dempster, A. P., "Generalization of Bayesian Inference", J. Royal Statistical Society, Series B, Vol. 30, 1968, pp. 205-247.

Duda, R., Gaschnig, J., and Hart, P., "Model Design in the PROSPECTOR Consultant System for Mineral Exploration", Expert Systems in the Microelectronic Age, Michie, D., (ed.), University of Edinburgh Press, Scotland, 1979, pp. 153-167.

Dutta, A., "Reasoning with Imprecise Knowledge in Expert Systems", Information Sciences, Vol. 37, 1985, pp. 3-24.

Edgar, G., and Petty, M., "Location of Multiple Faults by Diagnostic Expert Systems", Proc. SPIE Int. Soc. Opt. Eng., Vol. 485, 1984, pp. 39-45.

Fink, P.K., Lusth, J.C., and Duran, J.W., "A General Expert System Design for Diagnostic Problem Solving", IEEE Proc. Workshop on Principles of Knowledge-Based Systems, IEEE Comp. Soc., Dec., 1984, pp. 45-52.

Gabriel, R., "Lisp Expert Systems are More Useful", Electronics, August 7, 1986, pp. 65.

Gazdik, I., "Fault Diagnosis and Prevention by Fuzzy Sets", IEEE Transactions on Reliability, Vol., R-34, No. 4, Oct., 1985, pp. 382-388.

Genesereth, M.R., "The Use of Design Descriptions in Automated Diagnosis", Artificial Intelligence, Vol. 24, 1984, pp. 411-436.

Georgeff, M., and Bonullo, U., "Procedural Expert Systems", Eighth International Joint Conference On Artificial Intelligence, Vol. 1, 1983, pp. 151-157.

Goodman, I., and Nguyen, H., Uncertainty Models for Knowledge-Based Systems, North-Holland, Amsterdam, 1985.

Gonzalez, A.J., and Lowenfeld, S., "On-Line Diagnosis of Turbine Generators Using Artificial Intelligence", IEEE Transactions on Energy Conversion, Vol. EC-1, No. 2, June 1986, pp. 68-74.

Grillmeyer, O., and Wilkinson, A. J., "The Design and Construction of a Rule Base and an Inference Engine for Test System Diagnosis", International Test Conference, 1985, pp. 857-867.

Hartley, R.T., "CRIB: Computer Fault-finding Through Knowledge Engineering", Computer, March, 1984, pp.76-83.

Hayes-Roth, F., Waterman, D., and Lenat, D., (eds), Building Expert Systems, Addison-Wesley, New York, 1983.

Hayes-Roth, F., "The Knowledge-Based Expert System: A Tutorial", Computer, Sept., 1984, pp. 11-28.

Heckerman, D., "An Axiomatic Framework For Belief Updates", Uncertainty in Artificial Intelligence, Workshop AAAI 86, Philadelphia, Pa., 1986, pp.123-128.

Hofmann, M., Caviedes, J., Bourne J., Beale, G., and Brodersen, A., "Building Expert Systems for Repair Domains," Expert Systems, January 1986, Vol. 3, No. 1, pp. 4-12.

Jackson, A.H., "Expert Systems For Fault Diagnosis", MILCOMP '85: Military Computers, Software, and Graphics, London, England, 1-3 Oct., 1985, pp. 449-455.

Jacobson, A., "Lisp is Not Needed For Expert Systems", Electronics, August 7, 1986, pp. 64.

Kacprzyk, J., and Yager, R. R., "Emergency-Oriented Expert Systems: A Fuzzy Approach", Information Sciences, Vol. 37, 1985, pp. 143-155.

Kalme, C., "Decision Under Uncertainty In Diagnosis", Uncertainty in Artificial Intelligence, Workshop AAAI 86, Philadelphia, Pa., 1986, pp.145-1502.

Kanal, L., and Lemmer, J.,(eds), Uncertainty in Artificial Intelligence, North-Holland, Amsterdam, 1986.

King, J. J, "Artificial Intelligence Techniques for Device Troubleshooting", Computer Science Laboratory Technical Note Series, Hewlett-Packard Co., Palo Alto, August, 1982.

Kitowski, J., and Ksiazek, I., "Fuzzy Logic Applications for Failure Analysis and Diagnosis of a Primary Circuit of the HTR Nuclear Power Plant", Computer Physics Communications, Vol. 38, 1985, 323-327.

Laffey, T.J., Perkins, W. A., and Nguyen, T. A., " Reasoning About Fault Diagnosis with LES",First Conference On Artificial Intelligence Applications, 1984, pp. 267-273.

Liu, Chen-Ching, and Tomsovic, Kevin, "An Expert System Assisting Decision-Making of Reactive Power/Voltage Control," IEEE Transactions on Power Systems, Vol. Pwrs-1, No. 3, August 1986, pp. 195-201.

Masarie, F.E., Miller, R.A., and Myers, J.D., "INTERNIST-1 Properties: Representing Common Sense and Good Medical Practice in a Computerized Medical Knowledge Base", Computers and Biomedical Research, Vol. 18, 1985, pp.458-479.

McDermott, D., and Brooks, R., "ARBY: Diagnosis With Shallow Causal Models", AAAI-82, 1982, pp. 370-372.

Merry, Martin, "APEX 3: An Expert System Shell for Fault Diagnosis," The GEC Journal of Research, Vol. 1, No. 1, 1983, pp. 39-47.

Miller, R.A., Pople, H.E., and Myers, J.D., "INTERNIST-1: An Experimental Computer-Based Diagnostic Consultant for General Internal Medicine", New England Journal of Medicine, Vol. 307, 1982, pp.468-476.

Montgomery, G. "Artificial Intelligence Applied to Aircraft Integrated Diagnostics", Flight Dynamics Laboratory, Air

Force Wright Aeronautical Laboratories, Wright-Patterson AFB, May 23, 1986.

Moon, R. E., Jordanov, S. Z., Turksen, I. B., and Perez, A., "Human-Like Reasoning Capability in a Medical Diagnostic System: The Application of Fuzzy Set Theory to Computerized Diagnosis", Journal of Clinical Computing, Vol. 8, No. 3, 1979, pp. 122-151.

Negotia, C.V., Expert Systems and Fuzzy Systems, Benjamin/Cummings, 1985.

Nelson, W.R., "REACTOR: An Expert System for Diagnosis and Treatment of Nuclear Power Accidents", AAAI: Proceedings of the National Conference on Artificial Intelligence, Pittsburgh, PA, 1982.

Norton, S., "An Explanation Mechanism For Bayesian Inferencing Systems", Uncertainty in Artificial Intelligence, Workshop AAAI 86, Philadelphia, Pa., 1986, pp.145-1502.

Odryna, P., and Strojwas, A., "PROD: A VLSI Fault Diagnosis System," IEEE Design & Test, December 1985, pp. 27-35.

Palowitch B.L., and Kramer, M.A., "The Application of a Knowledge-Based Expert System to Chemical Plant Fault Diagnosis", Proceedings of the 1985 American Control Conference, June, 1985, pp. 646-651.

Pan, J. Y., "Qualitative Reasoning With Deep-Level Mechanism Models For Diagnoses of Mechanism Failures", First Conference on Artificial Intelligence Applications, 1984, pp. 295-301.

Pau, L. F., "Survey of Expert Systems for Fault Detection, Test Generation and Maintenance," Expert Systems, April 1986, Vol. 3, No. 2, pp. 100-111.

Pazzani, M., "Refining the Knowledge Base of A Diagnostic Expert System: An Application of Failure Driven Learning", AAAI-86: Proceedings of the Fifth National Conference on Artificial Intelligence, Philadelphia, PA, 1986, pp.1029-1035.

Pearl, J., "On the Logic of Probabilistic Dependencies", AAAI-86: Proceedings of the Fifth National Conference on Artificial Intelligence, Philadelphia, PA, 1986, pp. 339-343.

Pipitone, F., "The FIS Electronics Troubleshooting System," Computer, July 1986, pp. 68-76.

Powell, C., Pickering, C., and Westcourt, P., "System Integration of Knowledge-Based Maintenance Aids", AAAI-86: Proceedings of the Fifth National Conference on Artificial Intelligence, Philadelphia, PA, 1986, pp. 851-855.

Ralescu, A.L., and Ralescu, D.A., "Probability and Fuzziness," Information Sciences, Vol. 34, 1984, pp. 85-92.

Reggia, J.A., Nau, D.S., and Wang, P.Y., "Diagnostic Expert Systems Based on a Set Covering Model", Developments in Expert Systems, Academic Press, London, 1984, pp. 35-58.

Reggia, J.A., Nau, D.S., and Wang, P.Y., "A Formal Model of Diagnostic Inference. I. Problem Formulation and Decomposition," Information Sciences, Vol. 37, 1985, pp. 227-256.

Reggia, J.A., Nau, D.S., Wang, P.Y., and Peng, Y., "A Formal Model of Diagnostic Inference. II. Algorithmic Solution and Application", Information Sciences, Vol. 37, 1985, pp. 257-285.

Robach, C., Malecha, P., and Michel, G., "CATA: A Computer-Aided Test Analysis System", IEEE Design and Test, May, 1984, pp. 68-79.

Rowe, M.,Veitch, A., Keener, R., and Lantz, B., "An Adaptive/Learning Diagnostic System for Complex Domains: ET", TI Engineering Journal, Vol. 3, No. 2, 1986, pp. 67-72.

Ryan, P. M., and Wilkinson, A. J., "Knowledge Acquisition for ATE Diagnosis", IEEE AES Magazine, 1986, pp. 5-12.

Schwartz, S., Baron, J., and Clark, J., "A Causal Bayesian Model For The Diagnosis of Appendicitis", Uncertainty in Artificial Intelligence, Workshop AAAI 86, Philadelphia, PA., 1986, pp. 229-236.

Shafer, G., A Mathematical Theory of Evidence, Princeton University Press, Princeton, NJ, 1976.

Shortliffe, E.H., Buchanan, B.G., and Feigenbaum, E.A., "Knowledge Engineering for Medical Decision Making: A Review of Computer-Based Clinical Decision Aids", Proceedings of the IEEE 67, 1979, pp.1207-1224.

Shubin, H. and Ulich, J. W., "IDT: An Intelligent Diagnostic Tool", Proc. AAAI-82, 1982, pp. 290-295.

Snow, P., "Bayesian Inference Without Point Estimates", AAAI-86: Proceedings of the Fifth National Conference on Artificial Intelligence, Philadelphia, PA, 1986, pp.233-237.

Sticklen, J., Chandrasekaran, B., Smith, J.W. and Suirbely, J., "A Comparison of the Diagnostic Subsystems of MDX and MYCIN", IEEE Proc. Workshop on Principles of Knowledge-Based Systems, IEEE Comp. Soc., Dec., 1984, pp. pp. 205-212.

Talukdar, S. N., Cardozo, E., and Leao, L. V., "Toast: The Power System Operator's Assistant," Computer, July 1986, pp. 53-60.

Talukdar, S. N., Cardozo, E., and Perry, T., "The Operator's Assistant-An Intelligent, Expandable Program for Power System Trouble Analysis", IEEE Transactions on Power Systems, Vol. PWRS-1, No. 3, Aug., 1986, pp. 182-187.

Thandasseri, M., "Expert Systems Application for TXE4A Exchanges", Electrical Communication, Vol. 60, No. 2, 1986, pp. 154-161.

Thompson, T.F., and Wojcik, R.M., "MELD: An Implementation of a Meta-Level Architecture for Process Diagnosis", First Conference on Artificial Intelligence Applications, 1984, pp. 321-330.

Turksen, T. B., "Fuzzy Reasoning in Medical Diagnosis", Modeling and Simulation - Proceedings of the Annual Pittsburgh Conf., Vol. 14, 1983, pp. 1029-1033.

Umeyama, S., "The Complementary Process of Fuzzy Medical Diagnosis and Its Properties," INFORMATION SCIENCES 38, 1986, pp. 229-242, 1986.

Waterman, D.L., A Guide to Expert Systems, Addison-Wesley, New York, 1986.

White, C.C., "A Posteriori Representations Based on Linear Inequality Descriptions of A Priori and Conditional Probabilities", Systems, Man, and Cybernetics, Vol. SMC-16, No. 4, 1986, pp. 570-572.

Wilkinson, A. J., "MIND: An Inside Look at an Expert System for Electronic Diagnosis", IEEE Design and Test, Aug., 1985, pp. 69-77.

Williams, T.L., Orgren, P.J., and Smith, C.L., "Diagnosis of Multiple Faults in a Nationwide Communications Network," Eighth International Joint Conference on Artificial Intelligence, Vol. 1, 1983, pp.179-181.

Yadrick, R., Vaughan, D., Perrin, B., Holden, P., and Kempf, K., "Evaluation of Uncertain Inference Models I: PROSPECTOR", Uncertainty In Artificial Intelligence, Workshop AAAI 86, Philadelphia, PA., 1986, pp. 333-338.

Yen, J., "A Reasoning Model Based on an Extended Dempster-Shafer Theory", AAAI-86: Proceedings of the Fifth National Conference on Artificial Intelligence, Philadelphia, PA, 1986, pp.125-131.

Zadeh, L., "Fuzzy Sets", Inform. and Control, Vol. 8, 1965, pp. 338-353.

Zadeh, L., and Ralescu, A., "On The Combinality Of Evidence In The Dempster-Shafer Theory", Uncertainty in Artificial Intelligence, Workshop AAAI 86, Philadelphia, PA., 1986, pp. 347-349.

Zwingelstein, G., and Monnier, B., "Artificial Intelligence Applications to the Surveillance and Diagnostics of Nuclear Power Plants", Trans. Amer. Nucl. Soc., Vol. 50, 1985, pp. 515-516.

Three Artificial Intelligence Issues in Fault Diagnosis: Declarative Programming, Expert Systems, and Model-Based Reasoning

Jean-Marc David & Jean-Paul Krivine

Laboratoires de Marcoussis
Centre de Recherches de la C.G.E.
Route de Nozay 91460 - Marcoussis, France

ABSTRACT

Artificial Intelligence applications for fault diagnosis are increasing in an amazing fashion. The purpose of this paper is to characterize and classify these applications, in terms of which AI techniques could be applied.

We will first describe and compare three knowledge-based systems we realized or we are currently working on. These three projects will be used during the discussion as examples of three different approaches : Declarative Programming, Expert Systems, and Model-Based Reasoning.

1. INTRODUCTION

1.1. What is Diagnosis ?

Diagnosis : The act or process of deciding the nature of diseased condition by examination of the symptoms. [Dictionnary 84]

The diagnosis is a part of a more general maintenance process. Taking into consideration the term of this context, we will consider different kinds of *causes* of some misbehavior. For instance, in an electronic system, a cause of fault could be a component or a complete board depending whether the reparation consists of exchanging only the faulty component or the whole board.

Then the whole maintenance process defines what kind of faults hypothesis diagnosis uses.

1.2. The Diagnosis Process

If the diagnosis is applied in various domains such as medicine, electronics, mechanics, etc, we can nonetheless describe a typical diagnosis scenario. Three main steps will be distinguished. Each one could be more or less complex depending on the application.

■ **Recognize a misbehavior**

The process starts with the observation of some behavior recognized as a deviation from the expected or desired one. (*symptoms*). At this stage, it could be necessary to confirm some symptoms or to exhibit other ones.

■ **Candidate Generation**

The problem solver tries then to generate some hypotheses about the causes of the malfunction. These malfunctions are often expressed as changes in the structure of the device from the specifications (stuck at 1 or stuck at 0 in electronic components, beam rocking in Turbo-generator,...).
The change we consider at this stage depends on some general hypotheses given by the whole maintenance process as seen before.

This association between symptom and cause can be known (*diagnosis knowledge*) or can be generated from a schema or a model (knowledge about how the system works).
In this step, we associate hypotheses of fault with observation.

■ **Reduce the set of hypothese**

The process tries now to confirm or reject each hypothesis of fault. Indeed, the previous step in this scenario often produces more than one hypothesis that might explain the observed misbehavior. New observations have to be interpreted.
In some case, it could be necessary to *perform* and then *interpret* tests. Executing a test consists of performing some action and expecting some particular result.

As for the candidate generation step, the test and its interpretation may be directly known (diagnosis knowledge) or built from a description of the system (*tests generation*).

This step is repeated until an acceptable diagnosis is reached.

1.3. Different kind of knowledge

Generally, diagnosis process involves two kinds of knowledge :

• *Diagnosis knowledge :* It associates symptoms, observations and actions with cause or hypotheses. These association are the basis for the diagnosis task [Chandrasekaran, Mittal 83]. Such associations may be heuristic or founded on a deeper knowledge.

• *Knowledge about "how the system works" :* Because the diagnosis knowledge is not always available or sufficient, we often have to use knowledge about how the system works. From a schema or a model, we will try to generate diagnostic knowledge. For that reason, diagnostis association seen before is said to be "precompiled pieces of knowledge" [Sigart 85].

1.4. The Automation of Diagnosis

Maintenance becomes a crucial point in most industrial systems (cost of troubleshooting, cost of immobilization,...). Moreover, systems are becoming more and more complex.
For these reasons, people try to automate the diagnosis process. Then, they face a lot of difficulties. So Artificial Intelligence applications are increasing in an amazing fashion.

In the following, we will comment on three kinds of AI systems for diagnosis. All of these systems have to deal with the different steps described above (interpretation of symptoms, candidate generation, tests generation, tests interpretation, etc). In particular, we will discuss the use of diagnosis knowledge and model based knowledge.

2. THREE FAULT DIAGNOSIS APPLICATIONS

In this section, we will briefly introduce three applications we have worked on at Laboratoires de Marcoussis. These applications will be used to illustrate our discussion.

THE SIAD PROJECT

This application is part of an on-board diagnosis system (diagnosis of subway locomotives power components) [Mouginot, David 87] ; the problem consists of continuously monitoring sensor data to detect and diagnose failures. The diagnosis system has to face around 300 failures in the electronic circuit for power-supply.

Building a fault-tree for that device would not really be a problem. But this fault-tree may be frequently modified ; and since the device is quite new, there is not yet knowledge about failures. Despite a careful *a priori* analysis, some failures may have been forgotten, and failures may also occur in a different way than expected while functioning.

THE DIVA PROJECT

DIVA is a project for the diagnosis of turbine generator vibrations. The purpose of the system is to continuously monitor the main working parameters - vibrations included -, to detect and diagnose as early as possible developing faults, and to prevent serious and indeed catastrophic consequences [David, Krivine 87].

THE FLAG PROJECT

FLAG is an interactive system for the *First Step Maintenance* of an Aircraft Navigation System [David, Descotte 84], [Cornille, Meller, Rulah 87].
The Navigation System is divided into physical units, called LRUs (Line Replaceable Units). When the pilot has noticed something wrong during a flight, the ground technician must find and change the failed LRUs to get the aircraft in an operational state.

To localize failures, the technician can perform simple tests (observations in the cockpit), or he can use a specialized processor, which has to be connected to the aircraft ; this processor is a sophisticated but slow diagnosis tool.

The purpose of the FLAG system is to assist the technician in repairing the Navigation System quickly.

3. THREE AI APPROACHES

Knowledge-Based Systems are systems where the **knowledge** about the domain or the problem at hand is separated from the **inference engine,** the mechanism which selects and uses relevant knowledge to solve the problem.

This approach is underlain and justifyied by the following axiom : *In Knowledge lies Power.*

By expressing the most knowledge in a declarative way, and keeping the inference engine as simple as possible, most of the modifications of behavior or improvement of performances will be obtained by adding or modifying knowledge.
These modifications are kept easier because of the explicit representation of domain and problem knowledge.

Despite the three applications mentionned above are all Knowledge-Based Systems, they are representative of three different approaches, namely :

- Declarative Programming : there exists a diagnosis algorithm, which uses diagnosis knowledge represented in a declarative way ;

- Expert Systems : there is no diagnosis algorithm, but there exist heuristics and human experts who solve this problem ;

■ Model-Based Reasoning : there exists a diagnosis algorithm ; the system will extract its knowledge from a device model.

3.1. Declarative Programming

□ Generally speaking, Declarative Programming is more appropriate when there exists an algorithm to solve the problem (fault-tree, fault dictionnary, ...) but conventional programming techniques are inadequate (the problem solving process involves symbolic manipulations, diagnosis knowledge is continuously evolving, etc).

□ A typical example of this approach is the SIAD project. Indeed, it would not have really been a problem to build a fault-tree ; but this fault-tree would often have been modified. For these reasons, the fault-tree is not directly programmed by the experts ; their knowledge is expressed in a declarative way : associations between failures and manifestations are expressed in a production rule formalism. This knowledge is then compiled in a fault-tree.

This allows easy modifications : adding a fault or a rule, modifying a criteria can be done in an easy way. Further more, the system is able to detect inconsistancies between rules, or lack of diagnosis knowledge : so we have feed-back on the knowledge acquisition process.

3.2. Expert Systems

□ There is no consensus on the definition of Expert Systems ; we will call Expert System a Knowledge-Based System whose knowledge is *Expert Knowledge.*

Because expert knowledge is continuously evolving, Expert Systems always instantiate the Declarative Programming paradigm. The main difference lies in the nature of encoded knowledge : expert knowledge is often inexact, uncertain, and incomplete ; expert knowledge is *heuristic* by nature.

A heuristic is a rule of thumb, strategy, (...) which drastically limits search for solutions in large problem spaces. Heuristics do not guarantee optimal solutions ; in fact, they do not guarantee any solution at all ; all that can be said for a useful heuristic is that it offers solutions which are good enough most of the time [Feigenbaum, Feldman].

If there exist algorithms to solve the problem, or if there are no heuristics, there is no place for Expert Systems.

□ Expert Systems have been applied to various domains, but the earlier applications, and

Figure 1: Declarative programming

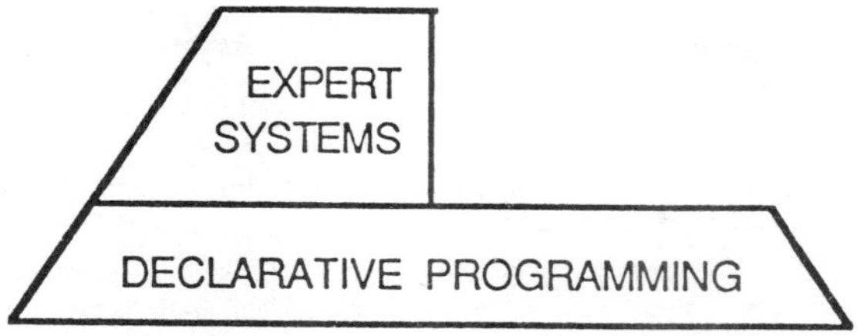

Figure 2: Expert System

may-be most of them, deal with diagnosis.

The DIVA project is a good example of Expert System ; attempts have been made to build decision-tree, but they all failed. On the other hand, there exists human experts.

Hence we formalize their knowledge, and their cognitive processes. The Knowledge-Base is organized in a hierarchy of "prototypes" [Aikins 83] ; each prototype is a description of a typical problem. DIVA makes use of a "hypothesize/establish/refine" strategy : when a situation is recognized among known problems (prototypes), DIVA evokes and tries to establish more specialized prototypes.
The knowledge is attached to the prototypes : rules to control information acquisition, rules to establish or reject the prototype, and rules to control the search of the next prototypes to consider. All these rules are heuristic knowledge.

☐ Difficulties while designing an Expert System are well established :

- it is difficult to "formalize" the problem : what knowledge is useful ? how should it be represented ? how should it be used ?
- it is difficult to acquire knowledge ;
- it is difficult to validate an Expert System.

First Generation Expert Systems also have many drawbacks :

- they are brittle, because they are unable to deal with atypical problems, or problems at the frontier of expertise ; instead, in such a case, we would prefer the system acknowledge its own incompetence.
- they have no real understanding of what they are reasoning on. They are unable to justify their knowledge ; their explanation capabilities are based on a trace of rules firing, without the possibility of formulating a deeper and more convincing explanation [David, Krivine 86-b].
- they are difficult to design, and they remain very specific to a given problem ; heuristic rules have no generality, of course, and a system designed for a particular device could not be used for another one, even though they both belong to the same domain.

3.3. Model-Based Reasoning

☐ Model-Based Reasoning is reasoning on a model of the behavior or the structure, and relations of domain objects. For diagnosis purposes, reasoning on the model will consist of extracting associations between actions, faults, and symptoms.

Examples of such models are :

- Qualitative Physics : the device behavior is modeled by means of *qualitative equations* between parameters [Sigart 85] ;
- Causal Network : a causal network is a kind of semantic network, where nodes represent states, actions, or observations, and links between them represent coupling and state causation [Sigart 85], [Rieger, Grinberg 77] ;
- Black-Box Network : the structural model describes how the boxes (modules) are connected, while the behavioral model describes how a particular module "behaves" (by stating a relation between output and input) [Davis 82], [Genesereth 82].

The model interpreter will use such device models to generate :

- associations between symptoms and fault hypothesis (candidate generation) ;
- associations between actions, fault hypothesis and observations (tests generation).

☐ One of the most promising techniques among these is "Reasoning from Structure and Behavior" [Davis 84]. There have been various implementations of that paradigm, but the basic idea is the same.

The device is described within a hierarchical network of black-boxes.
With such a model, the system is first able to generate expectations of the device behavior (simulation phase) ; when a discrepancy is detected between expected behavior and observed behavior, the model is then used to generate potential suspects (inference phase).

With such an approach, it is no longer necessary to list all symptoms, all faults and associations between them ; a symptom is a discrepancy between an expected and observed output ; a fault is a behavior which differs from the specified one.

☐ As an example of a Model-Based Reasoning approach, we will take the FLAG project.

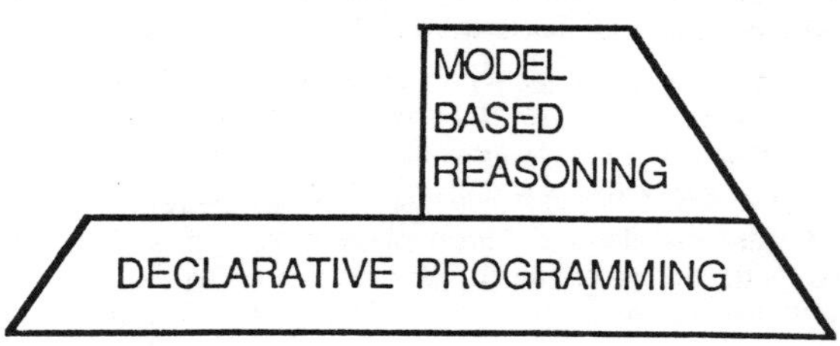

Figure 3: Model Based Reasoning

The first version of the system [David, Descotte 84], can be classified as a Declarative Programming approach :

- The diagnosis strategy was a very general one ; given the formal definition of a "repairing session", at each step , the chosen operation (test or replacement) is the one that minimizes the average probabilistic cost of the "repairing session".
- The Knowledge Base was made up of :

  i) knowledge about the LRUs (replacement cost, measure of fiability) ;

  ii) knowledge about the test (cost of operating, normal observation, how to proceed)

  iii) diagnosis rules in the following form : *conjunction of tests results* ⇒ *conclusion*

  where the conclusion part either suspects or discards LRUs failures.

The main problem with that system was the difficulty encountered by the experts in providing the diagnosis knowledge : tests and diagnosis rules. They found it easier to provide descriptions of the components and their dependancies.

HWnce we developed a new version [Cornille, Meller and Rulah 87] ; amongst the new developments, the system is now able to interpret the device model to generate tests, and to propose rules to interpret these tests.

□ The main difficulty when applying this approach is to find a good model (the good qualitative equations, the right modules, functions, and so on). This model has to be powerful enough to cover the faults we want to diagnose, but must remain tractable by the interpreter (no combinatory problems). (see [David, Krivine 86-a] for further discussion).

Usually, there are more than one model ; models differ according to fault assumptions [Davis 82]. We then need control capabilities to decide which assumption is the most plausible or valuable (which model is the most appropriate), and when to change or relax assumptions.

4. CONCLUSION

Our two main conclusions are : first of all, the emergence of a new programming style, we have called Declarative Programming ; and the convergence of two approaches, namely Expert Systems and Model-Based Reasoning.

We have seen that Model-Based Reasoning Systems need more and more control knowledge ; on the other hand, it would be interesting to provide Expert Systems with a better understanding of the device they are reasoning on. *Second Generation Expert System* will combine heuristic reasoning with deep reasoning.

But the definition of the role and nature of that cooperation between heuristic and deep knowledge is still an open problem.

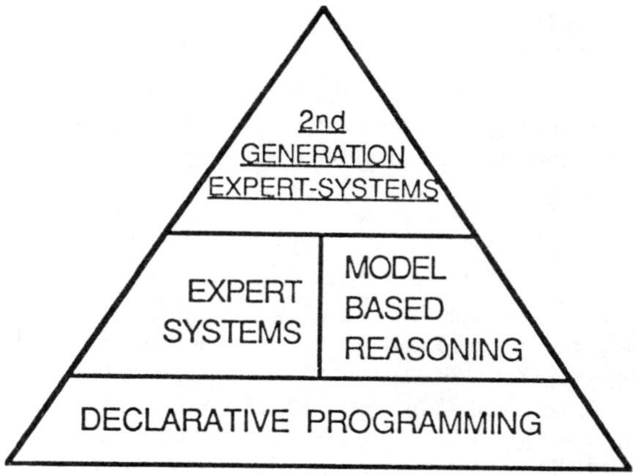

Figure 4: Second Generation Expert System

BIBLIOGRAPHY

[Aikins 83] J. Aikins, "Prototypycal Knowledge for Expert Systems" ; Artificial Intelligence 20 (2), pp163-210 ; 1983

[Chandrasekaran 83] B. Chandrasekaran, "Towards a Taxonomy of Problem Solving Types" ; AI Magazine pp 9-17 ; winter-spring 1983.

[Chandrasekaran, Mittal 83] B. Chandrasekaran, S. Mittal, "On deep versus compiled knowledge approaches to diagnostic problem solving" ; International Man Machine Studies 19 ; 1983.

[Chandrasekaran 86] B. Chandrasekaran, "Generic Tasks in Knowledge-Based Reasoning : High-Level Building Blocks for Expert System Design" ; IEEE Expert ; Fall 1986.

[Cornille, Meller and Rulah 87] J.M. Cornille, A. Meller et J. Rulah ; "Le système expert de maintenance PEDRO : Bilan et Perspectives " ; Septième journées internationales sur les systèmes experts et leurs applications. Avignon France, May 1987. (To appear. -in french-)

[David, Descotte 84] J-M. David, Y. Descotte, "Cost-Efficient Troubleshooting" ; The 6th European Conference on Artificial Intelligence, ECAI84 ; Pisa Italy.

[David, Krivine 86-a] J-M. David, J-P. Krivine, "Reasoning from Structure and Behavior :

Four Relevance Criteria" ; The 7th European Conference on Artificial Intelligence, ECAI86 ; Brighton UK ; July 1986.

[David, Krivine 86-b] J-M. David, J-P. Krivine, "What Happened ? Causal Reasoning in DIVA" ; First European Workshop on Fault Diagnostics ; Rhodos, Greece ; august 1986.

[David, Krivine 87] J-M. David, J-P. Krivine, "Utilisation de Prototypes dans un Système Expert de Diagnostic : le Projet DIVA " ; Septième journées internationales sur les systèmes experts et leurs applications. Avignon France, May 1987. (To appear -in french-)

[Davis 82] R. Davis, "Expert Systems : Where are We ? And Where Do We Go From Here ?" ; The AI Magazine 1982.

[Davis 84] R. Davis, "Diagnostic Reasoning Based on Structure and Behavior" ; Artificial Intelligence (24) ; 1984.

[Genesereth 82] M.R. Genesereth, "Diagnosis Using HierarchicalDesign Models" ; AAAI 1982

[Hamscher 83] W. Hamscher, "Using Structural and Functional Information in Diagnosis Design" ; Technical Report No 707, MIT June 1983

[Mouginot, David 87] Anne Mouginot and Jean-Marc David, "An Artificial Intelligence Programming Environment for the Diagnosis of Electronic Components" ; 2nd European Workshop on Fault Diagnostics, Reliability and Related Knowledge-Based Approaches ; April 1987

[Rieger, Grinberg] C. Rieger, M. Grinberg, "The Declarative Representation and Procedural Simulation of Causality in Physical Mechanisms" ; 5th IJCAI, Cambridge Massachusetts, pp 250-256 ; 1977.

[Robach and All 84] C. Robach, P. Malecha and G. Michel. "CATA : a Computer-Aided Test Analysis System" ; IEEE Design and Test. May 1984

[Sigart 85] Sigart Newsletter. Special section on reasoning about structure, behavior and function ; july 1985.

An Expert System Approach to Systems Safety and Reliability Analysis

A. Poucet†, S. Contini†, K.E. Petersen†† and N.K. Vestergaard††

†Commission of the European Communities,
Joint Research Centre Ispra
21020 Ispra (Va), Italy
††RISØ National Laboratory,
Systems Analysis Department,
Postbox 49
DK-4000 Roskilde, Denmark

ABSTRACT

A knowledge based approach to systems safety and reliability analysis to be implemented in an intelligent software tool (STARS: Software Tool for Advanced Reliability and Safety) is presented. The approach evolved from previous experience in the development and use of the CAFTS and RIKKE codes for automatic fault tree construction. This experience is shortly described.

The new tool will offer an intelligent and powerful aid in the logic modelling (fault tree, event tree construction) for large and complex systems.

Moreover, it will allow for a computer aided qualitative analysis performed to identify potential hazards or critical events of interest.

The tool is designed mainly for applications in the chemical process plant and nuclear industry.

KEYWORDS

Reliability; safety; fault tree analysis; hazard and operability analysis; expert systems.

INTRODUCTION

In systems safety and reliability analysis, the modelling of the system is an important task. This modelling task consists in general of:

1. A preliminary (qualitative) analysis in which the events and scenarios relevant for the safety and/or reliability are identified and ranked on basis of their consequences or hazard potential

2. The construction of (logic) models that describe the system behaviour with respect to the above mentioned events and scenarios either in a deductive or an inductive way.

While computerised methods for logic and quantitative analysis of the logic models (see e.g. Contini,1986) have already been introduced some decade ago, the modelling task is still mostly carried out without such aid. In the past, various attempts have been made to introduce

automatic or computer aided methods in system modelling, but their application has not been very wide.
A need for computer assistance in modelling is induced not only by the effort and cost involved in manual construction of logic models, but also by the wish to reduce the variability and the error proneness, and to increase the robustness and scrutiniseability.
This latter factors are becoming more and more important as probabilistic approaches are entering in the licensing framework.
Indeed, a Benchmark Exercise on systems analysis, organised by the C.E.E., has shown that the different assumptions and implicit procedures used in fault tree construction contribute significantly to the spread observed in the results and make the comparability of different analyses difficult (Amendola, 1985).

Several approaches for assisting the system modelling task have been proposed. They range from simple fault tree editing tools, that allow for computer aided drafting of trees but do not contain any domain knowledge, to computer codes for fully automatic fault tree construction.
In the sequel, two examples of computer codes for fault tree construction: RIKKE (Taylor, 1982; Olsen, 1984) and CAFTS (Poucet, 1985, 1987).
Besides the fact that the existing fully automatic fault tree construction codes have limited capabilities in real industrial applications, it is our opinion that computerised methods should aim to be tools rather than prostheses; that is: they should be designed to assist the analyst in his job rather than to replace him completely.
Indeed, the manual modelling involves an important process of learning-through-analysis about the potential system malfunctions which is lost in a completely automatic procedure.
Moreover, the collection of information, needed for complete and correct modelling of the system, is an iterative process which is partly driven by analysis. Therefore, it is difficult to provide the complete information in an a priori way as required in a fully automatic procedure.

The expert system technology offers the most promising perspectives for achieving a modelling support tool. Combined with graphics and analysis capabilities, it can provide a natural engineering oriented environment for computer assisted reliability and safety modelling and analysis.
Before presenting such an approach in further detail, a short overview will be given of two existing tools which, although not being designed as such, do use expert system concepts.
The experience gained with the development and use of the described tools has been used in the overall design of a new intelligent reliability and safety modelling and analysis tool called STARS.

FAULT TREE CONSTRUCTION IN RIKKE AND CAFTS

Overall approach

The fault tree construction procedures in both RIKKE and CAFTS are based on the use of a component knowledge base.
This knowledge base contains generic component models describing the behaviour of components irrespective of the system or process in which they are used.
Although CAFTS and RIKKE were not really designed as expert systems, the models in the knowledge base have the form of (IF-THEN) production rules written in first order logic.

The fault trees are constructed basically by backward chaining from given goal states to their

causes using the production rules and the topology of the system.
To illustrate this process consider the very simple sample system in figure 1.

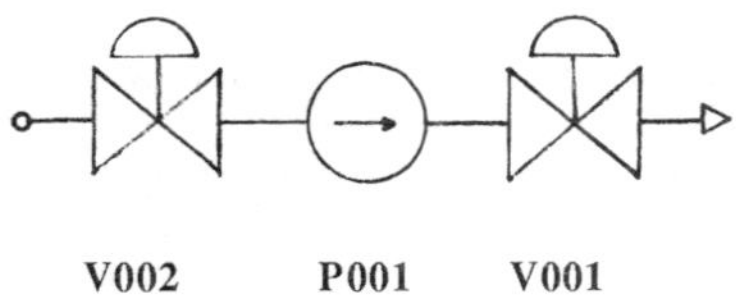

Fig. 1. Simple sample system.

Consider the following (simplified) rules for valves and pumps:

IF < valve *x* closed> **OR** < flow at output of component *y* immediately upstream *x* is zero>
THEN < flow at output of valve *x* is zero>

IF < pump *x* is failed> **OR** < flow at output of component *y* immediately upstream *x* is zero>
THEN < flow at output of pump *x* is zero>

By backward chaining from a goal state, e.g. flow at output of valve V001 is zero, the fault tree in fig. 2 can be obtained. The backward chaining begins by applying the valve rule to valve V001. It proceeds by consulting the topology (facts data base) to see that the component upstream is a pump and applies the rule for pumps. The process continues until the system boundaries are reached.

In real applications, the rules are obviously more complicated and the backward chaining process needs to be controlled in order to avoid loops and impossible branches.

Although the general concept is the same, RIKKE and CAFTS differ substantially in the realisation of it.

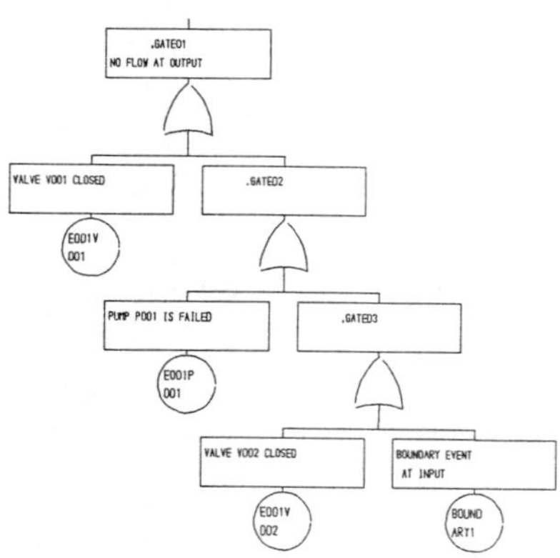

Fig. 2. Fault tree obtained using rules.

Approach used in RIKKE

The RIKKE code has essentially been designed as a fully automatic fault tree construction tool. The topology of the system to be analysed is entered in an interactive way by using a powerful graphics interface. Next a Top event is specified and the construction of the fault tree for that Top event is triggered. Once the construction process is started, the possibilities of interacting with the process are limited. The only interaction foreseen is the cut off of irrelevant or impossible branches in the tree (i.e. control of the backward chaining as mentioned in the previous section).
Since the only facts used in the process are the topology of the system and the Top event, and since the interaction during the construction process is limited to the one described above, it is not easily possible to enter information related to the process taking place in the system.
Hence, two systems with completely different processes but with the same topology will yield the same fault trees.

As a consequence, the fault trees generated by RIKKE can be overly complex since it may contain branches that are irrelevant for the system and process analysed. One could say that RIKKE generates a 'forest' of trees giving the causes for some Top event in all possible systems (system considered here as a set of interconnected components that fulfill a certain function) with a given topology.

As another consequence, the rules describing component behaviour can be quite complex since they have to represent all situations with respect to all processes.

Consider for example a heat exchanger: the rule for e.g. *'high temperature at the outlet of the primary side'* must take into account the different processes in which the heat exchanger can be used. In other words, it must take into account the situation in which heat is to be transported from primary to secondary side (as in a cooler) and the situation in which heat is to be transported in the opposite way (as in a heater).
It is however a merit of RIKKE that the component knowledge base is very exhaustive and detailed.

Approach used in CAFTS

The basic approach used to construct fault trees in CAFTS is an interactive one which is organised in two distinct phases.

First a high level macro fault tree is constructed. This construction is performed in an interactive process in which the *rules* of the component knowledge base are used and in which facts (topology but also other information) are prompted to the user as they are needed by the backward chaining. The control of the backward chaining process, necessary to avoid loops and/or running in impossible branches, is performed by using meta-rules on incompatible states and deviations.

In the second phase, the macro fault tree is expanded into a fully detailed one by using *frames* containing knowledge on generic component attributes (states, possible subcomponent circuits, support functions needed, failsafe states...) and detailed failure causes in function of these attributes. Once again the facts (concerning the attribute values of some particular instance of the component frame) are prompted as they are required during the expansion phase.

Even if the interaction possibilities are more extensive in CAFTS, and allow to enter other than topological data, the process-related information asked by CAFTS is also limited.
Moreover, the rules in the component knowledge base are rather simple and tend to describe the general case rather than the exception. It is supposed that the user manually intervenes to adapt the modelling in special situations, and ample provisions are foreseen to allow such manual intervention.
Despite the (relative) simplicity of the models in the component knowledge base, CAFTS is a very useful tool, not at least because of its powerful interface with the user (using interactive graphics, full screen data entry panels and menus).

Experience and insight gained

The experience gained with RIKKE and CAFTS was positive and the codes have been used on industrial systems in real applications .
It was felt that a combination of the best features of both codes would yield a powerful tool.

However, a major drawback, inherent to both CAFTS and RIKKE, is the fact that they have only limited ways of using knowledge and information about the *process* taking place in the system and base the modelling only on the *topology*.
This results in trees that may be overly complex in some respects (RIKKE) and not adequate or complete in other respects (CAFTS and RIKKE).

On the other hand, CAFTS and RIKKE can be used to construct logic models for events relevant for the safety or reliability of the system once these events have been defined, but they offer no assistance for identifying such events.
As a conclusion, it was felt necessary to introduce a front end part to perform a computer assisted preliminary (qualitative) analysis to identify critical events. The knowledge about the system and the process, acquired in this preliminary analysis, will be used, together with generic knowledge on component and subsystem behaviour, to construct the logic system models for the identified critical events.

QUALITATIVE ANALYSIS TOOL

The qualitative analysis tool will be used for two purposes:

1. the identification of potential hazards, events and event sequences leading to hazards;

2. the elicitation of *process specific* knowledge to be used in subsequent modelling;

The approach will be to emulate the reasoning and expertise applied by analysts in *hazard and operability* studies into an expert system.
In such studies, the plant is subdivided into systems and/or units each having well defined functions. The deviations from these normal functions and the way in which these deviations propagate in the plant are then analysed.

The expert system will use a knowledge base containing information on systems and units.

The knowledge base will contain:

1. a classification scheme of systems and units commonly encountered in process plants;

2. the characteristics of such units and systems;

3. properties of general classes of substances;

4. rules on possible deviations of process parameters and/or functions, their possible causes and effects;

5. generic critical events or event sequences;

6. checklists on potentially relevant facts to be prompted to the analyst.

The expert system will perform an inductive (event driven) reasoning by forward chaining through rules (see [4] above) that describe cause-effect relationships at the system/unit level. During this process, the analyst will be prompted to respond 'what happens if' type of questions. This allows to collect knowledge that is specific to the plant and process analysed, in contrast to the knowledge in the system/unit KB and the component KB which is generic. The knowledge will be represented in production rules (in 0 order logic or propositional calculus) to be used in complement to the generic production rules during the construction of the fault trees.
To illustrate this consider e.g. a mixer in which two products A and B are put together. Upon a question, *'what happens if the mixing ratio A/B increases (decreases)?'*, the analyst may respond that there is an exothermic reaction if the amount of A becomes too large. This process specific information is then translated into a rule that expresses the fact that high temperature at the outlet of the mixer can be caused by high (resp. low) mass flow in the input with product A (resp. B). This process specific rule is added to the component KB to complement the generic rule for high temperature at the outlet of a mixer.

FAULT TREE CONSTRUCTION TOOL

The fault tree construction tool will be largely inspired by the previous experience.
A frame/rule based expert system will be used, in which the deductive (goal driven) reasoning and the heuristics, applied during manual fault tree construction, will be modelled.

The construction will proceed in phases (as in CAFTS): first a macro fault tree will be constructed on the basis of the topological information of the functional flow diagram or P&I D obtained interactively in a plant description module, and of the process information obtained through the computer assisted qualtitative analysis. The construction of the macro fault tree will be triggered by specifying a state or some combination of states at some node(s) in the system diagram.
Afterwards, the macro fault tree will be further developed on the basis of facts provided interactively by the user for the components and the connections of these components with support or service subsystems. The process may be iterative in the sense that these latter subsystems can be modelled again in the same way.

The knowledge base used contains generic knowledge on components, knowledge specific for the

system and process analysed as collected during the qualitative analysis, and knowledge that allows to direct and control the reasoning process ('meta-rules' e.g. on incompatible situations).
The generic knowledge is represented in frames covering component characteristics, rules to be used for construction of macro fault trees and rules to develop the causes of the component states.
Whenever the frame is invoked for some component in the system, some slots will acquire values from e.g the system topology or the user will be prompted to enter values for some slots. The rules, which in general will depend on the slot values, can then be executed. The values of slots that deal with properties of some specific instance of a component type, are stored into objects (instantiations of part of the frame), one object for every component in the system.
The knowledge base is organised in an hierarchical way; some slots are defined at the level of a *class* of components (e.g. valves) and inherited down at the level of individual component types (e.g. motor driven valve). In this way it is possible to avoid information duplication and redundancy and to increase the maintainability of the knowledge base.

IMPLEMENTATION INTO VERSATILE PACKAGE

The methods described in the previous sections will be implemented in an integrated software package for safety and reliability analysis (STARS).
The overall structure of the package is illustrated in figure 4.
The package will allow to perform the modeling tasks, based on the methods described above, and also the more conventional tasks of fault tree drawing, editing and analysis.

To allow a versatile use, different ways of entry are foreseen: the normal way would be to start describing the system topology and other relevant information and then proceed to the qualitative analysis, the construction of the fault tree(s) and the analysis. However, at strategic points the possibility is foreseen to store information, retrieve previously stored information or enter information directly e.g. the topology of the system can be saved or some pre-existing topological data may be loaded, the values that are assigned to slots in the component frames may be stored for later use, the macro fault trees and expanded fault trees can be stored (this allows to use the package also for fault trees that existed already), the minimal cut sets of the fault tree can be stored so that they can be used later for quantifying event sequences.
Moreover, it is possible to use the package as a fault tree drafting tool if the user wants to manually construct a fault tree (e.g. for a special type of system for which no data exist in the knowledge bases and for which it is not worthwhile creating such data).
The analysis capabilities will include logical and probabilistic analysis based on either analytical techniques or simulation.

The package will be designed for running on personal workstations with the UNIX operating system. It will use graphics and windows to provide a powerful user interface.

CONCLUSIONS

An knowledge based approach has been presented for the modelling of complex systems. The approach allows to identify critical events and event sequences and to model the causes of these events or sequences in fault trees.
It has been shown that, alongside with topological information about the system to be modelled,

it is necessary also to consider process related information.
The elicitation of this process related information is performed by a computer assisted qualitative analysis, making use of knowledge bases containing system/unit knowledge and an event driven reasoning process.
The construction of fault trees is then performed by using generic component knowledge (in the form of rules in first order logic and frames), together with process specific knowledge (in the form rules in 0 order logic) into a goal driven reasoning process.
The approach will be implemented into a tool in which windowing, menus and graphics will be used to obtain an advanced user-interface, and in which analysis capabilities are included.

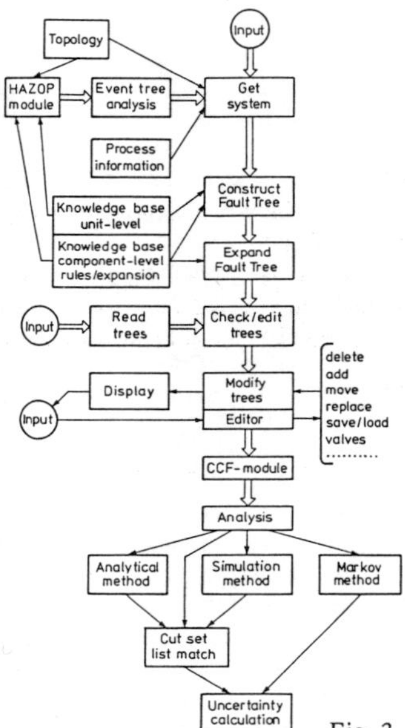

Fig. 3. Overall structure of STARS.

REFERENCES

Amendola, A. (1985). Systems reliability benchmark exercise, Final Report, CEC-JRC Ispra EUR 10696.

Contini, S., and A. Poucet (1986). European reliability computer code index, CEC-JRC Ispra SP.I.05.E3.86.90.

Olsen, J.V. (1984). RIKKE viewed as an expert system. RISO Electronics Dep., Internal note.

Poucet, A. (1985). CAFTS: Computer aided fault tree analysis. Proc. ANS/ENS International Topical meeting on Probabilistic Safety Methods and Applications, San Francisco, Feb. 24- March 1, 1985.

Poucet, A. (1987). CAFTS: Description and how to use. CEC-JRC Ispra EUR report (to be published).

Taylor, J.R. (1982). An algorithm for fault-tree construction, IEEE Trans. Reliab., R-31, Nr. 2.

An Artificial Intelligence Programming Environment for the Diagnosis of Electronic Components

Anne Mouginot & Jean-Marc David

Laboratoires de Marcoussis
Centre de Recherches de la C.G.E.
Route de Nozay, 91460 - Marcoussis, France

ABSTRACT

This paper describes an Artificial Intelligence approach for the diagnosis of subway locomotives power components. Our purpose is to show how Artificial Intelligence techniques can significantly help in solving some classical problems.

Our application is part of an on-board diagnosis project. The programming environment we developped helps the experts in expressing their diagnosis knowledge in a declarative way ; this knowledge base is then compiled in a fault-tree, which can be integrated on-board.

KEYWORDS :

Artificial Intelligence, Knowledge-Based Systems, Fault Diagnosis, Knowledge-Base Compilation, Diagnosis of Electronic Components.

1. INTRODUCTION

This paper describes an Artificial Intelligence approach for the diagnosis of subway locomotives power components. Our purpose is to show how Artificial Intelligence techniques can significantly help in solving some classical problems.

Our application is part of an on-board diagnosis project. The system we present has been realized within the framework of a collaboration between ALSTHOM, the Laboratoire d'Automatique et d'Analyse des Systèmes (LAAS), and the Laboratoires de Marcoussis.

The first section will introduce the context of the problem and the motivations ; in the second section, we will describe the system we realized ; in the last section, we will discuss the advantages of this approach.

2. THE PROBLEM

The increasing complexity of subway locomotive equipments (mechanical, electro-mechanical, and now the use of microprocessors) requires up to date maintenance tools.

Maintenance can no more be solely based on the subway driver complaints. On-board tools are needed to detect and diagnose failures when they occur ; the expected consequence will be a decrease of maintenance time.

The SIAD (Integrated System for Troubleshooting Aid) has been designed and realized to improve on-board fault diagnosis [Le Clech, Lepinay 84]. The SIAD is an independant external processor, which allows :

- ☐ the monitoring of sensor data ;
- ☐ the recording of relevant data when a fault is detected ;
- ☐ the diagnosis of the fault, based on these recorded data.

Our problem concerns the diagnosis of the electronic circuit for power supply. Designers of that device founded around 300 different potential failures ; we are concerned with the identification of these failures, based on the recorded data. But the on-board diagnosis system has to meet some requirements : the on-board processor is a small microprocessor with very few memory ; the diagnosis task stops the acquisition process : so, the diagnosis must be done as fast as possible ; there is no possible interaction with the user or with a human expert during the diagnosis : the recorded data must be sufficient to identify the fault.

The first solution we considered was to build a fault-tree and to implement it on-board. Building that fault-tree would not have really been a problem ; the search space is important but remains tractable. The drawback of that approach lies in the fact that this fault-tree may need to be frequently modified, for at least three raisons :

- despite the careful a priori analysis, some failures may have been forgotten ;
- failures may arise while running in a different way than expected : failures have been simulated or provoked, but their is no recording of real failures ;
- the device is planned to be used, for a long time : during this time, it will be modified, so the fault-tree will.

3. THE PROGRAMMING ENVIRONMENT

The expert will not directly build the decision-tree. He will rather give his knowledge in a declarative way ; this knowledge will then be compiled in a fault-tree.

We so have the advantage of beeing able to easily modify the knowledge-base, while the resulting compiled knowledge-base is compatible with the previous requirements.

We have developped a programming environment (figure 1) [Ghallab 86] [Mouginot 86], which is composed of :

- ☐ *the Knowledge Acquisition System (KAS),* which helps the expert to build and to debug the rule base in an interactive way.
- ☐ *the Knowledge Base Compiler (KBC),* which compiles the diagnosis rules in a decision-tree.

We will now describe the two main tasks performed by this environment.

3.1. Compiling Diagnosis Rules

• The diagnosis knowledge, namely associations between manifestations and failures, is represented in a production rule formalism : *IF <conditions> THEN <conclusion>*

The left-hand side of these rules is composed of conjunction of conditions : a condition is a predicate (equal, greater-than, member-of, ...) which tests parameter values, or may also refer to diagnosed faults.
Parameters are sensor data (uf, uci, ih, ...), or the result of a computation on these data (ih/uci, ...). Most of the parameters refer to fault manifestation at the "reference time" (the time the fault was detected) ; but some of them are concerned with the evolution of some values (mode-after-failure, ...).

The right-hand side of the diagnosis rules conclude about the initial failure, or the induced failures (consequences of the initial failure).

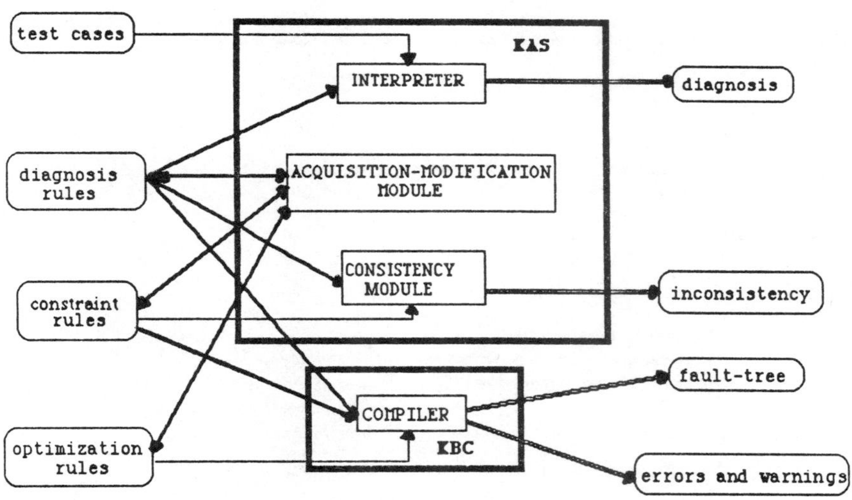

Figure 1: The Programming Environment

```
IF
   (eq configuration 'traction) (member-of mode '(vernier principal shuntage))
   (eq time-jmax 'short) (< i/j 0.4) (between fuci 0.001 314)
   (< ih/uci 0.4) (< um 300) (between ih-start 50 100)
   (between uci 1000 4000) (between uf 1000 2000) (> j 900)
   (eq ceq_after_failure 'open) (eq mode_after_failure 'inhib_clg_ferme)

THEN
   INITIAL-FAILURE ( 68 #| the sensor j is parasited |#)
```

If

- *the configuration is traction,*
- *the working mode [mode] is vernier or principal or shuntage,*
- *the courant j is higher than the courant i [time-jmax] during a short time*
- *the shuntage rate [i/j] is less than 0.4*
- *the frequency of reversing cycle [fuci] is belongs to the interval] 0.001 314 [hertz,*
- *[...]*
- *the working mode after the failure is inhib_clg_ferme*

then we can conclude that the sensor j is parasited (failure numbered 68).

```
IF
   (eq configuration 'freinage)
   (member-of mode '(vernier principal shuntage))
   (member-of failure '(122 39 49 40))

THEN
   INDUCED-FAILURE (AND 1 32)
```

If

- *the configuration is freinage,*
- *the working mode is vernier or principal or shuntage,*
- *we have already diagnosed one of the following failures : 122, 39, 49 or 40*

then we can conclude that failures 1 and 32 were induced.

Examples of Diagnosis Rules

• The diagnosis rules are then compiled in a fault-tree. To optimize the compilation process, the system will use the "constraint rules" and knowledge specially devoted to that task : the "optimization rules". The constraint rules can constrain some parameter values, while the optimization rules discard some parameters as no relevant for the diagnosis in a particular context ; these rules allow the compiler to prune its search space.

The system may face some problems during the compilation phase :

- a contradiction is detected between some rules (the suspected failures set is empty). This is an error : the contradiction has to be relaxed, i.e. the expert must modify the rules, and restart the compilation.

- the compilator find a case where no rule can be applied, or diagnosis knowledge is needed to further discriminate between suspected failures. This is not consider as an error ; but such information may warn the expert he misses some cases.

Some of these problems can be detected before running the compilation. Moreover, it is important to validate the knowledge-base before to compile it. The Knowledge Acquisition System is able to help the expert in building and debugging his knowledge-base.

3.2. Building and Debugging the Knowledge-Base

• To build the knowledge-base, the expert can use a dedicated acquisition module ; to test the so-built knowledge base, the interpreter will allow him to run cases. When he estimates the diagnosis rules are correct, he can run the compilation.

In order to optimize the test of the diagnosis rules, a library of test-cases has been constituted from simulation results or theoretical analysis. By running these test-cases, the expert can quickly be aware of the consequences of diagnosis rules modifications.

• All parameter values are not independant ; current laws, or device design impose relations or constraints between some of them. The *constraint rules* represent this domain knowledge. For instance, *if the working mode is "preexcitation" then the configuration is necessarily "traction" or "erre"*.

We have seen that these rules were used by the compiler to prune its search ; there are also used by the consistency module to detect "inconsistent" diagnosis rules. A rule is inconsistent (relatively to the constraint rules) if the context specified in its left-hand side would allow to fire constraint rules leading to a contradiction (different values for the same parameter).

Obviously, the more constraint rules the system has, i.e. the more domain knowledge he has, the more inconsistencies he will detect.

4. ADVANTAGES OF THE APPROACH

The programming environment we developped is an interesting approach. We have been able to compiled a fault-tree for the electronic circuit for power supply ; this fault tree has been programmed on a PROM, and integrated to the complete diagnosis system, which is currently under test on 20 locomotive prototypes.

While using a classical approach would have been possible, this AI approach has prouved to be more efficient.
The first knowledge base led to a criticism of the initial analysis : it appeared necessary to

change or add some criteria to further discriminate between failures. The need for such modifications was detected soon, and modifications were easy.

This is a very interesting property of that approach to permit feed-back, and to allow an incremental knowledge acquisition process.

REFERENCE

[Le Clech, Lepinay 84]
LE CLECH, B. LEPINAY
Les systèmes d'information d'aide à la maintenance embarqués sur le matériel RATP type MI 84 - Colloque ORE de Budapest - mai 84 -

[Ghallab 86]
M. GHALLAB
Environnement de développement d'un système de diagnostic embarqué : projet SIAD - Rapport LAAS 86312 - octobre 86 -

[Mouginot 86]
A. MOUGINOT
Système de diagnostic en électronic de puissance - Rapport final - Laboratoires de Marcoussis - Mai 86 -

From Fault-Defect Analysis to Knowledge Base Enrichment: An Epistemological Approach

F. A. Batzias

The Piraeus Graduate School of Industrial Studies
Department of Technology & Production Systems
Piraeus, Greece

ABSTRACT

The route from the first observation of a defect to the final disclosure of new knowledge is investigated through Fault Analysis. We found, by induction, that the following stages may be considered as mostly representative to describe this route:

1. Observation of a defect in a (semi)finished product or detection of a failure in a system.
2. Study of the symptoms surrounding the defect (or the failure).
3. Reproduction of the defect (or the failure) under controlled conditions by experiment and/or simulation.
4. Hypothesis about an intermediate (or ultimate) cause by fault analysis.
5. Verification (or attempt for falsification) of this hypotheses.
6. Collection of relevant data to unify and theorize the obtained results, by induction.
7. Incorporation of this knowledge into an existing Theory or a knowledge Base.
8. Application of the extended knowledge for the design of new products and methods.

A case example is analyzed to illustrate the fitting of the suggested route to a real situation and some methodological problems, concerning applicability of the above stages, are discussed.

KEYWORDS

Fault analysis; knowledge base; methodology; quality control.

INTRODUCTION

Fault detection may lead not only to remedial proposals but also to new products and methods as well as to deeper understanding of various physico-chemical processes. Of special interest is the mechanism according to which a discontinuous advancement in production technology originates from a fault occuring in the Laboratory or the pilot plant or even in the industrial plant. As an example we quote the well-known case of research on polymerization under high pressures, which led to the discovery/invention[1] of polyethylene: in the experimental procedure there

---

[1] There is a conceptual connection between discoveries and inventions, at least in

was a fault (due to a defect in the apparatus used), the pressure dropped and a small amount of white powder was found when the vessel of reaction was dismantled; this proved to be a turning point at the route to the final success.

A fault, considered as such according to some traditional pattern, may be proved to be the "secret" for an improvement in the industrial process or even for the avoidance of the production of a defective article. As an example, we quote the experience from the first application of the Float Glass process (in which a continuous ribbon of glass moves out of the melting furnace and floats along the surface of a bath of molten tin): after some months of initial successful operation of the industrial plant, it was found that the secret of success was a "fault" in the process, without which (i.e. with "corrected" conditions) a defective product was produced; so, the "fault" was studied and reproduced by the industrial inventor L. A. B. Pilkington and subsequently applied in the industrial plant with success; on the contrary, the initial measures taken for the correction of the "fault" were rejected (for further details, see Jewkes, Sawers and Stillerman, 1969).

The aim of this paper is to investigate, through Fault Analysis, the route from the first observation of a defect in a product to final disclosure of new knowledge, which is subsequently incorporated into an existing Theory or a knowledge Base.

METHODOLOGY

Initially, we analyzed fourteen case histories extracted from literature and materials science and from our own experienc in electrochemical engineering and chemical technology. The common characteristic in all cases collected was the observation of a failure in a subsystem of laboratory/industrial production or a defect in a product, which led through further investigation to new knowledge applicable to novel products and methods. Afterwards, Fault Tree Analysis (FTA) was used to thoroughly examine each collected case on an a priori basis, i.e. trying to think as if we didn't know the result of the investigation. This examination was aiming to the decomposition of each real case into common formal parts which then were synthesized to formulate by induction an applied research and development conceptual scheme; subsequently, the intermediate stages of a typical route, from the first observation of the defect in the product (or the failure in the process) to the final knowledge Base extention, were derived by deduction. These stages were tested for consistency in relation with the initial set of the fourteen case histories. Finally, seven new case histories were collected in a similar

---

Chemical Technology. We say that the properties of a substance are "discovered" but the special industrial process that makes use of them for the production of this substance is "invented". Respectively, we say "the discovery of polyethylene", meaning the finding of the result of certain existing properties when combined, while we say "the invention of polyethylene", meaning the finding of the result of certain artificial processes. Nevertheless, most industrial processes are imitating what happens or could happen in Nature. From this point of view, it is sometimes difficult to distinguish between discovery and invention (see also Batzias, 1986a). This is not the case in other disciplines of industrial applications, e.g. in Mechanical and Electrical Engineering Technology, where the materials used (e.g. metals, ceramics, etc.) are given (discovered or invented by Chemical Technology) and the findings are exclusively a matter of human creativity, i.e. invention. It is this reason that makes T. Edison himself to reject the pure empirical (Edisonian or empirical-inductive mode of thought, after Conant, 1964) method, which is closely connected with accidental findings, usually originating from a fault occuring during experimentation or production: "...but when it comes to problems of a mechanical nature, I want to tell you that all I've ever tackled and solved have been done by hard, logical thinking".

way and the typical route was tested again; corrections were performed and the deduction of intermediate stages was each time reexamined under the light of new evidence.

By omitting some sub-stages, which may differ from case to case, we concluded that the following stages were mostly representative to describe the route under investigation:

1. Observation of a defect in a (semi)finished product or detection of a failure in a system.
2. Study of the symptoms surrounding the defect (or the failure).
3. Reproduction of the defect (or the failure) under controlled conditions by experiment and/or simulation.
4. Hypothesis about an intermediate (or ultimate) cause by fault analysis.
5. Verification (or attempt for falsification) of this hypothesis.
6. Collection of relevant data to unify and theorize the obtained results, by induction.
7. Incorporation of this knowledge into an existing Theory or a knowledge Base.
8. Application of the extended knowledge for the design of new products and methods.

The continuously enriched knowledge Base (stage 7) can be considered as a mini "research programme", according to the meaning of the term given by Lakatos (1978), with (a) a conventionally accepted "hard core" protected from refutation by a "protective belt" of auxiliary hypotheses and (b) a positive "heuristic", i.e. a problem-solving machinery which digests anomalies, by the aid of user's intervention. Considering not only important scientific achievements but also common knowledge constructions on integrated scientific topics as "research programmes", we may distinguish, mutatis mutandis, progressing and stagnating knowledge Bases; the former can predict facts (e.g. on properties of materials, on expected results from combination of conditions,etc.) in the relative scientific domain, giving ground for an early building up of the 8th stage, while the latter can only be used for providing help for post hoc explanations either of chance discoveries or of facts anticipated by, and discovered in, a rival knowledge Base. Ceteris paribus, the latter knowledge Base can be used successfully for the realization of the 8th stage only in the absence of a rival Base; this condition is not uncommon in practice, as each knowledge Base is to some degree unique, at least temporarily. Last, but not least, it must be noticed that a knowledge Base is an a priori non-Popperian entity as it never includes "potential falsifiers" (see Popper, 1969); consequently, it cannot be considered as non valid but only as non competitive.

CASE EXAMPLE

A brief account of the stages carried out for the invention of $\gamma_{1,2}$-Al_2O_3 is following, to illustrate the fitting of the suggested route to a real situation extracted from the fourteen case histories collected initially (see previous chapter on Methodology). The branch of tree used for FTA is depicted in Fig. 1 and is obtained by extention of the main tree structure presented by Batzias and Kopsidas (1987). An inductive-deductive loop is established between stages 4 and 5; three main iterations (a, b, c) are presented below as stages but many sub-stages, including causal chains with stochastic variables, underlie beneath:

1. Observation of a characteristic defect near edges of articles made of aluminium industrially anodized to obtain thick anodic film (about 35 μm) in electrolytic bath of sulphuric acid.
2. The area in the vicinity of the defect was thoroughly examined before dyeing.
3. The defect was reproduced under similar conditions in a laboratory electrolytic cell of variable geometry. An intermediate surface zone was observed to exist between γ_1-Al_2O_3 and γ_2-Al_2O_3 named $\gamma_{1,2}$-Al_2O_3 . Stochastic models were developed for simulation.

4a. FTA suggests as possible cause "local overheating".

5a. Determination of temperature distribution verifies (actually does not falsify, after K. Popper) this Hypothesis.

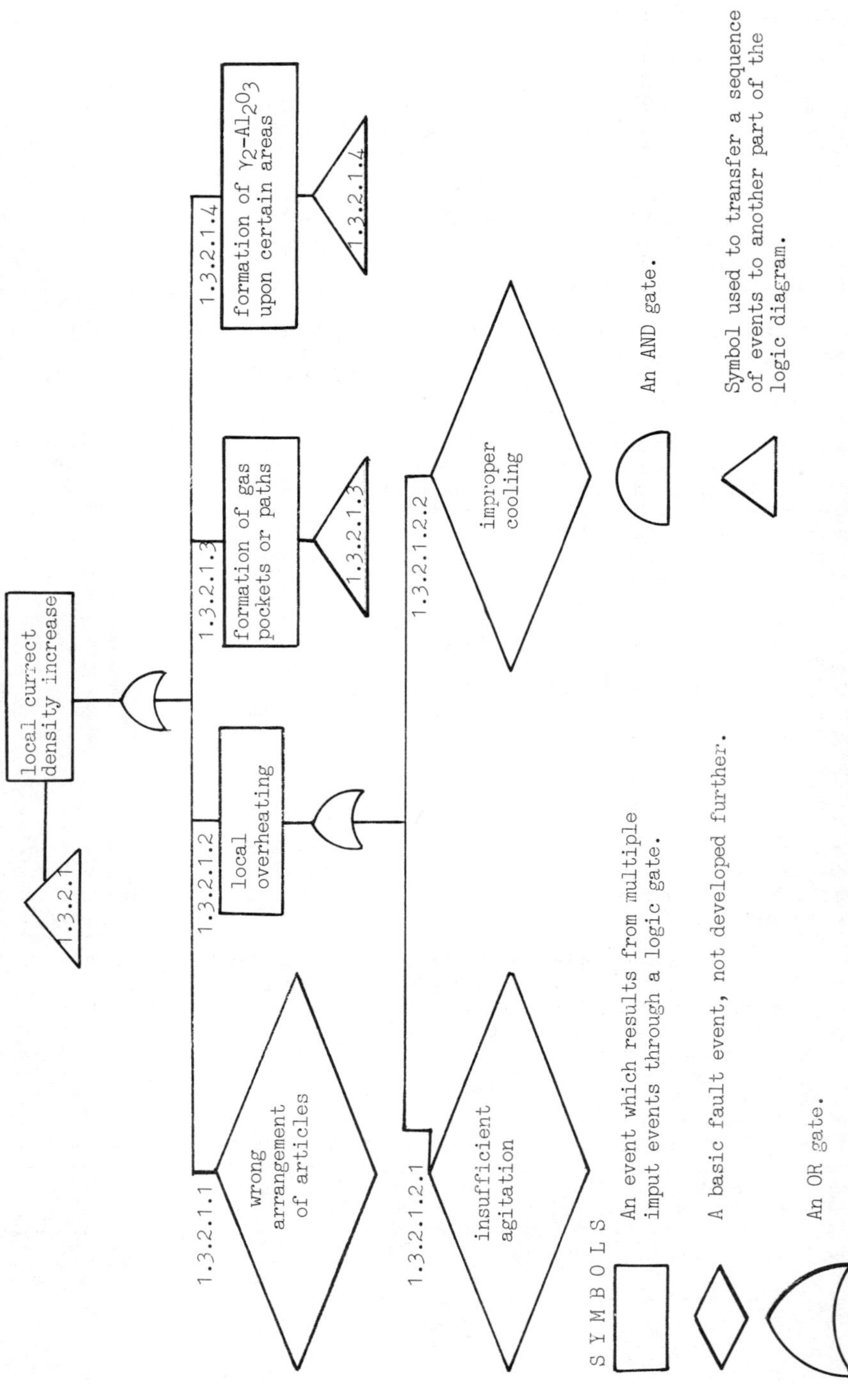

Fig. 1. A branch of the tree structure developed for FTA (stage 4).

4b. FTA suggests as further possible cause "insufficient agitation".
5b. Verification of this Hypothesis but the defect still remains though in a lesser extent.
4c. FTA suggests as an ultimate phenomenological possible cause the partial "formation of γ_2-Al_2O_3 upon certain areas" of γ_1-Al_2O_3, resulting in setting up a feedback mechanism favouring further development of γ_2-Al_2O_3 with a simultaneous surface propagation of $\gamma_{1,2}$-Al_2O_3.
5c. Verification of this Hypothesis and elimination of the defect by the aid of remedial proposals obtained from the knowledge Base of the diagnostic expert system ELECTROCHEM which has been presented recently (Batzias, 1986b, 1987).
6. Collection of relevant data and formation (by induction) of a unified Theory (actually extention of an already existing one) to include the new findings about the defect in the product and the fault in the system of electrochemical treatment.
7. Incorporation of this Theory (or Research Programme, after I. Lakatos) into the knowledge Base of the diagnostic expert system ELECTROCHEM.
8. Various implications were extracted by deduction, based on the enriched knowledge Base. Experimental verification, through an inductive-deductive loop finally led to the design of the new sorption medium $\gamma_{1,2}$-Al_2O_3, with potential catalytic properties (Skoulikidis and Batzias, 1978).

DISCUSSION

The study of the symptoms surrounding the top event (defect in the product or fault in the process) and of the effects associated with these events can be performed on different phenomenological levels. To me ion an example from electrochemical technology, research on defects/faults in the manufacture of semiconductor devices is usually carried out on at least two phenomenological levels: (a) statistical correlation between the presence of certain defects and the attendant electrical aberrations, and (b) determination of the structure of a particular defect on an atomic scale and attempt to better understand its interactions with p-n junctions; the latter phenomenological level is the deep one as it is formed by causal chain structures, aiming to answer questions like this: to what extent are the observed defects (stacking faults, dislocations decorated by metallic precipitates, etc.), which cut across p-n junctions, the cause of excessive reverse currents below the normal avalanche breakdown voltages of the junctions, emitter-collector pipes in bipolar transistors, etc.?

Evidently, between a shallow and a deep phenomenological level (or between shallow and deep knowledge, respectively) there is a continuum consisted of various levels combining shallow and deep knowledge, according to the personal attitude of the investigator and the state of knowledge in the particular mini "research programme". Therefore, the combination of results coming from studies carried out on different phenomenological levels faces some difficulties for its successful incorporation into an integrated knowledge Base.

Furthermore, setting up a Hypothesis about an intermediate (or ultimate) cause by fault analysis (stage 4) has sometimes to overcome contradictory views retrieved from the corresponding information Base. In the same example mentioned above, dislocations have been suggested as providing paths for locally enhanced diffusion and hence cause leakage between the emitter and collector of shallow diffused transistor; a contrary view point indicates that grow-in, or the so-called primary dislocations, are generally decorated with other impurities and are hence incapable of functioning as diffusion pipes during subsequent operations (for further details on this topic, see Ravi, Varker and Volk, 1973). In such cases of contradictory evidence and/or suggestions, the formulation of alternative hypotheses is preferable, so that in the next stage 5 testing of both hypotheses can be performed; thus, the most likely hypothesis is finally chosen on a comparative basis.

Last but not least, the collection of relevant data (stage 6) by automatic retrieval from the existing information Bases exhibits certain difficulties. Actually, the keywords used to describe defects or faults in the scientific/technical literature do not usually coincide with the (key)words used by industrial engineers/ technologists or professional journals/handbooks/manuals to denote the same top events. E.g., such (key)words usually met in professional readings on quality control of porcelain enameling and of other surface treatments are: alligatoring, blistering, boiling, crackledama, crawling, crinkled, curling, curtains, dimple, drain line, drying crack dry spray, fishscaling, flaw, fuzzy texture, gassing, gussy surface, glass eye, hairlining, lift, liver, lump, orange peel, peeling, pigskin, pinholing, pop-off, poppers, process fishscaling, ripple, sagging, scab, scaling, scumming, shiner, shorelines, sliding, spalling, spider, spongy enamel, spontaneous chipping, spray sagging, tearing, water mark, water spot, water streak. These characteristic (key)words are rarely used as conventional keywords in scientific/technical literature for purposes of classification although they usually appear in the corresponding texts of the scientific/technical publications. Instead, the respective terms which are used in scientific literature are usually referring to characteristics of the material structure and to the parameters of models describing properties and cause-effect relations. Moreover, the above mentioned (key)words currently used in industrial practice have not always the same denotation (e.g. scaling denotes either the process of forming scale or the spontaneous detachment of scale or both) or the same connotation (e.g. peeling refers to a surface feature sometimes considered a defect and sometimes not).[2]

The above discussed problems obstruct the communication between a multi-disciplinary large information Base and a knowledge Base dedicated to support a "research programme" or incorporated into an expert system for fault diagnosis in industrial processing. Nevertheless, the continuous enrichment of a small knowledge Base absorbing/reforming new information from large Bases can be improved by constructing a small semantic network for the transforamtion of scientific keywords to practical (key)words. This transformation can be achieved through the construction of a glossary of terms with a network resembling architecture (not a conventional indexer which transforms simple descriptions into a formal descriptor language); i.e. each term must be related to other terms lying on different semantic levels. This task presumes upon deep understanding of the linguistic peculiarities of every-day language used in industrial practice as well as upon the corresponding scientific terminology. Evidently, such a difficult task cannot be realized without a permanent human intervention, for continuous inter- and intra-facing (rearrangement of the nodes in the semantic network of the knowledge Base).

---

[2] The same semantic problem is also met in subsystem failure. E.g. when beta alumina solid electolytes are sujected to ionic charge transfer in Na/Na or in Na/S cells, degradation of the electrolytes may occur; but failure initiation definitions vary as scientists use the appearance of a specified level of nonfaradaic behaviour while battery manufacturers use the appearance of mechanical faults as a criterion for such a failure. Moreover, the connotation of the term "failure" is sometimes extended to include causal relations. E.g. failure, referring to a settling tank of an activated sludge system in wastewater biological treatment, may connote either settling failure or thickening failure.

CONCLUSIONS

In this paper, it is indicated that a typical route can be established by induction, which guides from the first observation of a defect in a (semi)finished product or of a failure in a system to the final disclosure of new knowledge, which is subsequently incorporated into an existing Theory or a knowledge Base. Several case histories with common characteristics were collected from scientific/technical literature and from the writer's research experience and Fault Tree Analysis (FTA) was used on an a priori basis, i.e. as if the final results of each case study were unknown, to decompose into formal parts the research path followed in each study. These parts were recomposed to form the route which is presented herein together with an example illustrating in brief the successful fitting of the proposed model to an actual situation (a fault observed in industrially anodized aluminium, which led to the design of a new sorptive medium with potential catalytic properties).

The continuously enriched knowledge Base is suggested that can be considered as a mini "research programme", after Lakatos; some methodological/semantic problems concerning its enrichment from, and communication with, larger information bases are discussed. The construction of a small semantic network as an interfacing mechanism intervening between multi-disciplinary large information bases and a small knowledge Base, dedicated to support a "research programme" or incorporated into an expert system, is finally proposed as a permanent solution to these problems.

REFERENCES

Batzias, F. A. (1986a). Intermediate Substances and Stages of Research and Development in Chemical Technology. Proc. IVème Colloque "Les Corps Intermédiaires", Athènes (1983). École des Hautes Études Industrielles du Pirée (Ed.). pp. 347-368.

Batzias, F. A. (1986b). An Expert System Design for Fault Diagnosis in Electrochemical Operations: A Qualitative Approach. In S. Tzafestas, M. Singh, and M. Schmidt (Eds.), System Fault Diagnostics, Reliability and Related Knowledge-Based Approaches, vol. 2. D. Reidel Publ. Co., Dordrecht, Holland. pp. 519-535.

Batzias, F. A. (1987). An Expert System for the Control of Industrial Dyeing of Anodized Aluminium. Proc. 1st Symposium on Paints. Nat. Tech. Univ. Athens (Ed.). pp.162-172.

Batzias, F. A. and G. Kopsidas (1987). An Expert System Design for Fault Diagnosis in Electrochemical Operations - A Quantitative Approach. Proc. 2nd European Workshop on Fault Diagnostics Reliability and Related Knowledge-Based Approaches. Pergamon, Oxford, UK (forthcoming).

Conant, J. B.(1964). Two Modes of Thought. Trident Press, New York.

Jewkes, J., D. Sawers, and R. Stillerman (1969). The Sources of Invention. Macmillan Co. Ltd., London. p. 280, 335, 53.

Lakatos, I. (1978). The Methodology of Scientific Research Programmes. J. Worrall and G. Currie (Eds.). Cambridge Univ. Press, UK.

Popper, K. R. (1969). The Logic of Scientific Discovery. Harper Torchbooks, Harper & Row, Publ., New York. pp. 78-92.

Ravi, K. V., C. J. Varker, and C. E. Volk (1973). J. Electrochem. Soc., 120, 533-539.

Skoulikidis, Th., and F. A. Batzias (1978). Structure Secondaire des Alumines Actives, Préparées Électrolytiquement. Proc. 4th Intern. Congress for the Study of Bauxites, Alumina and Aluminium (ICSOBA). Nat. Tech. Univ. Athens (Ed.). pp. 338-355.

Application of AI Techniques in CAR Environments

Paolo Camurati, Marco Mezzalama, Paolo Prinetto

Politecnico di Torino
Dipartimento di Automatica e Informatica
Corso Duca degli Abruzzi 24
I-10129 Torino TO, Italy

ABSTRACT

This paper presents the introduction of Artificial Intelligence techniques into the field of Computer-Aided Testing and Repairing. Such techniques are useful since it is increasingly important to improve ATE diagnostic capabilities and AI seems a suitable response. The overall goal is to set up Test/Repair environments which guarantee not only high diagnostic capability, but also high productivity. The system presented in this paper is based on a loop Tester --> Repair Station --> Tester. In a conventional approach, data coming from a Repair Station are not used to improve the Tester's performances. In the solution here proposed, data are first validated by the Tester and successively processed in order to gather new knowledge about real cases. Such knowledge is learned by the Tester to improve its symptom interpretative capabilities. Two methodological aspects are relevant in this approach: the knowledge the Tester receives from the Repair Station and the knowledge the Tester itself needs to learn from experience. It is consequently necessary to organize knowledge in suitable knowledge bases and to use it with a proper reasoning mechanism. AI techniques are used to achieve both goals: from one hand they allow us to formalize the problem in a rigorous way. From the other hand they facilitate the generation of an open system, in which the introduction of new diagnosis rules is easy and where a real feedback from the industrial testing environment is possible.

1 INTRODUCTION

The environment of testing, diagnosis, and repair plays a key role in the optimization of electronic board production. Within this environment the choice of the testing strategy strongly depends on factors such as production volumes, used devices, and available financial investments (Brendan, 1982), (Scheiber, 1985), (Scheiber, 1986), (Bennetts, 1982). Unfortunately, the advent of Very Large Scale Integration (VLSI) components and the introduction of new technologies for device mounting, e.g., SMT (Surface Mounted Technology) are making testing problems harder and harder. The same considerations apply for the diagnosis and repair phases. All these causes force the continuous growth of testing costs for devices, boards, and systems.

In realistic models of typical environments test and repair processes are represented by a queue based model. Boards enter the test and repair center at a given flow and join a queue of boards waiting for test. If a board passes a test, it moves to the output queue and leaves the center. Otherwise it moves to a conveyer that transfers it to a repair station. After repair, the board moves back to test, where it joins the other boards waiting for test. It is worth noticing that a failing board will cycle between test and repair until either it finally passes the test or it is scrapped. Different methods exists for estimating center size, throughputs, cycle time, number of test and repair stations, and queue lengths. The majority of them are based on stocastic process theory and are aimed at optimizing the production process (Howland, 1983). In addition, faults are often repetitive and a skilled operator is able to isolate the fault on the base of previous experiences. Generally speaking, diagnosis and repair problems are extremely difficult and time consuming, but they often are effectively solved by highly skilled people. These problems seem to be an ideal application domain for Artificial Intelligence solutions (Robinson, 1984), (Mullis, 1984), (Wilkinson, 1984).

Knowledge based systems implement today sophisticated reasoning schemes and provide empirical rules introduction facilities. Furthermore, they usually allow to modify these rules, updating the knowledge source. The domain of test generation and mainly the diagnosis and repair phases, seems to be an interesting area for the application of knowledge based systems. In this paper we present an attempt to apply conventional AI concepts to improve productivity in the test/repair loop.

2 THE REPAIR ENVIRONMENT

During the last decade a significant evolution in repair environments occurred, which increased rework efficiency and effectiveness and has reduced repair time. Essentially we moved from manual diagnostic/repair functions to computer based repair stations. Two peculiar facts determined this innovation: the use of graphics at the repair station and networking.

Traditionally, in the manual approach, the tester's operator attaches a tag, containing the fault list, to the PCBs that do not pass the test. In case of In-circuit testers, the fault ticket identifies the shorted traces and/or the failed components. With Functional testers using guided probe diagnostics, the failure message identifies a node (or a set of nodes) which will be assumed as a starting point to trace back. When a guided diagnostic is not applied, the message indicates one (or more) component(s), that has (have) been hypothesized to be failed by means of a diagnostic procedure which uses either a fault dictionary or similar information items. At the repair station the rework operator reads the tag and tries to identify the faulty component(s) and/or the short(s), with the aid of board documentation. When he/she has determined the fault location, he/she replaces the component(s) and/or eliminates the short(s). Failure messages are often ambiguous and it is the fundamental job of the operator to interpret them. The ability to identify and remove the faults on PCBs is essentially a function of "resolution and accuracy of the message, operator skill knowledge, and experience with the specific PCB" (Bateson, 1985).

Rework efficiency is usually expressed as a percentage of rework effectiveness. As an example, a 90% rework effectiveness, means that given 100 PCBs identified as faulty, after repair actions, 10 of them will still result failed when re-tested. Typical values of rework effectiveness are 90% for In-circuit testers and 85% for Functional ones. Strategies to improve rework effectiveness include Computer Aided Repair (CAR) stations, with graphic capabilities and, more recently, the adoption of AI techniques.

CAR stations consist of computer systems, normally connected to testers via point-to-point connections or LANs (Local Area Networks), with limited mass storage, and a color graphic display monitor. This, among other functions, provides a simpler way to identify fault locations on PCBs. For instance, when the failure message refers to a component, such component is outlined on the screen using special colors. In addition, at the bottom of the screen, a message usually appears to provide detailed information concerning the faulty component. The short term history (if any) of the board is also displayed.

Even if the experience has shown the effectiveness of the graphic support, we decided to move to AI, in order to investigate the possibility of transferring the knowledge from skilled operators to computers. It has in fact been noted that the rework effectiveness increases with the specific knowledge the operator has of a given board. After a sufficient amount of time, when repairing the same type of PCB, an operator is able to deduce and to learn the rule to interpret the message, avoiding ambiguities. The use of AI techniques is not limited to diagnosis: it can be effectively used in improving production planning and real time production monitoring, too. The latter possibility is of great importance in the loop repair environment, since it permits flexible production scheduling, together with the possibility of detecting any distortion, or error, in both the production line and in the test system.

The necessary condition to be able to provide this kind of facilities is to work in a networked environment. The system we have designed can be indifferently implemented by two distinct networking structures, both based on LANs. The first architecture implements a hierarchical structure. A minicomputer (host) works as data concentrator and job dispatcher. ATEs can be connected to the host via both a high speed data link and RS232 low speed connections. CAR stations use another high speed data link to communicate with the supervisor. Both the ATE LAN and the CAR LAN can work independently, thus guaranteeing a higher reliability to the system. The management functions of the test and repair area are concentrated in the host, which provides automatic logging of test data and repair activity, real time monitoring of ATEs and CAR stations and, more generally, system statistics. It also provides automatic downloading of test and repair programs. The second architecture is based on a traditional LAN, whose nodes are both ATEs and CAR stations. One of the computers connected to the LAN can also operate as a file server and performs some monitoring functions.

3 KNOWLEDGE BASES

As previously seen, two main problems must be considered in the management of the test and repair environment. The former is mainly related to the need for continuously monitoring the overall process, in order to detect and correct systematic errors at their very early occurrence. The latter comes out from the necessity of improving the repair phase, by trying to overcome the intrinsic symptom ambiguity of ATEs. Both the above mentioned problems require "intelligent" decisions, based on a powerful handling of both the "experiences" acquired during the process life and on the knowledge of the characteristics of the overall environment.

These considerations naturally lead us to improve the classical test and repair process the addition of a knowledge based system (Fig. 1).

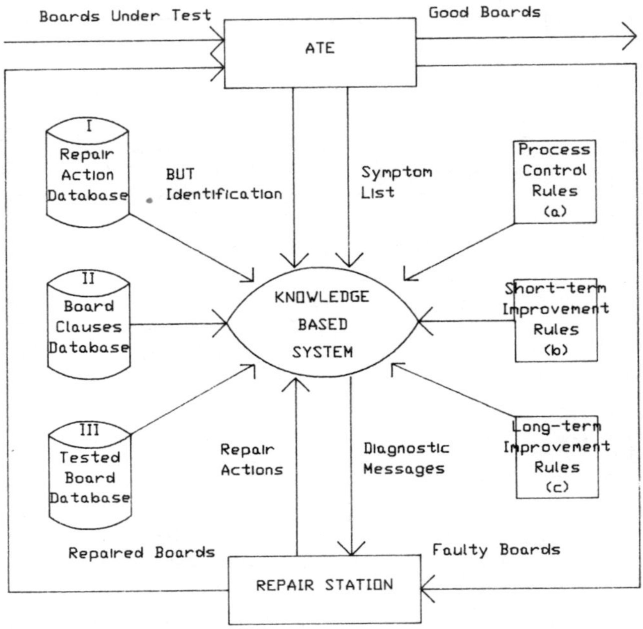

Fig. 1. Knowledge-Based System Organization

This system, taking into account both the structure of the board under test and the acquired experience (constituted by the list (if any) of the already performed repair actions and the history (if any) of the current board), provides the repair station with the list of repair actions to be taken. More in details it utilizes three data bases, and namely:

- repair actions data base:
  it contains information related to the effects of repair actions that have been (successfully or unsuccessfully) taken, in the past, in consequence of certain symptoms for each specific class of boards. This could be referred to as "long term experience" and it is extremely useful to improve the diagnostic capabilities of the system.

- board classes data base:
  it contains information related to the organization and to the structure of each class of boards that has to be tested.

- tested boards data base:
  it contains, for each tested board, a trace of the cycling of the board itself through the test and repair loop.

Data contained in the last data base could be thought of as the "short term experience" and are used both to watch for systematic errors in the overall process and to avoid endless loops of a given board in the test/repair environment. In practice, such a control is performed by the knowledge based system through a set of rules provided by the system manager. These are stored in the "process control rules" file. Rules to control the test and repair loop of each board and to detect systematic errors both in ATEs and in production lines are provided. An example of rule is:

```
if  ( this is the 3rd time that this board passes
      through the repair station )
then scratch it.
```

In addition to the above mentioned entities, the knowledge based system receives three other kinds of information, and namely:

- from the ATE, an item univocally identifying the board under test. Through it the knowledge based system can access the three data bases examined above.
- from the ATE, the symptom list.
- the list of repair actions performed at the repair station.

3.1 General Approach To Diagnostic Capabilities Improvement

As previously said, the diagnostic messages generated by the Testers are analyzed on the basis of the past history (learning process) in order to improve ATE's diagnostic capabilities. Two different types of learning are applied:

- short term learning
- long term learning.

Short term learning concerns data acquisition and the application of rules to each failed board, depending on the actions previously made on the single board, only. Long term learning involves the application of rules based on the experience acquired on the whole set of failed boards for a given board type.

Each learning phase is accomplished resorting to two information sources, and namely:

1. Data acquired from the repair station, concerning either the single failed board (in the case of short term learning) or the whole set of boards belonging to the same family (in the case of long term learning).

2. Rules provided to improve the diagnostic capabilities((b) and (c) in Fig. 1).

The logic flow of operations performed on each board is partially dependent on the kind of tester faulty boards come from. In fact, although the global organization remains the same, we must take into account the different logical and physical formats of diagnostic messages, which slightly change from tester to tester, the line protocol adopted in the communication with the tester, and the rules which are peculiar of each kind of tester.

Since the short term learning is almost the same for all kind of testers, we shall here focus on long term one, only. For each board, the tester produces a set P of symptoms corresponding to the diagnostic codes of those test patterns which provide wrong results. These symptoms are interpreted by the tester itself , in order to generate the set C of "faulty components". This second step corresponds to the application of a mapping function f(.), such that:

$$C = f(P)$$

The function f(.) strongly depends on the kind of ATE and the expert system must modify it on the basis of the acquired experience, i.e., on the basis of the results of the repair actions. To do this, the following entities must be known for each board:

- the set P,
- the set C,
- the set R, containing the result of the repair action on a single board.

the knowledge of the set P is essential, since the function f(.) is not biunivocal, in the sense that the same set C of faulty components may be obtained starting from different sets P of wrong test patterns.

The expert system is mainly composed of two blocks:

- a "filter", which, for each board, accepts in input C and P and provides a new set C* of faulty components;
- a module for the generation and the updating of the rules and/or of the function f(.), on the basis of the acquired experience.

Within the architecture of the global system, the implementation of the above mentioned blocks strongly depends on the kind of ATE, being related to the complexity of f(.) and of the module for the generation and updating of the rules. In the real world implementation of the system, two different solutions have been adopted, in dependence of the actual ATE type and, more precisely, on whether the evaluation rules of the ATE are accessible and modifiable ("dynamic tester") or not ("fixed tester").

3.2 Implementation Within An In-circuit Tester Environment

Since the repair operator must be provided with a list of diagnostic messages, i.e., a list of faulty components to be replaced, and the ATE provides a list of symptoms, a translator from symptoms to faulty devices must be adopted. It is here extremely useful to introduce a "Symptom Probability Matrix" (SPM), characterized by as many columns as the board devices and as many rows as the possible symptoms. SPM (j, k) is the probability that, having received the symptom Sj, the device Dk be faulty.

If we assume that the i-th test module Mi tests exhaustively and univocally the i-th device, the SPM matrix becomes a diagonal matrix. Seen from another point of view, this means that each test module is assumed to have no intersection with the other ones. Unfortunately, this is not the real case, since many factors cooperate to invalidate such assumption. Among the others, we simply remind the possibility that a given physical fault, such as a stuck-at on a wire of a bus, can induce malfunctioning in all the devices connected to the bus itself. These malfunctionings can cause symptoms to be generated for each of them. Obviously their replacement with new devices will not solve the problem, since the fault is elsewhere. Another important factor is the unreliability of the test module Mi: written to test the device Di, it actually tests it only partially, and, in addition, it may test another device Dj.

From a practical point of view, these considerations lead to the conclusion that the matrix SPM is neither a diagonal one, nor it contains probabilities all equal to one. In fact, a symptom Si can identify a fault in the device Di with a given probability P(Si,Di), a fault in the device Dj with a given probability P(Si,Dj) and so on, in such a way that:

$$\sum_{k=1}^{M} P(Si,Dk) = 1$$

This modification of the SPM must be carried out on the basis of the acquired experience. In fact, on the basis of the results obtained by the repair actions performed when, in the past, the same set of symptoms had been provided by the ATE for boards of the same class of the current one, it must be possible to select the new repair actions that have the highest probability of effectiveness. It is then possible, as a consequence, to modify the contents of the SPM matrix.

In the case of fixed ATEs, such as In-circuit ones, since it is not possible to introduce the SPM matrix within the tester, it is necessary to "filter" the symptom list provided by the ATE itself into the proper diagnostic message sent to the repair station. We have thus adopted the solution of a "real time filter". It operates on the basis of both the contents of the SPM matrix and of a set of filtering rules, stored in an ad hoc "filtering rule file". These rules include the strategy to properly "filter" the global symptom set provided by the ATE.

The knowledge based system takes care of two main tasks: the former is related with the dynamic modification and updating of SPM matrix, on the basis of acquired experience. The latter is concerned with offering the user the possibility of updating the set of rules contained in the "filtering rule file". A consistency check is performed anytime the file is modified.

3.3 Implementation Within A Functional Tester Environment

In the case of Functional testers, the function f(.) is more complex. In fact, the generation of the set C of faulty components is performed starting from the symptom dictionary generated through simulation, on the basis of the results P of wrong test patterns, and resorting to pattern matching functions, based on maximum similarity criteria. As a consequence, it is much better to use the knowledge based system to directly modify the symptom evaluation rules used by the symptom evaluator of the ATE. In this situation ("intelligent ATE") the "filter" does not work in "real time" on the symptoms generated by the ATE. Instead, it stores data at the end of each repair process, interprets repair results and, "off line", it modifies the rules to produce diagnostic messages.

The process of symptoms generation may thus be considered as a pattern matching process, aimed at finding the best match between the symptom list of a given board and the "patterns" stored in memory and created using the fault dictionary.

4 CONCLUSIONS

The methodologies presented in this paper have been applied in the management of the error messages generated from both WPS (Wiring Pattern Verifier for bare printed circuits) and ICT (In-Circuit Testers, for complete boards) equipments. In order to improve the productivity of the overall loop Test -> Repair -> After Repairing Test, "intelligent" computer-based repair stations have been implemented.

The adopted solution has proved to be valid, since valuable benefits have been gained in many areas, and specifically:

- Improvement of the diagnostic capabilities of the testers, on the basis of the global experience progressively acquired through all the technicians engaged with repairing.
- Possibility of transferring the acquired experience to the technicians just started in repair activities, without requiring long training periods.
- Lowering the mean number of recycles caused by failures in the repair actions.
- Consequent lowering of the mean repair time.
- Introduction of a real time feedback, used to improve the quality of both the global production process and specific test programs.

The system has been implemented resorting to three different programming languages, and specifically PASCAL, Assembler and BASIC (for graphics). Future trends include the introduction of guided fault isolation. Similar systems are the MIND system, developed by Teradyne and devoted to system testing (Wilkinson, 1984) and the TDMS (Test Data Management System) developed by GenRad (Williams, 1983).

5 REFERENCES

Brendan, D. (1982). "The economics of Automatic Testing", McGraw Hill.

Scheiber, S.F. (1985). "Economics of Testing: Deciding whether to Test and How Much It Will Cost", Test & Measurement World, February 1985, pp. 110-130.

Scheiber, S.F. (1986). "Cost effective PCB testing", Test & Measurement World, February 1986, pp. 31-40.

Howland, J.C., and Harding, P.T. (1983). "Estimating the required size of an automated test and repair system from subassembly volume and failure information", IEEE International Test Conference 1983, pp. 212-219.

Bennetts, R.G. (1982). "Introduction to digital board testing", Crame Russak.

Robinson, G.D. (1984). "Artificial Intelligence and Testing", IEEE International Test Conference 1984, pp. 31-40.

Mullis, R. (1984). "An Expert System for VLSI Tester Diagnostics", IEEE International Test Conference 1984, pp. 196-199.

Wilkinson, A.J. (1984). "A method for test system diagnostics based on the principles of artificial intelligence", IEEE International Test Conference 1984, pp. 188-195.

Bateson, J. (1985). "In-circuit testing", Van Norstrand Reinhold.

Williams, R.W. (1983). "Considerations While Introducing a Test Data Management System to the Factory Floor", IEEE International Test Conference 1983, pp. 220-225.

Simulating Operator Behaviour in Process Control by a Goal Oriented Approach and Fuzzy Logic Formalism

U. Bersini and P.C. Cacciabue

Commission of the European Communities,
Joint Research Centre Ispra 21020 Ispra (Va), Italy

ABSTRACT

In this paper, a model of plant operator behaviour is proposed as a contributing component to a larger Man-Machine study devoted to the analysis of accident. The model considers two levels of cognitive processes: a High Level Decision Making, whereby the reasoning about the plant as a whole takes place and a Low Level Decision Making, in which the actual actions of control are carried over. Accounts is also given for the mechanisms of error making and recovery.
Two basic assumptions are made:
1) the operator behaviour develops within a goal-oriented hierarchical structure; and
2) fuzzy-logic is the formalism to handle imprecise and uncertain understanding of the reality.
An example of management of contradictory information and error making during planning is presented as the sample case of the model.

KEYWORDS

Man-machine interaction; process control; human behaviour modelling; fuzzy-set theory; goal-oriented structures; knowledge-base systems.

INTRODUCTION

The model of a process plant operator behaviour, proposed and described in this paper, is a contributing component to a larger Man-Machine study project devoted to the analysis of accident, where a balanced simulation between man and machine activities is studied.
Modelling human behaviour in accidental conditions implies primarily the simulation of the primitive cognitive processes performed by human beings, accounting for the environmental constraints in which they are activated. Many existing technics focus separately on models of detection, planning or diagnosis with adequately different formalisms (Rouse, 1983; Sheridan, 1985). In the present model, the main tendency will be towards an integrated simulation allowing to tackle all the activities of the operator in the same framework (Cacciabue and Bersini, 1987).
Intentionality is fundamental in cognitive processes, then a hierarchical goal-oriented structure is proposed for the development as well as the execution of operator's strategies. Moreover the basic formalism adopted for handling human decision making is based on fuzzy logic which is a well suited theory for representing the approximate operator's knowledge and allows to create a semantic interface between the system simulation and the operator cognitive model. While the adequacy of fuzzy-set theory to represent imprecision seems to be generally accepted, the use of

multi-values logics, other than probability, for treating uncertainty, is still the object of debate (Garbolino, 1986; Dubois and Prade, 1982). In a pure descriptive approach, without considering, for instance, the problem of efficiency in normative decision theories, but regarding these specific theories as possible measures or modelling technics, the important advantage of possibility or plausibility on probability theory lies on their relaxed axiomatic, i.e. the monotonicity axiom. A first consequence of this is the existence of two dual measures in order to bound a real certainty measure. This interval allows a larger flexibility of modelling.
In the proposed model two cognitive levels are considered:

- the first one, High Level Decision Making (HLDM), allows the simulation to exploit operator's knowledge by continuously recognising situations and by building supervisory and control strategies. It represents pure mind work with no direct interaction with the actual control system;
- the second one, Low Level Decision Making (LLDM), is supported by the working and conscious memory dynamics, when the operator implements a preprogramed response or a planned strategy in order to satisfy a clearly defined intention.

The overall architecture of this model accounts also for the connection of these two levels with the mechanisms of error detection and recovery.

In the following the technics employed in order to formalise the model will be discussed and an example of managing conflictual information and of error making during the planning will be given.

HIGH LEVEL DECISION MAKING MODEL

A crucial aspect for the model formalism is the consideration of an overall and integrated human behaviour without a clear-cut separation between planning and diagnosis. Indeed the diagnosis is rarely a punctual process but, on the contrary, while the incident develops there is a continuous need for situation assessment (Woods, 1986). This entails diagnosing and planning in parallel: the operator, following a superficial and very imprecise first understanding of the situation can consider some initial possible recovery strategies, undertake urgent actions and only during the subsequent phases of the transient he reviews and trys to confirm the selected plan.
The planning is envisaged as the identification of intentions and the selection of actions or procedures to be carried over on the plant itself. Starting with a main intention or "top-goal", the result of this process is the construction of a hierarchical network where the different sub-goals and sub-tasks to be performed by the operator are schematically arranged in a "tree" type structure and are linked to each other by different gates such as "AND" and "OR". A "goal" is an element of the network at any level. An "act" is the last elementary goal of the network, i.e. the elementary action that the operator will perform as, for instance, "close a valve", "stop a pump", etc... A "task" is a set of acts that the operator has to perform in order to attain a certain goal at any level.
This high level planning process lies on two different types of frames, system and goal-frames, belonging to two interconnected knowledge-bases: the *Intentional Knowledge-Base* and the *System Knowledge-Base*. Moreover, as in classical expert system design, an inferential engine has been constructed in order to link the goals in the tree structure.

The Intentional Knowledge-Base

This knowledge-base contains all the goals that the operator has the possibility to attain. The formalization of the network is obtained by "a priori" analysis of the goals and tasks to be performed for the plant management, rather like it is done when developing a task analysis procedure. Each goal is characterized by a certain number of attributes and parameters.
The attributes define the boundary of a goal. They are: pre-conditions, continuous-requirements, expectancy, side-effects, sub-goals and super-goals.
The parameters regulate the unfolding of execution. They are: priority, membership, certainty

and satisfaction.
Depending on the initial assignment of the attributes, a goal can be :

- pre-compiled and specific : then its sub-goals are precisely defined "a-priori";
- pre-compiled and generic : the goal has a generic structure that has to be characterised by the system to which it refers. The attributes will be generated by exploring the system knowledge-base.
- not pre-compiled : the goal is isolated in the intentional knowledge base, without actual or potential links. The inferential engine must deduce the required sub-goals depending on planning attributes.

The attributes considered in the planning process are :

- pre-conditions : describing the necessary initial conditions for the goal attainment;
- continuous requirements: describing the necessary and continuous conditions that have to be present during the whole process of attaining the goal;
- expectancy : containing what should be the desired and expected effect of this goal;
- side-effects : describing the undesired but unavoidable consequences of the goal.

The physical variables behaviour are represented by linguistic terms, i.e. words of natural language such as "high", "low", etc.., which are interpreted as fuzzy restrictions in the form of appropriate membership functions. Each membership function associates each linguistic term with a value in the interval [0,1], representing the compatibility with the fuzzy restriction. A trapezoidal form for the membership function is currently used for the evaluation of any linguistic expression. For the system variables, two kind of possible description are allowed: a static one, as "high" or "low", and a dynamic one, as "increase " or "remain constant".

The executional parameters which regulate the navigation through the network are :

- the degree of priority (**GDP**), which expresses the measure of hierarchical sequence between goals;
- the degree of membership (**GDM**), which defines the measure of the dependency between a goal and its directly superior goals;
- the degree of satisfaction (**GDS**), which represents the correlation between the result of a goal and the operator's expectancy; and finally
- the degree of certainty (**GDC**), which represents the measure of the attainment of a goal.

The System Knowledge-Base

The system knowledge base contains the structural representation of the plant. This knowledge base could be considered as a kind of operator's internal representation of the system. When a goal is only generically specified, it is associated with a particular routine or demon, which interacts with the system knowledge-base in order to fill some of the goal attributes.
This knowledge base has been built with the KEE software (Fikes and Kelher, 1985) in an optics similar to the EPRIKEE development. Three large units have been implemented : the components, the line-elemenents and the indicators.
Each of these units contains some specific attributes. A component is characterized by its position, its state (close/open), its availability and the associated indicators. A line-element is characterized by its input and output, the components that it contains and the associated indicators. The hierarchical formalism and the associated inheritance technics of KEE have been substantially used.

The Goal-Tree Automatic Construction

By means of the inference engine, starting from a Top Goal, the tree is constructed by a recurrent exploration of the two knowledge bases. For each goal, the attributes are beeing recursively examined in order to define the related sub-goals and the type of link. The problem of

matching between attributes is solved by means of a fuzzy pattern matching mechanism (Dubois and others, 1986). The recurrent process terminates when a level composed of all elemetary goals has been reached. Then the problem of interaction among goals is treated assigning the parameters **GDP** and **GDM**.
If all goals recurrently considered in the automatic construction are pre-compiled and specific, the tree is automatically preconstructed and no real planning process is activated. In this case, it can be considered that the operator possesses an adequate, unique and precise executional frame.

LOW LEVEL DECISION MAKING MODEL

The tree of goals selected by the HLDM process is actually carried over within the LLDM process which interacts with the simulation of the plant and its interfaces.
The navigation through the network is regulated by the values assumed by the parameters of the goals. In particular :

- travelling, in a top-down way, through the network is governed by **GDP** which allows the model to select the acts to be executed. At any level the operator attends to the goal of highest priority.
- the attainment of a goal is measured by the **GDC** of the goal. This is related to the **GDS** of the goal itself and to the information gathered from directly dependent sub-goals, in terms of their **GDCs** and **GDMs**. If the calculated value of **GDC** of a goal is greater than a pre-established threshold, then the goal is considered as attained and the simulation proceeds to tackle the other goals in order of priority; otherwise the goal is not attained and an error is detected. In this case, as well as when a new disturbing event occurs,
- the recovery process is activated which implies two possible ways of action :
    1. the first one consists in exploring the tree under the same Top Goal in order to find possible alternatives to the current unattained goal; this corresponds to look for the presence of an "OR" gate in the goal structure;
    2. the second one implies the return to HLDM in order either to review the original strategy, or to create some new goals in order to front the novel situation.

Fuzzy Feedback Mechanism

The use of the fuzzy logic as well as possibility theory for representing the operator imprecise and uncertain understanding of the reality must be accounted in a descriptive perspective.
While **GDP** and **GDM** have been assigned by the HLDM model, **GDS** and **GDC** are evaluated during the dynamic evolution of the man-machine interaction by applying fuzzy relations to the perceived and the expected results of a goal.
GDS is the result of matching the goal expectancy and the real behaviour of the related indicators, expressed by means of a trapezoidal membership function:

$$\mathbf{GDS(goal) = f_{trapez}(a,b,c,d)} \qquad (1)$$

GDC of a goal is evaluated in terms of **GDS** of the goal itself and **GDC** and **GDM** of the connected sub-goals.
For example if the operator is not "satisfied" of a certain goal attainment then **GDS = 0**; however the degree of certainty of this goal, i.e the measure of the attainment of the goal, will take into consideration also the fact that, to reach the goal, the operator had to perform some other sub-goals which could have been fully attained, and thus he has to balance between these two contradictory information. This is taken into consideration by two weighing factors (x for **GDS** and y for the sub-goals **GDC** and **GDM**) and by using the fuzzy logic dual concepts of "necessity" (**N**) and "possibility" (Π).

Under these assumptions, the expression of **GDC** of a goal is:

$$\Pi = \max[\min(x, GDS(goal)), \min(y, GDC^*(sub-goals))] \quad (2)$$

$$N = \min[\max(1-x, GDS(goal)), \max(1-y, GDC^*(sub-goals))] \quad (3)$$

$$N \leq GDC(goal) \leq \Pi \quad (4)$$

where :

$$GDC^*(sub-goals) = \max_{j=1,k} \{ \min[GDC(sub-goal_j), GDM(sub-goal_j)] \} \quad (5)$$

in case of an "or" gate connecting the goal with its sub-goals; or :

$$GDC^*(sub-goals) = \min_{j=1,k} \{ \max[GDC(sub-goal_j), 1-GDM(sub-goal_j)] \} \quad (6)$$

in case of an "and" gate connecting the goal with its sub-goals.

By this aproach it is possible to model various degress of confidence experienced by the operator during the management of the accidental sequence.

Resolving Conflictual Information

Possibility theory allows to take naturally into account and to model a problem of lack of understanding and of decision uncapabilities in cognitive processes.
Having executed a certain action, the operator activates his fuzzy feedback mechanism in order to verify some related goal attainment.
Given eq. (4) and assuming a threshold value **D** for the **GDC** parameter, three different situations can occur :

-1- $N > D$: the operator is then satisfied and carries on with the execution of the tree;

-2- $\Pi < D$: the operator detects a problem in the tree execution;

-3- $N \leq D \leq \Pi$: this situation is symptomatic of conflictual information; for instance, the operator is certain of his previous actions (**GDC*** = **1**) but doesn't verify the expectancy of the current goal (**GDS** = **0**).

When condition -3- is encountered, the operator is caught in a dead-end condition and no further progression in the goal structure is made. Indeed the operator waits for some new information to pop up in order to help him to decide whether to neglect one of the two previous information (x or y = 0), or to give more credit to the possibility value **Π** (accounting for the redundance of the information and keeping the most satisfactory one) or the necessity value **N** (accounting for the coherence of information). This stalemate remains until condition -1- or -2- is reached.

ERROR MECHANISM

The almost generally accepted subdivision of this very special category of human undertakings considers two types of errors : the mistakes and the slips, according to whether the error has been made during planning or alternatively during the execution phase. In the conceptual framework of our human model, this distinction has been kept by accounting for the possibility that errors are made during the HLDM as well as the LLDM processes.
The mechanism of generation and the forms of errors are accounted for in the model being developed.
The *error generation mechanism* is exploited by a further interaction of HLDM and LLDM with a model of the psychological primitives which describe the operator decision making. The Lens model (Rasmussen,1986) or the GEMS approach (Reason,1986) are being studied for this purpose.
The *forms of errors* are already directly linked to the goal oriented model and to the mechanism

of navigation through the network. Indeed, as example, at the high level, a wrong structural knowledge representation of the plant would lead to the formulation of an erroneous planning of actions. Similarly, at the low level, errors can take the form of an inversion of priority or of a wrong expectation or of an omission of action etc.

A SAMPLE CASE OF OPERATOR-SYSTEM INTERACTION

In order to illustrate what has been described in the previous chapters, a sample case is presented in which the operator deals with the relatively simple system described in the Fig.1.

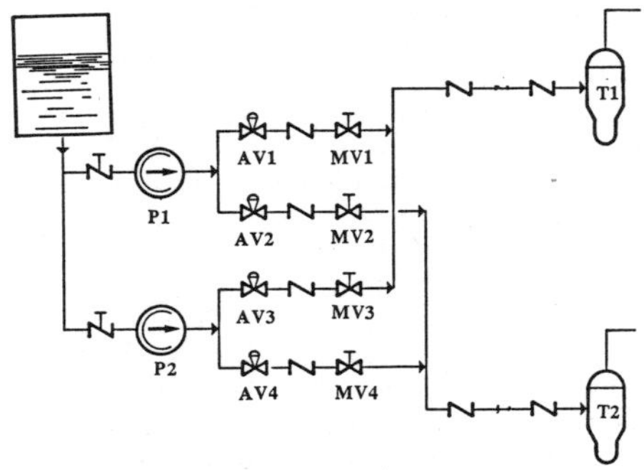

Fig. 1. Reference feedwater system

The actual goal of the operator is to fill the tank T2 only, and not tank T1.
During the HLDM process, the *System Knowledge-Base* supports the structural representation of the system and the *Intentional Knowldege-Base* contains generic goals such as : "start pump", "stop pump", "open automatic valve", "open manual valve", "close automatic valve", "close manual valve", "isolate tank", "regulate flow in tank"... The Top Goal activation, i.e "fill tank T2 only", demands some *continuous requirements* : "T1 isolated" and "pumps running" (Fig. 2).

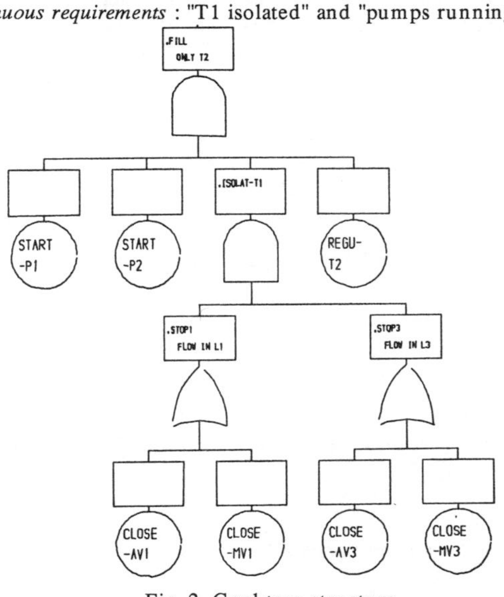

Fig. 2. Goal tree structure

The priority distribution takes into account the interaction between the *pre-conditions* and the *expectancy* of the goals which result in the construction of the final goal tree.
This has to be executed during the LLDM process. Two interesting cases can be examined: the first one concerns the managing of contradictory information and the second one the behaviour following a mistake in planning.
In the first case, it is assumed that: a) the automatic valve 3 (AV3) is stuck open, b) its indicator does not represent the real behaviour of the component but it follows the actuator positions and c) all the other components and indicators work correctly.
Following the goal tree, the operator starts the pumps, closes AV1, in order to isolate T1, and checks his actions. Since the valve indicator shows "valve closed", according to Eq. (1), one has: **GDS (AV1) = 1**. Then he closes AV3 and checks it: **GDS (AV3) = 1**, because the indicator shows success of action even if, in reality, the valve is stuck open. The operator verifies the achievement of his isolation task, and observes that the inlet flow indicator, contrary to his expectancy, carries on indicating a certain value. This implies that: **GDS (isolate-T1) = 0**.
According to the fuzzy formalism previously described and assuming that the two contradictory information, i.e. valves being closed but still presence of flow in the line, are weighed 1 and 0.7 respectively (x= 0.7 and y= 1.), the certainty of the goal attainment can be evaluated: from Eq. (6), **GDC*(valves-AV1-AV3) = 1** and from Eqs. (2) and (3), with **GDS (isolate-T1) = 0**, it results that **Π = 1.** and of **N = 0.3**. The **GDC** parameter is thus comprised between 0.3 and 1, while the treshold value **D** is assumed 0.7. The operator is now confronted with a contradictory information and the way to resolve this conflict is to wait for some new relevant information to come up. Indeed from the rising T1 level indicator, the operator will receive a new information coherent with the previous one. This would modify the weights x and y in x= 1 and y= 0.5, which represents the increased belief of the operator towards the information received from the tank T1 indicators vs. the valves indicators, and would lead to the new values of **N = 0** and **Π = 0.5 < D** (condition -2-).
A detection process will occur inducing a first recovery consisting in closing the manual valves. This reestablishes **GDC** to values above the treshold limit.
In Fig.3 the values of **Π** and **N** vs. the time sequence of actions are shown.

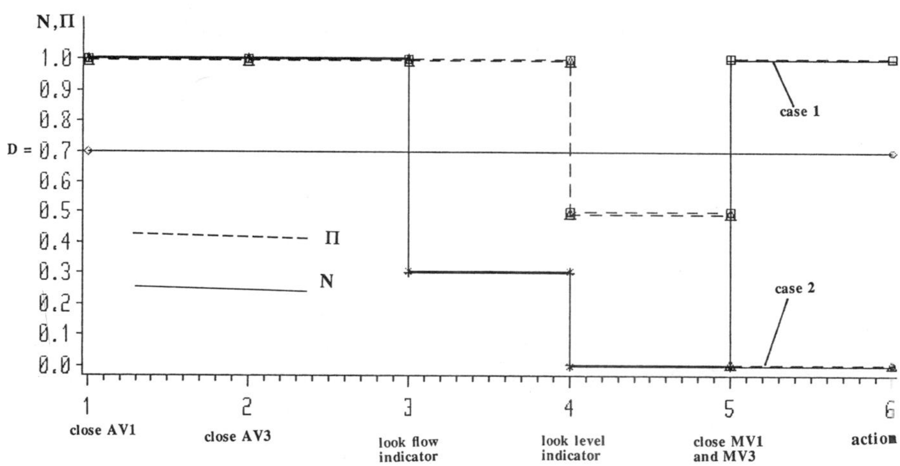

Fig. 3. Operator action time

In particular it can be observed that, between steps 3 and 4, the treshold value **D** lies in between **Π** and **N** implying that there is no certainty on the result of the conflict, while, elsewhere, **D** is either smaller or greater than both values of possibility and necessity leading to a definite selection of alternative.

In the second case, a possible mistake at HLDM can be simulated by modifying the *System Knowledge-Base*, for instance exchanging line-2 with line-3 as input line of T1. In this case the execution of the tree develops in a very similar manner than the previous case, up to the point when manual valves are being closed. Given the intrinsic error made by the operator, even after the manual closure of the valves, the flow meter would still indicate a non-zero value and this would lead to $\Pi = \mathbf{0}$ and $\mathbf{N} = \mathbf{0}$.
At this stage an error assessment mechanism, for the identification of possible slips or mistakes, and a recovery model would be activated.

CONCLUSIONS

This paper attempts to formalise two fundamental characteristics of cognitive processes: the intentionality and uncertainty.
Being aware that from the beginning, one of the challenge of AI is to represent in a normative way these two phenomena, our approach has been more apprehended in a reliability perspective, regarding with the same acuity erroneous and optimal behaviours.
Only the planning process has been dealt with here, but it seems to be psychologically acceptable that the diagnosis can be simulated by a similar dynamics. Indeed, a goal-tree construction can represent the development of a diagnostic strategy.
The future of the work will be oriented towards a deeper development of High Level Decision Making Processes and the error making model focusing on the simulation of the diagnosis and of the associated frames.

REFERENCES

Cacciabue, P.C. and U. Bersini (1987). Modelling Human Behaviour in the context of a simulation of Man-Machine Systems. In J. Patrick and K. Duncan (Eds.), *Human Decision Making and Control*, North-Holland, Elsevier, Amsterdam.

Dubois, D., and H. Prade (1982). On several representation of an uncertain body of evidence. In M.M. Gupta and E.Sanchez (Eds.), *Fuzzy Information and Decision Process*, North Holland, Amsterdam.

Dubois, D., H. Prade, and C. Testemale (1986). Weighted Fuzzy Pattern Matching. Proceedings of Journee Nationale sur les Ensembles Flous, La Theorie des Possibilites et leurs Applications. Toulouse, 27 june 1986.

Fikes, R. and T. Kelher (1985). The role of frame-based representation in reasoning. *Communications of the ACM*, September 1985, vol.28, No. 9

Garbolino, P. (1986). A comparison of some rules for probabilistic reasoning. NATO-ASI Conf. on Intelligent Decision Support in Process Environments. Ispra, Italy, 11-14 November 1986. To be published in *Journal of Man Machine Studies*

Rasmussen, J. (1986). Simulation of Human Decision Making in accident scenarios. CEC-JRC Ispra Contract No. 2529-84-11 ED ISP DK.

Reason, J. (1986). Recurrent errors in process environments: some implications for the design of Intelligent Decision Support Systems. In E. Hollnagel, G. Mancini and D. D. Woods (Eds.), *Intelligent Decision Support in Process Environments*, NATO ASI Series, Springer-Verlag, Berlin.

Rouse, W. (1983). Models of human problem solving: detection, diagnosis and compensation for system failures. *Automatica* 19, 6, 613-625.

Sheridan, T.B. (1985). Forty-five years of Man-Machine systems: history and trends. Keynote Address. Preprints of 2nd IFAC Conf. on Analysis, Design and Evaluation of Man-Machine Systems, Varese, Pergamon Press.

Woods, D.D. and E.M. Roth (1986). Models of cognitive behaviour in nuclear power plant personnel. NUREG/CR-4352.

A Rule Based Adaptive Blood Pressure Controller

M.W.N.M. den Brok *) and J.A. Blom **)

\) Elektroson Engineering Consultants, Liempde, The Netherlands
***\) Eindhoven University of Technology, Department of Electrical Engineering, Eindhoven, The Netherlands*

ABSTRACT

A rule based expert system was built to realize a robust, adaptive blood pressure controller. The expert system's knowledge consists of PID-controller tuning rules, knowledge about pattern recognition for the detection of frequently occurring artefacts in the blood pressure signal and knowledge about the system's probable non-linearity, its time variance and the maximum allowed drug dose. Simulations were carried out to test the expert system based controller and its performance was compared with that of an existing controller in clinical use [Sheppard, 1976]. The performance of the new controller is excellent, especially in extreme cases.

KEYWORDS

Blood pressure control, Sodium NitroPrusside SNP, robust adaptive PID-control, knowledge based adaptive control, expert systems.

INTRODUCTION

During some types of surgical procedure the patient's blood pressure needs to be controlled at a lower than normal value. Usually the drug Sodium NitroPrusside (SNP) is infused to lower the blood pressure. An earlier study [Blom, 1982] into the dynamic and static characteristics of this drug, as well as clinical experience, show that application of the drug is often difficult due to little a priori knowledge of the patient's sensitivity, the change of of the patient's sensitivity in time, a pronounced non-linearity and a (possibly changing) dead time in the control loop. Manual control is often difficult and requires close attention to the patient's response. Existing closed loop controllers do not cope well with non-average patients. An automatic controller thus far is not a good alternative for manual control, since those cases which are the most difficult to treat manually are impossible to control automatically. The goal of this research was to realize an unconditionally stable automatic controller.

THE MODEL

Blom [1982] showed that the patient, viewed as a 'system' with input SNP and output Mean Arterial Bloodpressure (MAP), can be characterized by

- a second order dynamic response with significant dead time;
- a non-linear static response curve;
- large differences in sensitivity (a factor 80);
- initially unknown, and time-varying parameters.

A system with these characteristics is known to be difficult to control.

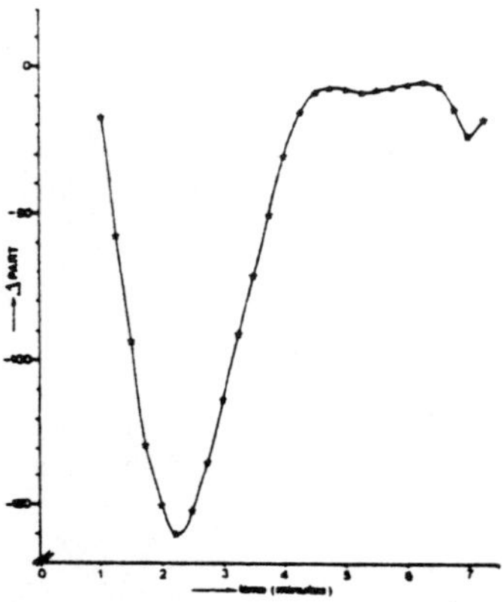

Fig. 1. Dynamic response

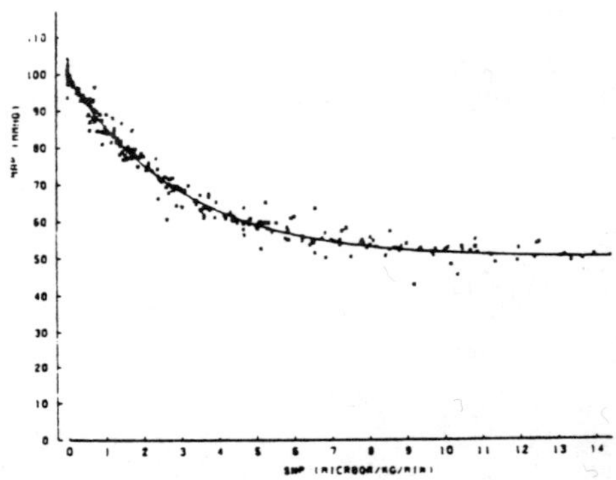

Fig. 2. Static response

The patient's behaviour can be modelled in several ways: Sheppard [1976] gives a Laplace-domain transfer function; several authors [Arnsparger 1983, Stern 1985] use an ARMAX model. These authors have the following in common:

- They use a model with a fixed structure representing the patient, with parameters tuned for an average patient. By making the controller adaptive (usually the patient's characteristics are estimated by some kind of least squares method) the controller is adjusted for non-average patients and the time variance is taken care of.
- The controllers generally work acceptably for average patients, but for patients who are exceptionally sensitive or insensitive the controllers experience problems with their stability and accuracy.

Our own experience shows that the blood pressure signal is contaminated with much 'noise' due to spontaneous fluctuations and transient disturbances, making it unlikely that a parameter estimation method will consistly produce reliable estimates. Besides this, Rohrs [1982] showed that none of the known adaptive algorithms, amongst which the earlier mentioned model-reference adaptive control algorithms, are reliable enough to be used in critical environments. Stability can only be garantueed if the number of zeroes and poles is exactly known, and/or if the time delay in sampled systems is an integer multiple of the sample period. These stability conditions are almost never met in real systems.
Clinical experience shows that automatic controllers are not very effective: only patients that are easy to control manually are suited for automatic control.
In view of the problems of conventional modeling and estimation methods for

this particular problem, we decided to have a better look at the (rather successful) clinical practice of ad hoc adjustment of the infusion flow, and to investigate the use of a simple, but robust controller combined with an expert system to monitor the controller's behaviour and adjust its parameters if necessary. Robustness is the key issue.

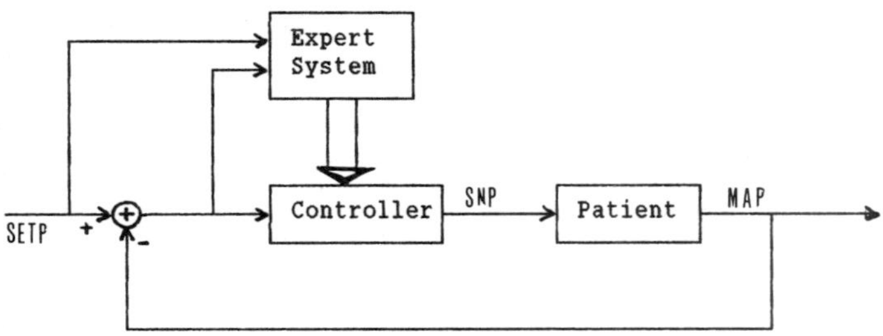

Fig. 3. Expert system based closed control loop.

THE KNOWLEDGE BASE

The mean arterial pressure (MAP) of the patient needs to be controlled as accurately as possible by adjusting the SNP flow, but stability is even more important. The expert system based closed loop control, depicted in figure 3, uses a PI-controller.
The PI-controller was chosen because:

- PI(D)-controllers are well known, reliable and are applied generally;
- PI(D)-controllers are very robust;
- PI(D)-controllers are easy to tune;
- other types of controllers (dead beat-, cancellation controllers, etc.) are not sufficiently robust.

Because of the significant dead time in the system and the large amount of noise in the blood pressure signal no D-element is used in the controller. This results in a slightly lower quality of control, but this is compensated by a greater robustness.

Three different kinds of knowledge were implemented in the expert system: tuning rules to adjust the PI-controller, pattern recognition knowledge to detect large disturbances, and some general knowledge.

Tuning rules. Tuning rules for PI(D)-controllers are well known. Tuning can be based on the system's response after a step input, or it can be based on induced oscillations. The second method cannot be used in this application, except possibly after unwanted oscillations, which should not normally occur. Step changes of the input are frequently available, however. From the response after a step input the the system's dead time, gain and integration time constant can be determined. Errors in these estimates are likely because of the bad quality of the blood pressure signal; therefore the controller's integration time constant was chosen to be fixed and only its gain is adjusted by choosing one of a few discrete gain values.
The integration time constant is chosen to be optimal for a dead time of three

sample periods, which is the expected value of the time delay. For extreme cases of the dead time (very short or very long) this means that the controller will be slightly too fast or too slow respectively.
The variability of the sensitivity was found to be a factor 80. This means that the controller's gain should also be adjusted over this interval. The gain of our controller can have one of five different values.
The patient's reaction to a step input is qualified as follows:

1. gain correct;
2. gain too low, systems responds slowly;
3. gain much too low, hardly any reaction;
4. gain too high, system responds with oscillations.

After a step input the system determines in which class the response belongs and chooses the appropriate gain factor.

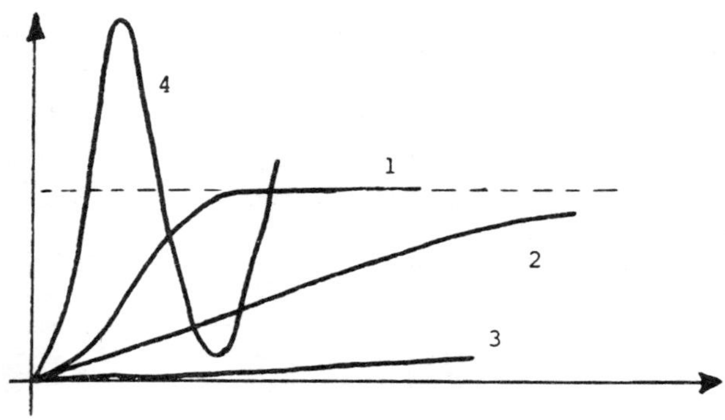

Fig. 4. Classes of responses to a step input.

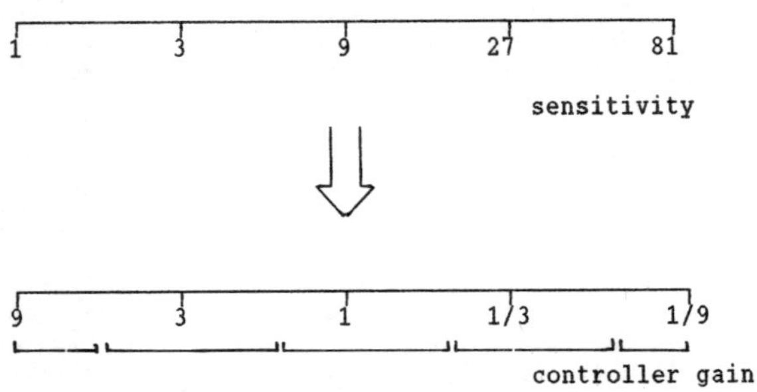

Fig. 5. Sensitivity scale with gain factor.

Only five different gain factors can be chosen. A more accurate setting of the controller's gain is not meaningful and, in order to guarantee the robustness of the controller, undesirable.
A step response will occur whenever the medical staff changes the setpoint value of the MAP, after which the system can check the correctness of the gain

factor. Positive and negative setpoint changes (relative to the last setpoint value) require a different approach because of the non-linearity: as the sensitivity can increase when changing to a higher setpoint value, this case requires extra care.

Pattern recognition. The system should recognize large, transient disturbances in the blood pressure signal. A transient is detected because the blood pressure measurement does not fall within the 'noise band' around the previous blood pressure samples. Again positive and negative transients should be distinguished.
A much too low MAP is life threatening, and the system reacts to a negative transient by shutting off the SNP flow. When the blood pressure signal exceeds the setpoint value again, the SNP flow is set to the value it had just before the occurrence of the transient, and this SNP flow is maintained for a time significantly greater than the system's time constant, to prevent oscillations. Due to the switching off of the SNP flow a large overshoot in the blood pressure signal will occur, but this is inevitable for safety reasons.
In the case of a positive transient the SNP flow is maintained at the level it had just before the occurrence of the transient, but released as soon as the transient ends. In this way undershoot in the blood pressure signal is minimized.

General. This category includes:
- the maximum allowed flow rate;
- negative flow rates: as it is not possible to withdraw SNP from the patient, the controller's output is set to zero when the computation shows that it ought to have a negative value;
- excessive offset cancellation: after long periods without setpoint changes the controller's gain can become too low due to the time variance. This is concluded from an increased offset.

THE EXPERT SYSTEM SHELL

A rule based, forward chaining expert system shell was built. The system allows the user to edit the rules in a pre-formatted but readable form, using a text editor. The system is not interactive, but has a tracing facility to check the performance of the rules. The programming language used is Fortran 77.

SIMULATIONS

In order to test the expert system based blood pressure controller on a series of 'worst case' patients, a simulation program was written. This program simulated the closed control loop controller and a second order system with adjustable characteristics: for each simulation run the system parameters, the degree of time variance, the degree of non-linearity, the amplitude of white measurement noise and the amplitude of system noise with various disturbances could be freely chosen. To compare the performance of the controller, the same simulation program was used with a well known, clinically used blood pressure controller in the loop [Sheppard, 1976]. In figures 6 to 8 simulation results are shown for a very insensitive, an average sensitive and a very sensitive patient. The graphs show the SNP flow rate, the blood pressure and the setpoint with both controllers. These simulations had a sensitivity time variance of a factor 4 and a pronounced sensitivity non-linearity. Figure 9 shows the blood pressure signal and SNP flow with the expert system based controller (for a patient of average sensitivity) together with graphs of the variation in system dead time during the simulation and the system noise.

CONCLUSIONS

The expert system based blood pressure controller performs very well in these simulations, especially in cases with extreme sensitivities. There is little difference between the two controllers in the average case. The expert system based controller reacts faster on large disturbances and has almost no undershoot after positive transients; after positive transients it stabilizes sooner at the setpoint. Overshoot after negative transients is inevitable for safety reasons; negative transients are usually artefacts, but not always.
Simulations allowed us to test the new controller extensively, for cases that are probably more difficult to control than will ever be encountered clinically.
We found that an expert system based blood pressure controller is a meaningful and realistic approach. Our system will soon be exposed to a clinical test.

REFERENCES

Arnsparger J.M. Adaptive control of blood pressure. IEEE Trans. Biomed. Eng., Vol. BME-30, no.3, March 1983, p.168-176.

Blom J.A. Peroperative estimation of sodium nitroprusside sensitivity. Proc. IEEE Southeastcon 1982, p.564-566.

Rohrs C.E. Robustness of adaptive control algorithms in the presence of umodeled dynamics. Proc. 1982 IEEE Conf. Dec. Control, p.3-11.

Sheppard L.C. Correlation analysis of blood pressure responses to vasoactive drugs, with particular reference to clinical surveillance of the post-surgical cardiac patient. Ph.D. Thesis, University of London, 1976.

Stern K.S. The self-tuning controller: comparison with human performance in the control of arterial pressure. Annals of Biomed. Eng., Vol. 13, 1985, p.341-357.

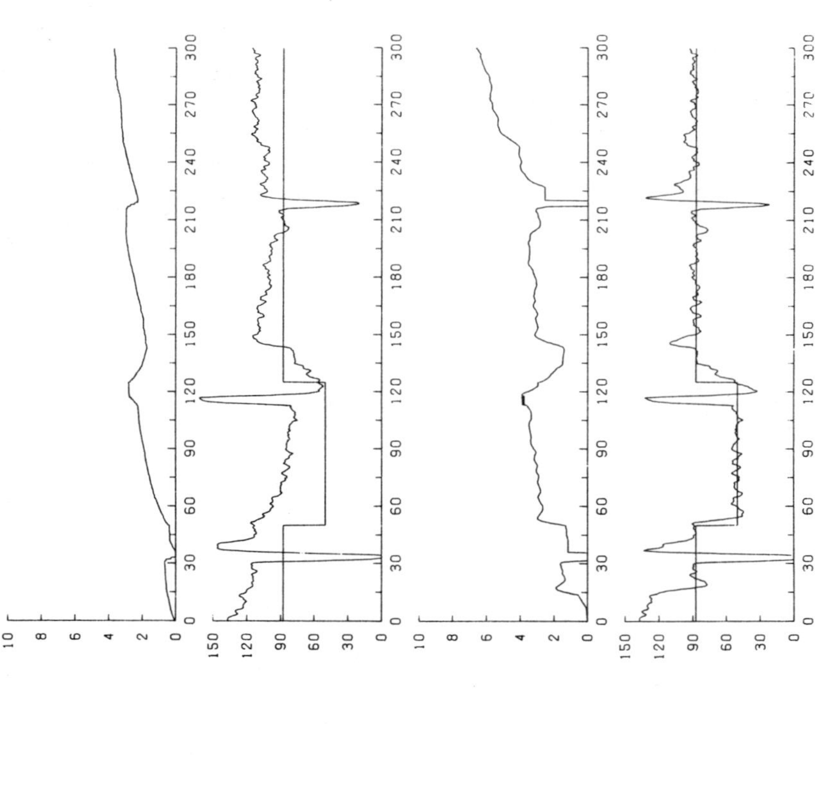

Fig.5 Simulation with Sheppard controller and rule based controller; patient sensitivity = 1.

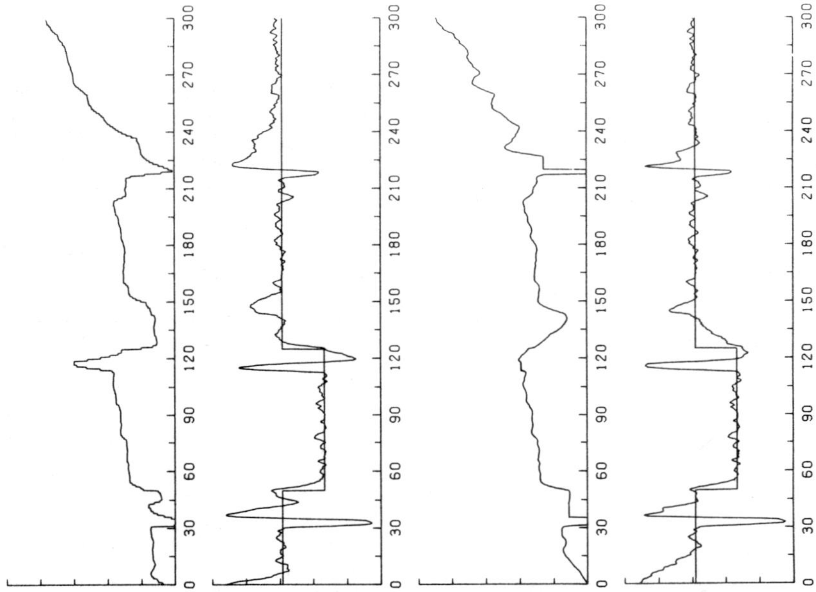

Fig.6 Simulation with Sheppard controller and rule based controller; patient sensitivity = 9.

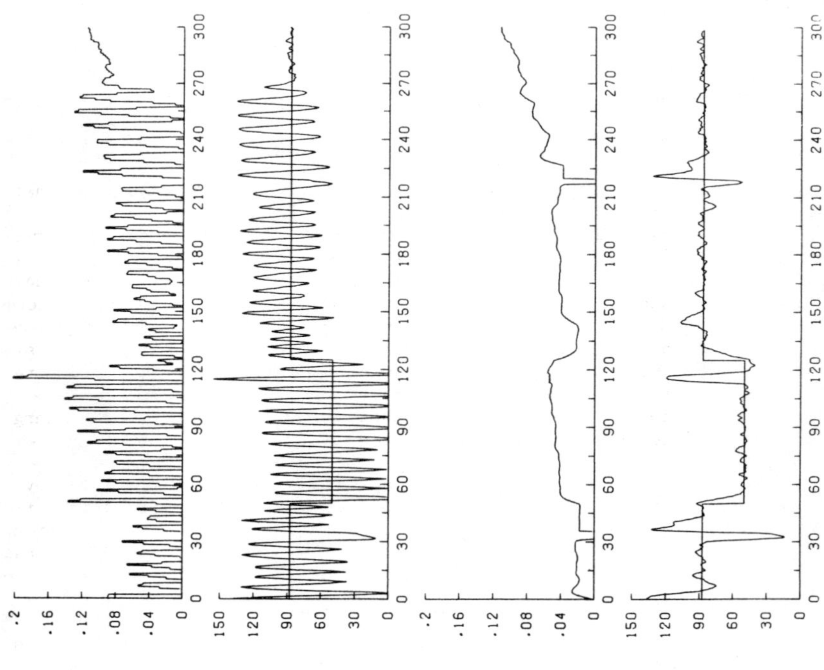

Fig.7 Simulation with Sheppard controller and rule based controller; patient sensitivity = 81.

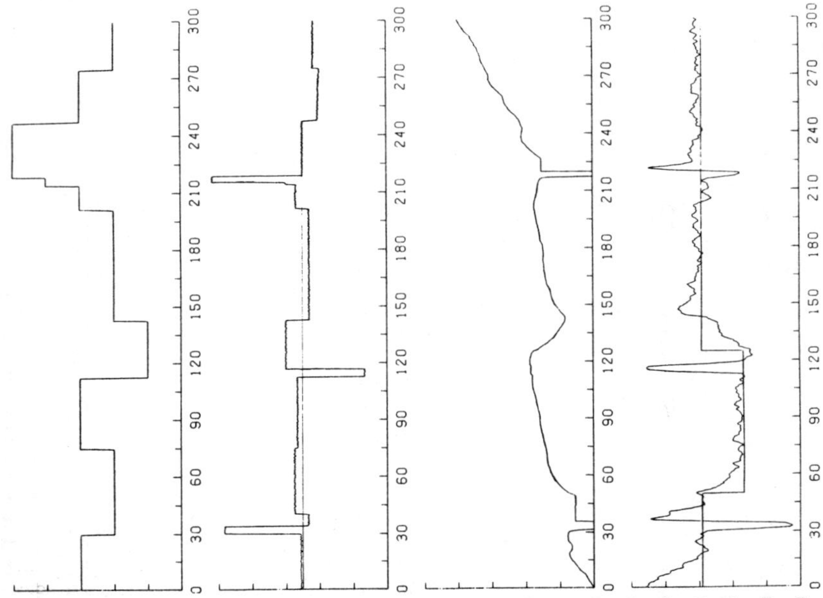

Fig.8 Simulation with rule based controller, with graphs of system noise and variation in dead time.

An Expert System for Optimal Adaptive Control Automation

M. de la Sen and J.J. Miñambres

Departamento de Física
Facultad de Ciencias, Universidad del País Vasco
Campus de Leioa (Bizkaia), Spain

ABSTRACT

This paper describes an expert system to optimize the use of adaptive control algorithms for purposes of achievement of high quality performances during the adaptation transient. The developed expert system is of Diagnosis Via Causal reasoning type and it has the capability of taking decisions such as switching between adaptive control algorithms via error prediction, and to up date an to modify the free parameters of the above algorithms according to optimal rules.

KEYWORDS

Adaptive Control, Expert Systems.

I. INTRODUCTION

The National Aeronautic and Space Administration (NASA) has begun a program that will result in a permanently manned, fully operational United States space station by the mid-1990s. The space station is intented to support scientific and commercial endeavors in space, stimulate new technologies and enhance space-based operational capabilities. The role of the space station in a national space infrastructure is multiple. The station can be used as a base by orbital maneuvering vehicles and orbital transfer vehicles. The first one is a remotely controlled vehicle that can be controlled from several hundred miles away by a crew member on the shuttle or station. The second one would extend man's presence for beyond the range of the remotely controlled platform by being able to travel tens of thousands of miles from the station to geosynchronous orbit and then to the moon, transporting large platforms, a crew of several astronauts or another orbital transfer vehicle [3-6].

In a recent study, SRI International examined automation concepts postulated by four NASA contractors to determine the required research and development in artificial intelligence to attain these concepts, by ending with demonstrations of a previous plan leading. The final report [1] , [15] describes the demonstrations. The plan leading needs to attain the expert system technology for the evolving space station.

By its complexity, this is one of the most sophisticated possible developments of expert systems [8] with possibility of real-time decision actions. The term "expert systems" was originally used [2] to denote systems that utilize a significant amount of expert information about a particular domain to solve problems in that domain. Because of the important role of knowledge in such systems, they have also been called knowledge-based systems. However, the term has since been applied to so many diverse systems that its original meaning has been largely lost. There are essentially two uses of the term that need to be differentiated. First, the term is used to describe any system constructed with special Kinds of "expert system" programming languages and tools. These include production systems, rule-based systems, frame-based systems, "blackboard" architectures, and the PROLOG language. The second use of the term expert system is to describe any system that "reasons" about a problem in much the same way humans do. Some of the features distinguishing these systems from standar applications programs follow ([1], [15]).

(a) Knowledge. Each contains a data base of knowledge (usually in a spezialized area) represented in a relatively natural form that allows some sort of reasoning to be carried out. The knowledge representations are usually symbolic, reflecting the qualitative nature of much human reasoning. Early expert systems used relatively simple rule-based or frame-based schemes for representing this knowledge. Recent systems have augmented these approaches by more sophisticated knowledge-representation formalisms.

(b) Extensibility. The representation of knowledge is such that modifications of or additions to the knowledge base do not require extensive modification of the entire system. Thus, the systems are extensible, degrade "gracefully" rather than catastrophically as elements are removed, and can evolve without excessive rewritting. This requires highly modular systems, in which the semantics of each module can be specified independently of other module.

(c) Explanation. Many systems can retrace the reasoning sequence employed and explain what was done at each step and why. Its usefulness is closely related to the similarity between the expert reasoning processes.

(d) Incomplete or Inexact Data. Many of these systems can carry out reasoning process on incomplete, uncertain, or inaccurate data. For example, the effects of a given action may be incomplete, the conclusion of a diagnosis may be uncertain, or there may be errors in sensory information. At present, expert systems do not acquire their expertise through experience but that information is supplied by the team expert / programmer.

The terminology "expert system" is included in the wider field of Artificial Intelligence. A detailed list of references about particular items on these system is cited in this paper (See [3] - [15]).

II. EXPERT SYSTEM FOR OPTIMAL ADAPTIVE CONTROL

Many of the applications of expert systems have been orientated to static (non real-time) applications such as manufacturing designs. The most important cathegories lie in the types of Interpretation and Diagnosis and Design Systems. In real-time applications the design of expert systems is more problematic since the processing of the data through the rules to activate the decision maker takes some time which depends on the amount of information and rules and may be incompatible with the concrete real-time application.

In adaptive control the problem is to generate an input to a non completely known industrial process (some knowledge is usually available and some "a priori" knowledge is required from the adaptive thoery in order to asymptotically achieving reference tracking).

The problem which usually occurs which is an important drawback for applicability is that transients are very poor with large tracking / error deviations and high input demands. Adaptive control is usually implemented by implementing in a digital computer analytic equations which have been theoretically deduced. It has been observed that the performances are greatly dependent on the algorithm which is used and on the free and initialization parameters. Although some practical idea is usually available about like to cope with these problems, analytic mathematical studies are quite incomplete or even inexistent in many situations. The software Lisp is used because of its programming flexibility and because of its special ability to be used by expert Rule-Based Expert Systems.

2.1. Adaptive Control problem definition

The objective is to correct transients based upon predictions and supervisor levels in adaptive control.

In each problem, the following elements are present which constitute the basis for development of a mixed expert system for Interpretation, Diagnosis and Design.

(A) PROCESS TO BE CONTROLLED (P). Non fully known Required "a priori" knowledge: upper-bounds of the dynamics order, sampling period domain. (If this knowledge is "erroneous" and leads to failures, it can be monitored). The process must belong to the class of admissible processes (AP), i.e., $P \in \{AP\}$.

(B) ADAPTIVE ALGORITHM (A). It must belong to the class of admissible algorithms for the class AP, namely AA (AP) ; i.e., $A \in AA\ (AP)$. (Note that given $P' \in AP'$; $P \in AP$, it may occur that $AA\ (AP) \neq AA\ (AP')$ but some particular subsets in AP are in AP' (i.e., $(AP)_{(i)}\ (\subset AP) = (AP')_{(j)}\ (\subset AP')$ with $AP \neq AP'$ some positive integers i,j denoting subsets of AP, AP'. It becomes apparent that

$\forall P \in AP\ ,\ \forall P' \in AP'$: (i) $AA\ (P) = AA\ (P') \Longleftrightarrow AP = AP'$

(ii) $AA\ (P) \cap AA(P') \neq \phi \Leftrightarrow AP_{(i)} = (AP')_{(j)}$, some $i,j \in z$ with $(AP)_{(i)} \subset AP\ ,\ (AP')_{(j)} \subset AP'$

(iii) Equivalently $AA(P) \cap AA(P') \neq \phi$ if $AP \subset AP'$ or $AP' \subset AP$

(iv) $AA(P) \cap AA(P') = \phi$ if $AP \cap AP' = \phi$

- The classes AP ($\mathcal{P}$) of the set of possible processes $\mathcal{P}$ are determined via experimentation (which characterizes "a priori" the process into cathegories (for instance, possibley stable / unstable, of possible inverse stable / unstable of degree of dynamics $\delta \leq \delta_o$, etc.).

- The classes AA (AP ($\mathcal{P}$)) are known from mathematical stability studies depending on the prespecification of $\mathcal{P}$.

The adaptive algorithms consist of two modules:

(B1) Initialization module(IM). Initial parameter vector for the adaptive algorithm. It contains initial values of the adaptation matrix (AM(0)), of the free parameters of the algorithm $FP_k(0)$, $\forall k \in Z_o \subset Z^+$ which must belong to a real domain of admissibility defined by the algorithm A from its theoretical analysis.

(B2) Execution module (EM). It consists of two basic actions:

(A.1) Rule of adaptation of the adaptive controller (updating its parameters and adaptation gain, if any).

(A.2) Rule of generation of the input to the plant so that the output asymptotically tracks the reference sequence.

The same adaptive controller is usually used by specialists to describe (IM + EM).

At a first step of research, a method to optimize the values $f_k \in F\ P_k(t)$ at each sampling point t was developped in [16] to be used in real-time. The optimal choice of the $f_k(.)$ modifies the numerical values obtained in Actions 1-2 of EM so that the process is monitored. The optimization action is based on building an equivalent linearized system of equations, which is of the linear difference equation type. Its (easily computable) optimal control needs low computation time / memory requirements. Several prediction methods are alternatively derived, implemented and used (according to prior registered performances) in order to smooth the optimization process with better results, namely:

- The Heuristical Prediction Method (HPM) which predict the process output using extrapolation of its previous values.

- The reference model prediction Method (RMPM) which predicts the process output via use of the reference model. It is required for applicability that the process parameters be close to those of the reference model.

- The mixed reference-prediction method (MRPM) which is a weighted combination of the two above ones. It may be useful in douftful cases about the method to be applied.

The prediction methods are implemented in the

(B3) Prediction module (PM). The following modular structure unes (B3).

(B4) Optimization module (OM). It is activated from PM so that $f_k \in F\ P_k(t)$ takes a new value and EM is activated again.

At a second research level (See [17]) another optimization procedure consisting of using Numerical methods for acceleration of the convergence were used for substitution of (A.1) on in EM for another sequence (A.1) which converges faster to the same limit; i.e., we have

(B5) Modified optimization module (MOM) = EM(with (A.1) $\longrightarrow$ (A.1)$'$)

Now, the problem is defined at different levels by a most three elements:

LEVEL 1 : (IM + EM).- BASIC ADAPTIVE ALGORITHM

The result in practice is that the process output asymptotically tracks a given reference sequence and that all the signals within the system remain bounded.

LEVEL 2: $\begin{cases} (IM + EM + S)\ ;\ S_1 = PM + OM \\ (IM + S)\ ;\ S_2 = MOM \end{cases}$, or where $S_{(,)}$ is a supervisor which consists of PM + OM or MOM according to the philosophy in [16] or, respectively, [17] . This solution level maintains the results of level 1 and, furthermore diminishes the transient error deviations and the process input (energy) demand if it is conveniently designed (by including fault diagnosis and by respecting the stability domain of the FP when the supervisor is applied.

It becomes apparent that a LEVEL 3 maintaining the performances of LEVELS 1-2 and improving them would be useful in the following way would be suitable.

LEVEL 3 . It includes the use of

- A Basic Adaptive Algorithm as LEVEL 1
- A Supervisor as LEVEL 2

Furthermore, for a family $\mathcal{P}$ of processes P :

- It selects the best AA (P) for each $P \in \mathcal{P}$ from a finite AA(P) family.
- It selects the supervisory level S from a family $\mathcal{S}$ including fault diagnosis at lower levels.
- It is able of commuting from a an element to AA(P) to another or from a supervisor S to another (and even between the submodules of the supervisor as, for instance, HPM, RMPM and MRPM above
- It performs in real time for each given P of $\mathcal{P}$,

It is obvious that for an adequate implementation of such a Level 3, it is necessary to design an expert system having a "backtracking" and a prediction capability to allow several choices to be made if each current decision is not a fruitful one.

2.2. Expert System

A typical expert system such as developped in [1] is shown in Fig. 1. In our current case related to the problem [16]-[17], the following (standard) technical abilities are needed in order to implement the general functions of Interpretation, Diagnosis an Designs.

. Knowledge Representation :

Formalisms to represent time, space, actions, processes, mechanisms and other complex objects need to be developed including procedural knowledge — that is, where expert knowledge involves reasoning about sequences of test and actions. This includes.

- CONTROL PROCESS REPRESENTATION

(a) Via mathemtical equations at a design level.
(b) For verification purposes in real-time applications.

- ADAPTIVE CONTROLLER ALGORITHMS $AA_i(P)$, $i \in Z_o \subset Z^+$.
- SUPERVISORS (S) $\supset$ predictions : PM, OM, MOM, etc..

All these modules have an activator $\mathcal{A}$ which enables them to perform, a task T to perform and a set of transfer actions to other modules $M_{(.)}$ $(ST(M_i) \longrightarrow ST(M_j))$.

- Time, space etc. are represented in the Tasks themselves (for instance, in discrete applications, the left-hand-side of certain equations is related to the current sampling instant and the right-hand-side to previous ones).

-TESTS for instance

- Is the stability domain of the FP violated by the OM? If so does the system performs well according to predictions?
- Is the prediction correct?
- Is the input level / tracking error admissible? Its prediction?
- Is the optimization size horizon correct?
- Is the dynamics order correctly upper-bounded?
- Does the $AA_{(.)}(P)$ performs well according predictions and observations?

In order to have a good design the above tests have a priority sequence and an action to be done. The "commonsense" knowledge the type of knowledge a person uses in dealing with the problem is incorporated when stating the above priorities.

. Reasoning

The reasoning system must be able to reach conclusions on the basis of information about the current situation and the knowledge contatined in the knowledge base. Much work is required at a design Level to cope with this problem, especially, for real-time applicability purposes. In our case the reasoning procedure must lead to two types of results:

(a) To select which Level 1, 2, 3 is suitable.

(b) To select which options at each Level are suitable prediction method type/Optimizer type / Supervisory type / Algorithm type / FP type (if options are present) etc.

The main (sets of) rules are now ordered according to a priority order.

— **RO** : Rules for determining the fiability of measurements or generated signals.

— **R1** : Level 1 is selected if test says system performs well according to predictions. Combine with a subset of **RO.**

— **R2**: One of Levels 2/3 is selected otherwise. One of the $AA_{(.)}(P)$ is selected according to off-line "a priori" knowledge on P. Level 2 is prioritary on Level 3 and $AA_{(i)}(P)$ is prioritary on each $AA_{(j)}(P)$, $\forall i \neq j \in Z_o(P)$ if satisfies better the test on admissibility of input level/tracking error. Combine with a subset of **RO.**

— **R3**: If Level 2 has performed well for $\forall t \leq t_o$ (current sample), Level 2 is prioritari on Level 3 at (t_o+1). If the stability domain of the FP has been violated slightly (according to a prior quality measure), the test of horizon size is prioritary on that of the algorithm, that of the used prediction method (if it fails activate change of prediction method) and that of use of the second (or nth) supervisor. Combine with a subset of **RO.**

— **R4**: If stability domain of the FP does not perform well, then change it if the fail goes on, then modify the FP. If the fail goes on, then change supervisory $ST(S_i) \longrightarrow ST(S_i+1)$. If the fail goes on yet, then reapply second part of the set of rules **R2** (i.e., excluding the priority of Level 2 on Level 3). Combine with a subset of **RO.**

There are also "structural regles" (programming and real-time applicability rules) which are general "a priori" requirements such as absence of infinite cycles when evaluating the rules and process modelling test. The last one may be repeated periodically since dynamics of real physical processes is highly dependent on the supplied input signals.

The reasoning has a mixed qualitative and quantitative reasoning, the first one being supplied by the expert experience while the second one is used mainly to evaluate the decisions to be taken from the tests.

. Knowledge Acquisition

It is a general fact that it is a major problem in obtaining, representing, and debugging expert knowledge about a particular domain. Methods are now being developed for dealing with that acquisition and these methods should reduce the time system takes to build new systems.

In our current system, the data adquisition is mainly concerned with the database and the supervisory modules.

The database is organized in such a way that it is able to admit new "a priori" parametrical values (estimation of the nominal controlled process) if the modelling inaccuracies test says that system is underdimensioned and so its dimension (and then that of the controller) should be increased. The supervisory models may be of different types as, for instance, those developed in [16]-[17]. However, the problem remains open to new methods which must first be theoretically experimented. The same considerations may be made about the prediction methods. For these purposes the memory is organized in a direct accessibility way according to an "a priori" data test.

. Verification and Explanation capabilities

Since the system may be inconsistent in its knowledge of rules, it is important that manual and eventually automatic verification techniques be developed. However, the problem of formally verifying a knowledge base constructed with an expressive representation scheme is, in general, intractable.

In our expert system the verification is only at a design level. The quality tests allow several repetitions of the reasoning with different optional parts (see the above rules) within a "backtraking" capability context.

The explanations consists of giving the used "a priori" data, P definition, sampling period, $AA_{(.)}(P)$, supervisor and the sequence of Activations followed from the applications of the rules. As complementary information, there are also a set of premotecnic sentences which summarize the followed reasoning.

. Learning Capability

It is an important task to determine which rules need modification when the expert system is not performing up to an expert's standards. In our case this capability development is yet being searched. At the present stage, it is trhought (and some experience has been obtained) that the priority of tests, prediction methods and supervisors are highly dependent on the set (Reference sequence, process, adaptive algorithm, deviation from true parameters from the initialization values of the algorithm) and so rules may be modified accordingly. Exhaustive off-line implementation of this capability is being developed. Little theoretical study is available so that experience on each particular class of processes {P } of type P must be taken.

2.3. (Summarized) Example

It has been noticed that for third order discrete control processes [16-17] with a zero and two conjugate complex poles, a step functions with several values is first test followed by Landau-Lozano's algorithm (until 20-30 samples about, sampling period = 1 sec., and then by Lozano's algorithm with S_1 of { 2.1 modified optimization module (Level 2), until complete adaptation is performed.

III. CONCLUSIONS

Dynamic analysis and control design for flexible adaptive control problems involves variable-order dyanimic systems with sets of inputs and outputs for which a little "a priori" knowledge is available to the designer. Computers tools are essential for such analysis. This paper has described the adaptive control (information structures) supervisors interaction analysis (with several options at different elemental modules) of an "expert system" valid for off-line design which some possible projections in real-time applications. We thiak the proposed application is interesting especially due to the fact that applications in real-time of "expert systems" software design is not abundant nowadays. Because various subsystems are operating simultaneously, it is important that the representation be sufficiently rich to enable reasoning about concurrency and subsystem interaction and that procedures for scheduling and synchronization be developed. These points have been carefully taking into aceount in the presented design and they are closely related to the mathematical structures of the describing equations of both plant and controller and to the allowed sampling period domain which is defined from the expert's experience. The functions of interpretation, diagnosis and design are realized by the proposed system.

. Interpretation-Diagnosis : From the database and the reasoning capability the system interprets results and behaviour predictions via knowledge rules including several tests (stability, optimization horizon, signal levels, failures, etc.), and makes a diagnosis which is output via its explanation capability. The Diagnosis implies in many cases a set of actions to be taken.

. Design : The decision maker based upon the database and knowledge base (previous and predicted) takes the decision of activating some module (task) so as to drive the process via generation of a control action

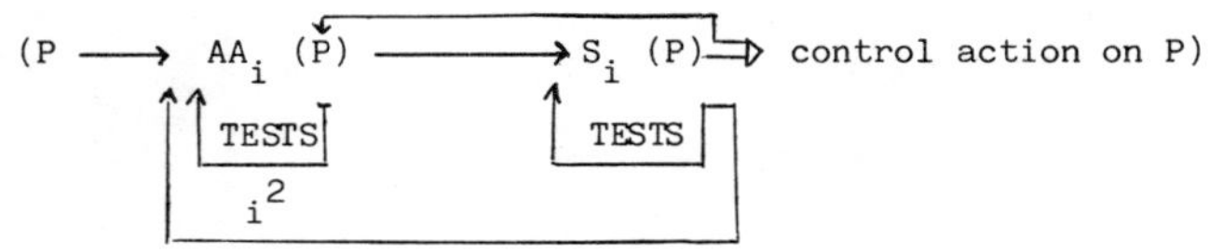

REFERENCES

[1] M. Kárny, A. Halousková, J. Böhm, R. Kulhavý, P. Nedoma, "Design of Linear quadratic adaptive control. Theory and algorithms for practice", Kybernetika, Vol. 21, suppl., 1985.

[2] H. Wertz, Intelligence artificielle. Aplications á l'analyse de programmes, Masson, Paris, 1985.

[3] Nasa Space Station Automation: "AI-Based Technology Review", (Final Report SRI Project 7268), Menlo Park, CA, SRI International, Mar. 1985.

[4] "SSAS: Automation Requirements Derived from Space Manufacturing Concepts", (Final Report NR 740-9), Valley Forge, PA, General Electric Space Systems Division, Nov. 1984.

[5] "SSAS: Automation Study for Space Station Subsystems and Mission Ground Support", (Final Report F5713), Los Angeles, CA, Hughes Aircraft Company, Nov. 1984.

[6] "SSAS: Satellite Servicing" , (Final Report Z 410.1-84-160), Redondo Beach, CA, TRW Space and Technology Group, NASA, Contract NAS-8-35081, Nov. 1984.

[7] B.G. Buchanan and E.A. Feigenbaum "DENDRAL and Meta-DENDRAL: Their Applications Dimensions", Artifical Intelligence, 11, pp. 5-24, 1978.

[8] S.M. Weiss and C.A. Kulikowski, "EXPERT: A System for Developing Consulation Models", IJCAI, pp. 942-947, 1979.

[9] B.G. Buchanan and E.H. Shortliffe, "Addison-Wesley series in Artifical Intelligence. Rule-Based Expert Systems", Reading, MA: Addison-Wesley, 1984.

[10] J. McDermott and C. Forgy, "R1: An expert in the computer Systems Domain", AAAI, pp. 269-271, 1980.

[11] R. Davis, "Diagnosis Via Causal Reasoning: Paths of Interaction and the Locality Principle", Proceedings of the 3rd National Conference on Artificial Intelligence William Kaufman, Inc., 95 First Street, Los Altos, CA, Washington, DC, pp. 88-94, Aug. 1983.

[12] R.R. Cantone, F.J. Pipitone, W.B. Lander, and M.P. Marrone, "Model-Based Probabilistic Reasoning for Electronics Troubleshooting, "Proceedings of the 8th International Joint Conference on Artificial Intelligence, William Kaufman, Inc., 95 First street, Los Altos, CA; Karlsruhe, West Germany, pp. 207-211, Aug. 1983.

[13] D.R. Antonelli, "The Application of Aritifical Intelligence to a Maintenance and Diagnosic Information System (MDIS)", Joint Services Workshop on Artifical Intelligence in Maintenance, Volume I: Proceedings, Air Force Human Resources Laboratory, Brooks Air Force Base, San Antonio, TX; Boulder, CO, 1983.

[14] M. Georgeff and U. Bonollo, "Procedural Expert Systems", Proceedings of the Eigth International Joint Conference on Artificial Intelligence, Karlsruge, West Germany, 1983.

[15] M. Georgeff and O. Firschein, "Expert Systems for Space Station Automation", IEEE Control Systems Magazine, Vol. 5, pp. 3-8.

[16] M. de la Sen, "On-line optimization of the free parameters in discrete adaptive control systems", Proc. IEE-D, Vol. 131, No.4 (1984) pp.146-157.

[17] J.J. Miñambres and M. de la Sen, "Acceleration of the convergence in Adaptive Control. The one-dimensional case". Computers and Mathematics with Appl. (in press).

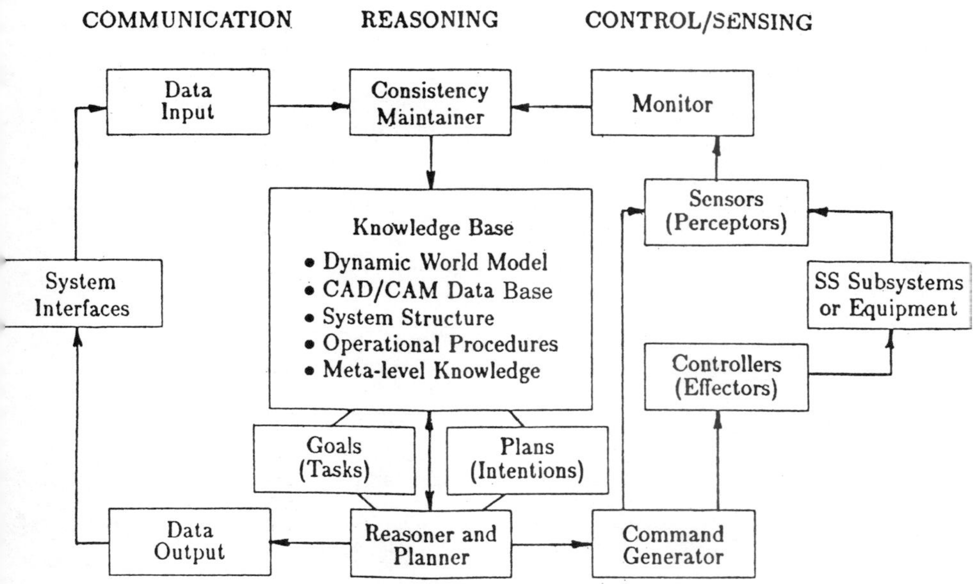

Fig. 1. Expert system

Decision Support Systems in the Design of Process Control Systems

Lj. Vlačić and B. Matić

Institute of Control and Computer Science - IRCA
Energoinvest, 71000 Sarajevo, P.O.B.158, Yugoslavia

ABSTRACT

The dramatic increase of availability of computers put them down from the pedestal, in their proper place as very powerful and sophisticated tools, but stil tools that serve humans. This has motivated a rapid growth of research on and applications of decision support systems - computerized systems of processing information, knowledge and models that help in preparation of decision made by humans. An automatic control system is build in order to relieve humans from tedidus and repetitive decisions while the construction of such a system requires other, more complex human decision that cannot be fully automatised but can be helped by decision support.

KEYWORDS

Decision support systems; systems engineering; multicriteria decision-making; control systems design.

INTRODUCTION

Approximately twenty years before the end od twentieth century, the rapid developments of microelectronic technology brought about a revolution in the availability of microcomputers - relatively inexpensive and small in size but with enough computational power and memory to support most diverse needs and applications. This revolution will have most profound impacts in the organization of human societies; now, at the end of the first decade of this revolution, we stil observe only the most fundamental effects of this transition to the so-called information society. Among many trends related to this transition, extremely important are shifts in the relation between the user of a computer and the computer itself.

In early days of the development of computer technology, when computers were scarce, they were put on a pedestal by their users: for example, when we considered computer applications in decision making, such applications were typically named "computer-made decisions". The dramatic increase of availability of computers put them down from the pedestal, in their proper place as very powerful and sophisticated tools, but stil tools that serve humans. This has motivated a rapid growth of research on and applications of *decision support systems* - computerized systems of processing information, knowledge and models that help in preparation of decision made by humans. Various aspects of the theory or rather theories that underly the construction of such systems have been long developed before; but the changes in the philosophy of computer use that accompanied the revolution in availability of computers have also heavily influenced the development of decision support systems and motivated new theoretical approaches.

A decision support system should not outguess the human decision maker, replace him in the final responsibility od taking decisions--except in cases of repetitive, customary decisions with established decision rules where a human decision maker would find repetitive decision taking too tedious. However the decision situations from the latter class have been long studied and resolved by the techniques of a automatic control; the new challenge of developing decision support systems relates to nonstandard and changing situations of strategic character or to very complex decision situations where a large amount of information has to be processed and aggregated during the decision making process.
Here, the basic characteristics of a decision support systems and the main results in application of SCDAS decision support system in the domain of systems engineering of industrial plants are explained.

DECISION SUPPORT SYSTEMS

Decision support system is a system including computerised procedures of decision support as well as models and data basis relevant for the decision problem. The fundamental goal is to improve the

quality of the decision[1] and to support the user in making a decision by providing quick and easy access to data and models relevant and applicable to that decision.
Decision support systems alow the user to confront a problem in a flexible, personal way by providing the ability to manipulate the data and models in a variety of ways while progressing through the decision making process. The user, through the dialog system chooses relevant data and models and directs their use to explore and analyze the problem and possible solutions.
Four levels of flexibility can be identified (Ford, 1985):

(1) giving the user the ability to approach a problem analysis in a flexible, personal way,

(2) allowing modification of the configuration so that the decision support systems can handle a changing set of problems,

(3) allowing changes extensive enough to require a completely different decision support system,

(4) providing a means for evolution in response to changes in the basic nature of the technology on which decision support systems are based.

A detailed analysis leads to the conclusion that the application of decision support systems depends primarily on the character of multicriteria decision situations (Sprague, Carlson, 1982).
Hence, in some cases with repetitive decision making procedures, DSS could be intended for a complete replacement of human behaviour, and could render the process of decision making completely automatic.

---

1) The fundamental goal of expert system is basically the same, but the objective is to provide to the user a conclusion or decision that is correct all the time. An expert system is a problem solving program that achieves good performance in a specialized problem domain that generally requires specialized knowledge and skill. Less attention is concentrated on the rationality of interaction between the user and the system. But, how to make a big data base usefull for the user and how to represent knowledge to the user are the basic characteristics. Practically, the objective of an expert system is to provide the user with a conclusion or decision significantly better than the user could reach.

Though the industrial plant oriented human decision making appears to be a repetitive process, it does not mean that it should be automated. Hence, DSS for the support of human decision making within industrial plant systems engineering can not be considered a replacement, but merely a support with respect to

- classification of aggregated information,
- generation of alternatives,
- multicriteria choice.

A few examples of existing systems are: Geodata Analysis and Display system (Sprague, Carlson, 1982), Portfolio Management system (Keen, Scot Morton, 1978), Capacity Information System (Keen, Scott Morton, 1978), Expert choice (Decision Support Software Inc., 1983), LIGHTYEAR (Lightyear Inc., 1984), Selection Committee Decision Support System, (Lewandowski, Johnson, Wierzbicki 1985) etc. Here, the basic characteristics of SCDAS support system are explained and, also, its applicability in the domain of systems engineering of industrial plant.

THE SCDAS DECISION SUPPORT SYSTEM

The theoretical background is related to the concept of an order-consistent achievement function (Wierzbicki, 1985) based on aspiration and reservation levels specified by a decision maker. This achievement function can be interpreted by fuzzy set theory, but it is used in this application as a cardinal utility function explicitly dependent on the contextual information (reservation and aspirations) supplied by the committee members (Lewandowski, Wierzbicki, Toth, 1985).

Basic features of SCDAS are the following:

- the method is dynamical (non-statical);
- aspiration levels are formed adaptively, if necessary;
- aspiration levels and achievement functions are needed instead of weighting coefficients and utility functions in setting the full set of efficient solutions;
- the ranking of alternatives is exclusively under the control of decision maker, i.e. depends on specified aspiration levels;
- the generation of additional alternatives is possible in the course of decision making;
- with evaluation of a discouraging large set of alternatives, the method permits testing of each single alternative in order to

evaluate the most promising alternative;
- the procedure of decision making, based on SCDAS system, recognize the influence of alocated aspiration levels upon the choice of the alternative; it also permits the checking of rank sensitivity of alternative to the extent of uncertainty in alocation of attributes values.

The complete procedure of SCDAS method, adapted to the cases of decision making with hierarchically grouped decision attributes, is shown, Fig.1.

The stages include: step 1. - defining the procedure, step 2. - setting and scaling attributes, step 3. - initial setting of aspiration levels, step 4. - initial survey of alternatives, step 5. - alternative - attributes assignement, step 6. - individual ranking, step 7. - individual analysis, step 8. - group ranking.

Since the problem addressed involves a hierarchical structure of attributes, the SCDAS methodology can be applied twice: once for aggregating the analytically modelled lower-level attributes, and the second time for evaluating overall system performance. Detailed description of this application was given by Vlačić (1986).

A MULTICRITERIA NATURE OF THE DESIGN OF PROCESS CONTROL SYSTEMS

Design of process control systems, regardless of the fact whether it is realised within the phase of systems engineering of industrial plants, is characterised by the problem of design of user-oriented process control system, where the term user is understood to mean both the automated process and the suppervising personnel (operators in the control room, technology specialists, control and instrumentation specialists, system engineers, etc.).

By evaluation of the efficiency of the system through multiple criteria, the design procedure for process control systems acquires the character of a multicriteria decision-making problem. Thus, design of process control systems (in the sense of defining the system structure, system associated functions and total system performances) represents a typical example of multicriteria decision problem. Since, the design of process control system is primarily the result of work of a team of experts the SCDAS decision support system is reading applicable.

The results, obtained for the case of hierarchically grouped decision

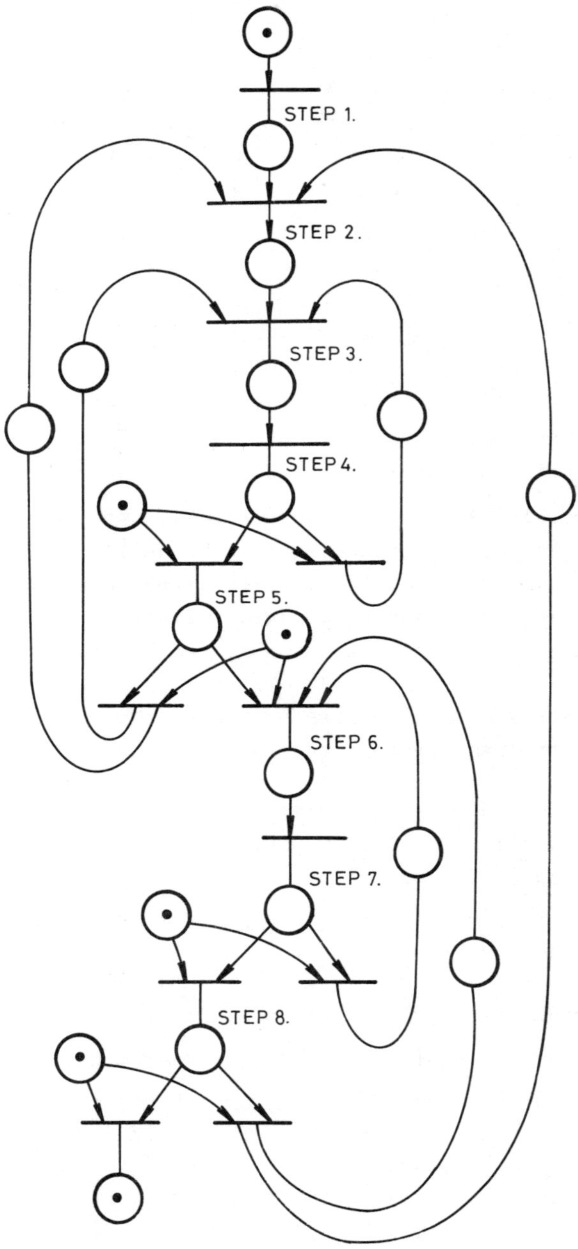

Fig. 1. Petri-net of SCDAS methodology

attributes (Vlačić, Wierzbicki, Matić, 1986), show that the SCDAS methodology leads toward a satisfactory solution.
Employing the SCDAS approach, a rational and theoretically-based methodology of decision making for a design of process control system is made available.

ACKNOWLEDGEMENT

The authors would like to express their gratitude to Professor A. Wierzbicki for his advice.

REFERENCES

Decision Support Software Inc. (1983). EXPERT CHOICE., Decision Support Software Inc., USA.

Ford, F.N. (1985). Decision Support Systems and Expert Systems: A Comparison. Information&Management 8., 21-26.

Keen, P.G.W., and M.S.Scott Morton (1978). Decision Support Systems: An Organizational Perspective. Reading, Mass., Addison-Wesley.

Lewandowski, A., S. Johnson and A.P.Wierzbicki (1985). A Selection Committee Decision Support System: implementation, tutorial example and users manual. Mimeograph, IIASA, Laxenburg, Austria.

Lightyear, Inc. (1984). LIGHTYEAR-The Decision Support Software, Lightyear, Inc., USA.

Sprague, R.H. and E.D.Carlson (1982). Building Effective Decision Support Systems. Englewood Cliffs. N.J., Prentice-Hall.

Vlačić, Lj. (1986). A multicriteria approach to determination of topology of distributed systems for control of thermo-power and petrochemical plants. Doctoral dissertation (in Serbo-Croation), University of Sarajevo, Yugoslavia.

Vlačić, Lj., B. Matić and A.P. Wierzbicki (1986). Aggregation procedures for hierarchically grouped decision attributes with application to control system performance evaluation. Proceedings International Conference on Vector Optimisation, Technical University of Darmstadt, F.R.G.

Wierzbicki, A.P. (1985). Negotiation and mediation in conflicts, II: Plural rationality and interactive decision processes. In M. Grauer, M. Thompson, A.P. Wierzbicki (editors), Plural Rationality and Interactive Decision Processes. Proceeding, Sopron 1984, Springer Verlag, Berlin.

An Expert System Design for Fault Diagnosis in Electrochemical Operations - A Quantitative Approach

F. A. Batzias* and G. Kopsidas**

**Technology & Production Systems Dept., The Piraeus Graduate School of Industrial Studies, Piraeus, Greece*
***Chemical Engineering Dept., National Technical Univ. Athens, Greece*

ABSTRACT

This paper deals with the design of an Expert System (ES) for fault diagnosis in electrochemical operations and especially in electroplating - anodizing. Fault tree analysis is used as a searching technique within the ES while the corresponding knowledge base includes standard quality control methods and simulation models. A part of the tree structure and a sample dialogue of the user with the ES are presented and some shortcomings concerning semantics of the keywords, used for communication between user and ES, are discussed.

KEYWORDS

Electrochemical operations; expert system; fault diagnosis; fault tree analysis.

INTRODUCTION

The aim of this paper is to contribute in the design of an Expert System (ES) for fault diagnosis in Electrochemical Operations and especially in electroplating - anodizing. Such an ES can be used (a) in an Electrochemical Institute offering technical support to small or medium size electrodepositors or anodizers and (b) in large electrochemical industrial units.

The searching technique adopted is Fault Tree Analysis (FTA), in a quite simple form, as there exist some problems in the synthesis of fault trees for metals processing plants and generally for plants with a major physico-chemical process. According to Lapp and Powers (1977), the most important of these problems are:

1. The variables have a continuous range rather than two states.
2. Failures may propagate both upstream and downstream from a disturbance.
3. Time delays are significant.
4. The equipment may have a continuous range of degradation rather than two states of operation and failure.

In contrast, electrical/electronic systems have a number of special characteristics that facilitate the application of FTA. According to Martin-Solis, Andow and Lees (1982), these special characteristics are that (a) the components have only two states, (b) time delays are not significant in system dynamics and (c) failure data are comparatively easy to obtain and are consequently well defined.

Electrochemical operations form a link between physico-(non electro)chemical processes and conventional electrical/electronic systems. The main characteristic of this link is the important role of electric current, which is the guiding force of the physico-chemical transformations that take place during electroplating or anodizing. Consequently, these electrochemical systems have some common points with pure electrical systems. On the other hand, the values of the parameters referring to (a) the properties of the product obtained after the electrochemical surface treatment and (b) the conditions of treatment, vary within a continuous range; therefore, from this point of view, the electrochemical treatment comes closer to conventional chemical processing.

As the observation of a defect in the product is usually the cause (used also as a top event in FTA), that gives rise to further investigation, the corresponding physicochemical parameters appear to play the most important role; consequently, the critical intermediate stages in searching through FTA exhibit the same problems in tree synthesis that are common in process plants.

METHODOLOGY

As it is widely accepted, current ES technology mainly employs the shallow knowledge technique. Typically, a production rule representation is used so that an ES's knowledge about fault detection is simply a series of IF - THEN statements that embody the rules-of-thumb from a group of experts in this field. The rules usually indicate how this sample of experts respond when a certain category of problems arise, without giving an explanation for this response. Shallow knowledge can be extended towards several domains of corresponding deep knowledge, bringing the latter to a phenomenological level where the depth of knowledge decreases in behalf of its width, i.e. its practical usefulness for a greater variety of cases. But acting so, deep knowledge lacks its explanatory power and is reduced to be a kind of shallow knowledge, frequently less practicable than rules-of-thumb. On the other hand, maintaining deep knowledge in its original dimensions, we are not certain that we can always introduce an inference engine that will keep the system in self-consistency and in a ready-to-act state. Even if the ES with deep knowledge incorporated into it, exhibits these merits, we cannot be certain that addition of new knowledge (or rules) will not have unforeseen consequences on the overall functioning of the ES.

Therefore, we decided not to use two distinct kinds of knowledge but only one, suitably modified to form a continuum from what has been referred to as shallow (Hart, 1982), surface (Nau, 1983), compiled (Chandrasekaran and Mittal, 1982), or low-road (Michie, 1982) knowledge to what has been referred to as deep or high-road knowledge. The architecture of our knowledge base has a tree structure suitably formulated for searching by an industrial user (with the necessary scientific background) other than the human interface who is responsible for the initial construction of the base and its enrichment during the first stages of its operation.

The process flow diagram of the electrochemical plant is converted to an integrated block diagram of successive stages, where the first stage S0 represents metallurgy of the base metal, though this industrial operation is not included in the electrochemical plant. This addition is necessary for attributing defects occuring before entering of the metallic articles into the stage of preparation for electrochemical surface treatment. In the basic part of the tree structure for anodizing, depicted in Fig. 1, there are five successive stages: S0 = metallurgy of the aluminium and construction of the article to be anodized, S1 = mechanical, chemical and electrochemical preparation of the surface to be anodized, S2 = anodizing, S3 = dyeing, S4 = sealing.

The common symbols used are shown in Fig. 2. The OR gates are normally inclusive OR (IOR) gates, but exclusive OR (EOR) gates can be included if necessary. Though published fault trees derived from synthesis algorithms often contain few, if any,

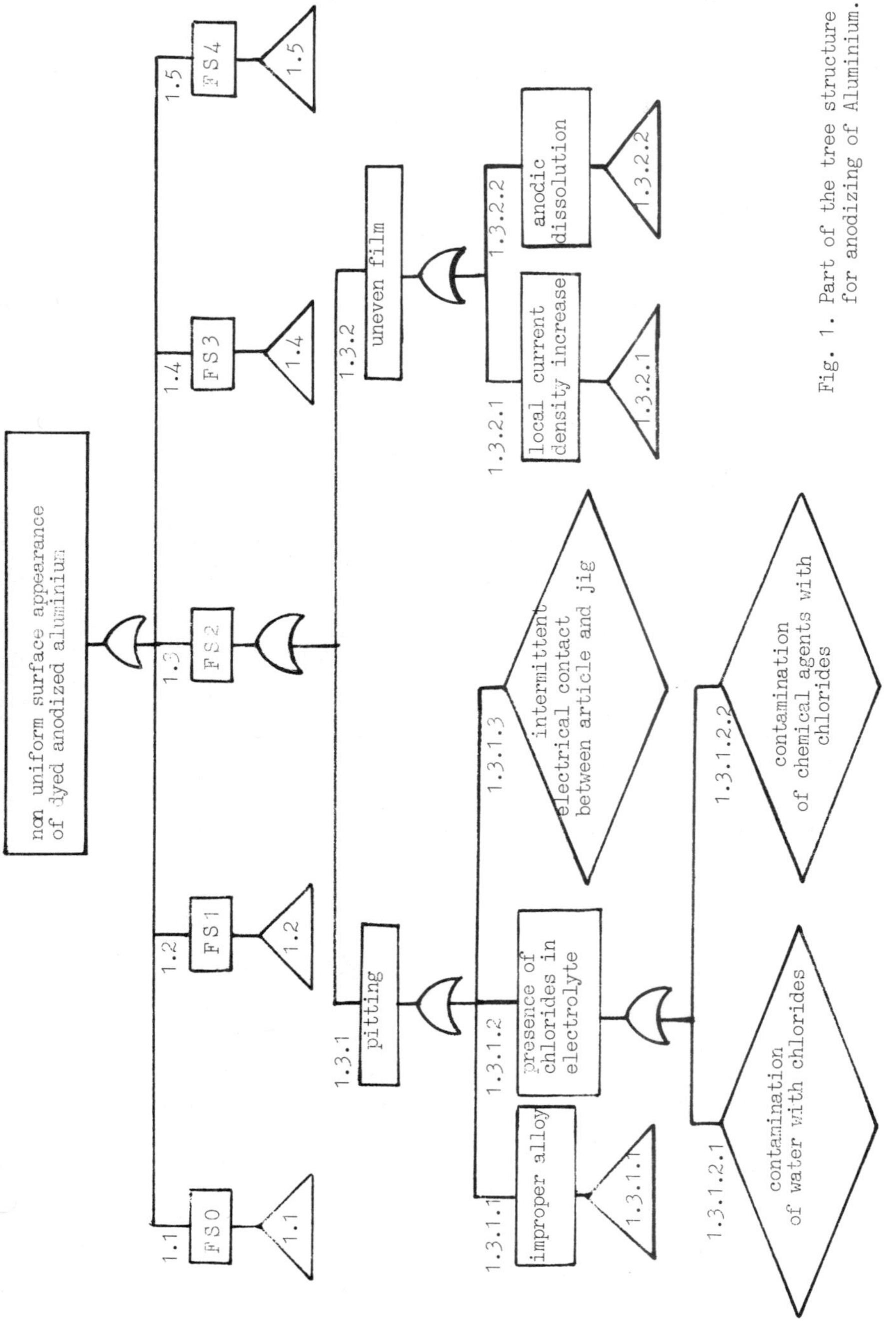

Fig. 1. Part of the tree structure for anodizing of Aluminium.

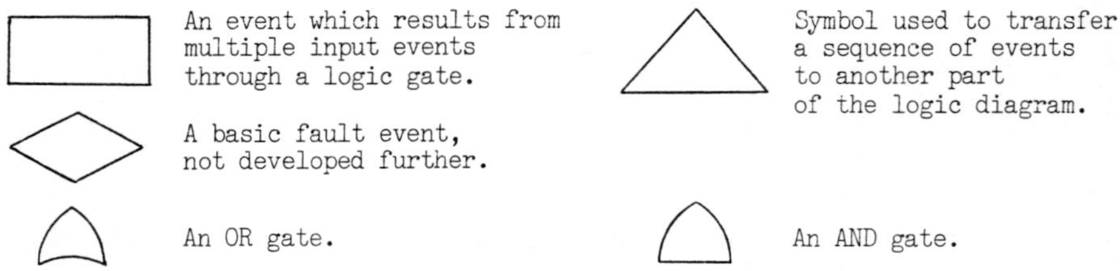

Fig. 2. Common symbols used for Fault Tree Analysis.

AND gates, we have successfully introduced AND gates by analyzing the probable causes of failure to their very elementary components and by synthesizing these components to form mini fault trees, as it is shown in Fig. 3.

The search method we have used is "available first" or "most likely first" rather than "depth first" or "breadth first" (semantics of the latter two terms and an illustrative example are quoted by Andow, 1985) and it is comparable to the well known in the operations research "branch and bound method", which is available in several algorithmic versions. The method used for searching is heavily based on the availability of quality control tests performed mainly at the end of each intermediate stage; for this reason, the first generation nodes FSi (standing for Fault observed or revealed at the end of Stage i, i = 0,1,2,3,4) determine the five main branches of our tree structure, serving as operationally grouped (by qua-

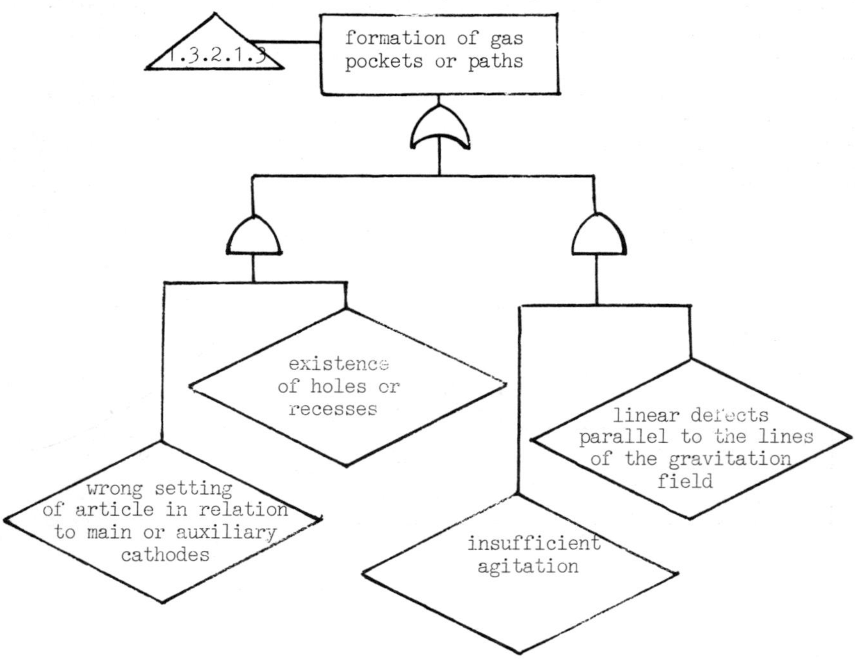

Fig. 3. Introduction of AND gates into the tree structure, by the aid of mini fault trees formation.

lity control tests) phenomenological attributes rather than as role indicators (for the meaning of the latter term in comparison with other terms of the same semantic family, see Paice, 1977).

A collection of quantitative relations, retrieved from the scientific/technical literature has been assigned to each relevant node of the tree structure. The only criterion for this assignment is the existence of at least one parameter within the quantitative relation, that may influence directly or indirectly (by setting up a deviation leading to a fault in the process) the appearance of a defect in the product. On the other hand, a collection of corresponding quality control tests, standardized by the well-known Organizations ASTM, BS, DIN, AFNOR, ISO, etc. or extracted from industrial practice, has been assigned to the same nodes. By scanning the tree structure with the "available first" or "most likely first" technique, a continuous combination of parameters is performed (between a standardized quality control parameter and a normalized quantitative operational model parameter, each time) giving rise to the construction of a combinatorial matrix, which serves as a decision making mechanism. The evaluation of these combinations, as far as fault analysis is concerned, is based on a set of weights extracted from a static set of rules, which obtains dynamic character only by intervention of the human interface, i.e. the industrial user.

The distinct values of the various combinations included in this matrix range between +2 and -2, with the following meaning: +2 = combination accepted, without any further recommendation; +1 = nearly good combination, but some improvement of product characteristics can be achieved, by the aid of correction of industrial treatment conditions; 0 = combination accepted with caution to sample statistics, e.g. the estimator of covariance; -1 = combination rejected because of the failure of one parameter to meet the quality control standards - correction is relatively easy to be achieved; -2 = combination absolutely rejected, because of the failure of two or more parameters to meet the quality standards - correction is relatively difficult to be achieved and the design of the corresponding industrial operation must be reconsidered.

A quantitative relation need not be explanatory in order to be used. Moreover, a remedial proposal may result as an empirical consequence of an evidence provided by a quality control test, as indicated in the following conceptual framework for integrated fault searching, where a bypass route guides directly from COMBINATION to REMEDIAL PROPOSAL. Nevertheless, an explanation, even on a surface phenomenological level, can predict an upstream or the ultimate cause and suggest the corresponding remedy, which may differ from the remedy retrieved by the empirically designed bypass route.

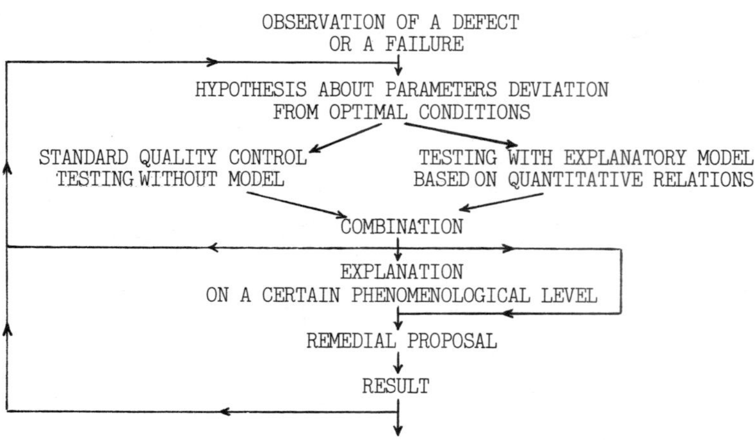

CASE EXAMPLE

The following case example represents a sample dialogue between the user and the ES, which uses inter-document search in a knowledge base of quality control standards, as we have described recently (Batzias, 1986a). The problem under investigation is "non-uniform surface appearance of dyed anodized aluminium"; the choice of this general defect for FTA is not arbitrary but is based on the qualitative approach of the same problem presented recently (Batzias, 1986b). In the following extract, the system output is in upper case while the user's responses are in lower case.

WELCOME TO THE ELECTROCHEM DIAGNOSTIC SYSTEM
THIS IS THE ANODIC OXIDATION FILE
STAGE 0 = METALLURGY/CONSTRUCTION OF THE ARTICLE, STAGE 1 = PREPARATION FOR ANODIZING, STAGE 2 = ANODIZING, STAGE 3 = DYEING, STAGE 4 = SEALING.

1. ENTER REAL DATA FOR EACH STAGE ACCORDING TO THE FOLLOWING PATTERN...
   ..... (user's answer)
2. DEVIATIONS FOUND : ... WHAT IS THE PROBLEM?
   non-uniform surface appearance of dyed anodized aluminium.
3. IS THE APPEARANCE OF THE ANODIC SURFACE UNIFORM AT THE END OF STAGE 2?
   no.
4. MEASURE THE THICKNESS OF ANODIC COATING IN THE DEFECTIVE REGION AT THE END OF STAGE 2 BY MICROSCOPICAL EXAMINATION OF CROSS SECTIONS, ACCORDING TO ASTM STANDARD METHOD B 487-85. DO YOU OBSERVE UNEVEN COATING?
   yes.
5. MEASURE THE THICKNESS OF ANODIC COATING IN THE DEFECTIVE REGION AT THE END OF STAGE 2 USING A LIGHT-SECTION MICROSCOPE, ACCORDING TO ASTM STANDARD METHOD B 681-81. DO YOU OBSERVE UNEVEN COATING?
   yes.
6. WHAT IS THE MEAN THICKNESS OF THE ANODIC COATING IN THE DEFECTIVE REGION?
   10 micrometers.
7. MEASURE THE THICKNESS OF ANODIC COATING IN THE DEFECTIVE REGION AT THE END OF STAGE 2 USING THE NON-DESTRUCTIVE VERSION OF THE DOUBLE-BEAM INTERFERENCE MICROSCOPE TECHNIQUE, ACCORDING TO ASTM STANDARD METHOD B 588-75. DO YOU OBSERVE UNEVEN COATING?
   yes.
8. MEASURE THE THICKNESS OF ANODIC COATING IN THE DEFECTIVE REGION AT THE END OF STAGE 2 USING AN EDDY-CURRENT INSTRUMENT ACCORDING TO ASTM STANDARD METHOD B 244-79. DO YOU OBSERVE UNEVEN COATING?
   no.
9. MEASURE THE WEIGHT OF ANODIC COATING IN THE DEFECTIVE REGION AT THE END OF STAGE 2 BY THE CHEMICAL STRIPPING METHOD, ACCORDING TO ASTM STANDARD METHOD B 137-45. DO YOU OBSERVE UNEVEN COATING?
   yes.
10. CAN THE CURRENT DISTRIBUTION BE SIMULATED TO A CHANNEL MODEL, IN THE DEFECTIVE REGION?
    yes.
11. IS THE FLOW LAMINAR?
    yes.
12. IS THE PECLET NUMBER LARGE?
    yes.
13. IS THE THIN DIFFUSION LAYER APPROXIMATION VALID?
    yes.
14. IS THE UNIFORMITY OF THE CURRENT DISTRIBUTION SATISFACTORY, ACCORDING TO F CRITERION?
    What is F CRITERION?
15. F CRITERION IS GIVEN BY $F = FO - (1/(2*PI*K))*\int_0^L (Ic(Z)*LN(SINH^2(PI*(X-Z)/(2*H)) + SIN^2(PI*(Y-H)/(2*H))) - Ia(Z)*LN(COSH^2(PI*(X-Z)/(2*H)) + SIN^2(PI*Y/(2*H))))\, dZ$, WHERE FO IS THE REFERENCE POTENTIAL OF SOLUTION, IN THE ABSENCE OF CURRENT, RELATIVE TO GROUND (V), K IS THE CONDUCTIVITY OF BULK SOLUTION (mho/cm), L IS THE LENGTH OF ELECTRODES (m), Ic(Z) IS THE CURRENT DENSITY ON THE CATHODE (A/

m^2), TAKEN TO BE NEGATIVE, X IS THE DISTANCE DOWNSTREAM FROM THE LEADING EDGES OF THE ELECTRODES (m), Z IS THE REFERENCE VALUE OF X IN THE LOCALITY (m), H IS THE DISTANCE BETWEEN ELECTRODES OR THE WIDTH OF THE CHANNEL OBTAINED BY SIMULATION (m), Y IS THE DISTANCE NORMAL FROM THE ANODE TOWARDS THE CATHODE (m), Ia(Z) IS THE CURRENT DENSITY ON THE ANODE (A/m^2), TAKEN TO BE POSITIVE. STANDARD EXPERIMENTAL METHODS AND VALUES FOR TESTING THE UNIFORMITY OF THE CURRENT DISTRIBUTION ARE GIVEN IN FILE E28. IS THE UNIFORMITY OF THE CURRENT DISTRIBUTION SATISFACTORY, ACCORDING TO F CRITERION?
yes.

16. IS THE UNIFORMITY OF THE CURRENT DISTRIBUTION SATISFACTORY, ACCORDING TO E CRITERION?
what is E CRITERION?
17. E CRITERION IS GIVEN BY E = 1/(1+LN(B*Eo*T+1)/(B*Eo)), B IS THE FIELD COEFFICIENT, Eo IS THE ANODIZATION ELECTRIC FIELD UNDER CONSTANT CURRENT CONDITIONS, T IS THE NORMALIZED TIME OF ANODIZATION (DIMENSIONLESS). STANDARD EXPERIMENTAL METHODS AND VALUES FOR TESTING THE UNIFORMITY OF THE CURRENT DISTRIBUTION ARE GIVEN IN FILE E30. IS THE UNIFORMITY OF THE CURRENT DISTRIBUTION SATISFACTORY, ACCORDING TO E CRITERION?
no.
18. IS THE NORMALIZED CAPACITANCE OF THE ANODIC FILM UNIFORM IN THE DEFECTIVE REGION?
no.
19. - - - - - - - - - - - - -

The dialogue goes on, until the fault is found to be a kind of porosity developed along roll marks on the basis metal. According to standard specification ASTM B 580-79, defects in the coating that result from such conditions should not be cause for rejection; i.e. the detection of fault in a stage previous to electrochemical sub-stages leads either to a remedy in stage S1 (pre-etching, electrobrightening, desmudging) or to a prevention action in the rolling room. The quantitative relations presented in the above sample dialogue, have been extracted from scientific literature (Caban and Chapman, 1976; Albella and co-workers, 1986), and their use indicates the way that explanatory models can be incorporated into the knowledge base. This incorporation extends the shallow knowledge without entering a source of deep knowledge, in which case an integrated diagnostic model, like the one presented by Fink, Lusth and Duran (1985), would be necessary to be used.

The evaluation of results obtained by different quality control tests is a helpful device for keeping the tree structure away from its corresponding network structure. E.g. a standard measurement of admittance (i.e. the electrical conductivity for AC) is related with the thickness of the anodic coating; according to Henley (1982), an admittance not exceeding 500 micro-sec/d, where d is the thickness of coating, is considered acceptable but the addition of bloom-preventing agents (see sample dialogue in Discussion) to sealing solutions and the adverse effects of phosphate and silicate in hot-water sealing (stage S4) are not always reflected by a high admittance value. Consequently, it is possible to obtain good results by the admittance test on unsatisfactorily sealed coatings (unidentified defect in FS4 branch of the tree structure shown in Fig. 1). To avoid this confusion, we must relate concentration of some substances in the sealing bath with the sealed film admittance and with results of thickness measurements obtained by an eddy-current instrument according to ASTM standard method B 244-79. Such a relation creates closed loops within the tree structure, transforming this structure to a directed multigraph and, subsequently, to a network, when weight values are assigned to each arc.

The evaluation of results obtained by different quality control tests contributes decisively to the avoidance of such a distortion of the original tree structure; in the above quoted example, the results of measurements of the thickness of sealed anodic coating by microscopical examination of cross sections, according to the ASTM standard method B 487-85, are not related with concentration of certain substances in the sealing bath; therefore, these results are considered to be more reliable and the corresponding mini fault tree (referring to the identification of uneven coatings) maintains its independence in relation to other branches.

DISCUSSION

The formal language initially used was PASCAL UCSD but later on we changed to PROLOG, which contains search and pattern-matching features that are well-matched to the computational requirements of the application. PROLOG does however lack some useful features of more conventional languages. A system combining PROLOG and procedural features, like POPLOG, has been suggested by Andow (1985) as the best way forward. Special algorithmic methods and computer programs for FTA have also been developed since 1970, when Vesely and Narum published two computer codes for FTA, PREP and KITT.

When the values of some parameters of the quantitative relations used in the tree structure are out of the recommended range, a defect in the product may occur. E.g. close temperature control is important in anodizing, because the current density and the rate at which the film is dissolved by the electrolyte alter markedly with temperature; these factors in turn affect the properties of the anodic film and, in particular, the "blooming" of the coating, which in some forms may not occur until a long period after anodizing.

The architecture of the knowledge base of the ES must include the links connecting such a defect with corresponding quantitative relations containing parameters closely related with the conditions of setting up the defect. The enrichment of the knowledge base with new information retrieved from data/information bases meets with severe difficulties as keywords representing defects have a different meaning, depending on the sub-stage where the defect has been observed. In the above mentioned example, "blooming" refers to a common defect occuring in stage S2 but under the same name another defect occuring in stage S4 is also known. By introducing the latter defect as a top event in the ES, we obtain the following remedial proposals at the end of the corresponding dialogue between the user and the ES:

```
    - - - - - - - - - - - -
 2. DEVIATIONS FOUND : ...  WHAT IS THE PROBLEM?
    blooming.
 3. IS THE APPEARANCE OF THE ANODIC SURFACE UNIFORM AT THE END OF STAGE 2?
    yes.
 4. - - - - - - - - - - - -
    - - - - - - - - - - - -
50. FOLLOW REMEDY R1 OR R2 OR (R1 AND R2). R1 = USE SPECIAL ADDITIVES TO THE SEAL-
    ING BATH TO MINIMIZE OR PREVENT BLOOM FORMATION. R2 = REMOVE BLOOM IN AN ACID
    DIP, PREFERABLY NITRIC ACID; SUBSEQUENTLY RINSE AND DRY THE ARTICLE. GIVE THE
    CODE OF REMEDY FOLLOWED.
    R2.
51. DO YOU OBSERVE BLOOMING AGAIN?
    - - - - - - - - - - - -
```

By introducing the first defect (blooming at stage S2) as a top event in the ES, we obtain the following remedial proposals at the end of the corresponding dialogue between the user and the ES:

```
    - - - - - - - - - - - -
 2. DEVIATIONS FOUND : ...  WHAT IS THE PROBLEM?
    blooming.
 3. IS THE APPEARANCE OF THE ANODIC SURFACE UNIFORM AT THE END OF STAGE 2?
    no.
 4. - - - - - - - - - - - -
    - - - - - - - - - - - -
50. FOLLOW REMEDY R1 OR R2 OR R3 OR (R1 AND R2) OR (R1 AND R3) OR (R1 AND R2 AND
```

R4). R1 = DECREASE TEMPERATURE INDICATION OF THERMOSTATIC CONTROL OF THE COOLING SYSTEM. R2 = INCREASE THE RATE OF AGITATION. R3 = DECREASE THE CURRENT DENSITY. R4 = REARRANGE AGITATION SOURCES TO ACHIEVE UNIFORM TEMPERATURE FIELD AND TO AVOID BLIND AREAS. GIVE THE CODE OF REMEDY FOLLOWED.
R1.
51. DO YOU OBSERVE BLOOMING AGAIN?
\- - - - - - - - - - -

In these cases, the ES can follow the proper branch in the tree structure and distinguish the cause of blooming from the answer to query 3; but in other cases, where the defect keyword is not referred to separate branches of the tree structure, some confusion may arise resulting in decrease of the effectiveness of the ES or in wrong fault detection and, consequently, in wrong remedial proposal.

CONCLUSIONS

In this paper, we have presented some basic points for the design of an expert system for fault diagnosis in electrochemical operations and especially in electroplating-anodizing. We have indicated that dfault tree analysis can be successfully used as a searching technique in this domain. The main characteristic of this expert system is a continuous combination of standardized quality control parameters with normalized quantitative operational model parameters; the so obtained combinatorial matrix serves as a decision making mechanism. Moreover, explanatory models can be incorporated into the knowledge base as an extention of the shallow knowledge, without entering sources of deep knowledge. It is also indicated that some confusion may result in searching through the tree structure as a result of usage of a keyword with various semantic content. This confusion can be minimized by setting up proper queries assigned to strategic nodes but cannot be eliminated, especially when a keyword obtains its different semantic content within the same branch of the tree structure.

ACKNOWLEDGEMENT

This work is an extended part of the authors' contribution to the "Study of the Greek Electrochemical Industry", a programme sponsored by the Ministry of Industry, Research and technology, especially concerning organization and management of an Institution for technology transfer in the Electroplating - Anodizing Industrial Subsector.

REFERENCES

Albella, J. M., I. Montero, O. Sanchez, and J. M. Martinez-Duart (1986). J. Electrochem. Soc., 133, 876-879.

Andow, P. K. (1985). Fault Diagnosis Using Intelligent Knowledge Based Systems. Inst. Chem. Eng., Symposium Series No.92, Pergamon, Oxford. p p. 145-156.

Batzias, F. A. (1986a). Design of an Interlaboratory Network for Fault Detection in Industrial Production Processes. In S. Tzafestas, M. Singh, and G. Schmidt (Eds.), System Fault Diagnostics, Reliability and Related Knowledge-Based Approaches, vol. II. D. Reidel Publ. Co., Dordrecht, Holland. pp. 648-661.

Batzias, F. A. (1986b). An Expert System Design for Fault Diagnosis in Electrochemical Operations: A Qualitative Approach. In S. Tzafestas, M. Singh, and G. Schmidt (Eds.), System Fault Diagnostics, Reliability and Related Knowledge-Based Approaches, vol. II. D. Reidel Publ. Co., Dordrecht, Holland. pp. 519-535.

Caban, R., and T. W. Chapman (1976). J. Electrochem. Soc., 123, 1036-1041.

Chandrasekaran, B., and S. Mittal (1982). Deep versus Compiled Knowledge Approaches to Diagnostic Provlem-Solving. Proc. AAAI-82. pp. 349-354.

Fink, P. K., J. C. Lusth, and J. W. Duran (1985). IEEE Trans. Pattern Anal. Mach. Intell., PAMI-7, 553-560.

Hart, P. (1982). Directions for AI in the eighties. SIGART Newslett., no. 79.

Lapp, S. A., and G. J. Powers (1977). IEEE Trans. Reliab., R-26: 2.

Martin-Solis, G. A., P. K. Andow, and F. P. Lees (1982). Trans. Inst. Chem. Eng., 60, 14-25.

Michie, D. (1982). High-Road and Low-Road Programs. AI Mag., 3, No. 1, 21-22.

Nau, D. (1983). Expert Computer Systems. Comput., Feb. 1983, 63-85.

Paice, C. D. (1977). Information Retrieval and the Computer. Macdonald and Jane's Publ. Ltd., London. p. 11.

Vesely, W. E., and R. E. Narum (1970). IN-1349, Idaho Nucl. Corp., Idaho Falls, Idaho.

II. Fault Detection of Technological Systems

Critical Assessment of Various Leak Detection Techniques for Gas Pipeline Networks

G. Lappus and G. Schmidt

Lehrstuhl für Steuerungs- und Regelungstechnik
Technische Universität München
D-8000 München 2, Arcisstraße 21, Bundesrepublik Deutschland

ABSTRACT

Conventional leak detection techniques for gas pipelines or networks are based on special hardware and manual inspection. Although these methods are still required, model-based on-line leak detection techniques can provide additional security. Over the last ten years various methods have been developed. Most of them are based on transient flow models and state estimation schemes. This paper discusses and compares some of these techniques recently presented in literature.

KEYWORDS
Leak detection, gas pipelines, gas networks, on-line model, observer, Kalman filter, nonlinear distributed parameter system.

INTRODUCTION

All over Europe large gas pipeline networks are operated in order to transport and distribute natural or industrial gases from production to consumption areas. The safe operation of these systems is of significant importance due to the serious and costly effects that may result from faulty operation. Leaks are of particular interest since they can lead not only to considerable product losses but also to situations imposing danger for the environment. The causes of leaks may be ascribed to several factors, such as unintentional pipeline violation by construction equipment, corrosion, defects in the pipeline joints, inadequate design and intentional sabotage. Pipeline segments having leaks need to be detected and isolated for repair. This is a difficult task in the case of compressible, transient gas flow and many different types of leak detection techniques and devices have been developed, accordingly. This paper discusses some model-based leak detection methods which have recently been suggested in literature. It is organized as follows: in the next two sections, we first outline the process under consideration and the methodologies of leak detection. Section 4 then discusses various model-based leak detection techniques. A summary of this comparison concludes the paper.

GAS TRANSMISSION NETWORKS

Here, we consider high pressure pipeline networks being used for the transportation and distribution of natural gas or industrial gases, such as hydrogen, ethylene, oxygen, etc. These networks are operated at pressures of between 10 and 80 bar. Gas flow is always transient due to the time varying supplies, offtakes, and network control actions, such as switching of compressor stations or valves. The average gas flow through a pipeleg is typically in the range of 10^2 to 10^6 m^3/h. Such a gas transmission system may be a meshed distribution network or a tree-structured transportation system with a total pipeline length of between 50 km and some 1000 km, see Fig. 1. The common subsystem which we will use for our comparison, is a homogeneous, long pipeline.

The modeling of transient gas flow in transmission networks has been widely studied (Fincham and Goldwater, 1979). Though some authors choose to neglect minor physical effects which others include and despite the different numerical solution schemes employed, it is sufficient for our discussion to review only one model in order to show the basic model properties. We selected the GANESI simulation model of Weimann and Schmidt

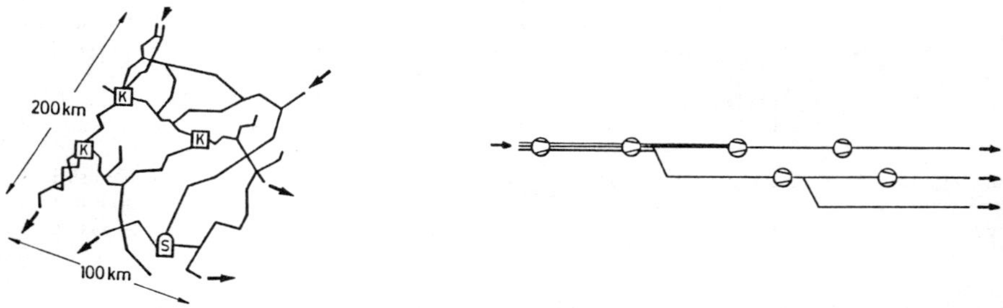

Fig.1. Typical gas distribution network and gas transport system.

(1977) which is widely accepted by the European gas industry.

For modeling purposes, the transmission network is decomposed into distributed parameter subsystems (long pipelegs) and lumped parameter subsystems (compressor, regulator, and valve stations). The essential dynamics in a network are due to the gas flow through the pipelegs, which is characterized by transportation lags and transient settling times, varying from several minutes up to several hours. Under normal operational conditions we can assume one-dimensional, isothermal and compressible gas flow. Using the state variables pressure p(z,t) and mass-flow q(z,t), gas flow through a pipeleg is represented by a nonlinear, hyperbolic-type set of two partial differential equations

$$\begin{aligned} \partial p(z,t)/\partial t &= -a_1(p)\cdot\partial q/\partial z \\ \partial q(z,t)/\partial t &= -a_2\cdot\partial p/\partial z - a_3(p)\cdot|q|\cdot q/p \end{aligned} \qquad (1)$$

z is the spatial coordinate along the pipeleg, and t is time. The coefficients a_1, a_3 and the constant a_2 follow from the geometrical and physical data of the pipeleg and gas. Boundary conditions at $z=0$ and $z=l$ are provided by coupling conditions between adjacent subsystems and/or by pressure or flow defining functions of time. Initial conditions may be specified by suitable spatial profiles of pressure and flow.

The spatial extensions and the dynamics of compressors, regulators and valves are negligible when compared with a long pipeleg. These elements are included in the network model as stationary, lumped parameter subsystems represented by nonlinear algebraic relations which describe the steady state relationship between the inlet pressure, the outlet pressure, the flow and a control variable.

The mathematical model of the total network is constructed by collecting and coupling all subsystem models. The coupling conditions at the connection points, i.e. at the network nodes, follow from mass and force balance equations. Based on a graph-oriented description of the network topology (input data), the network equations are generated by the program and then solved numerically by means of a modified Crank-Nicholson scheme, Newton-Raphson iteration and sparse matrix techniques.

GENERAL ASPECTS OF LEAK DETECTION

The variety of leak detection techniques which are available or presently under development demonstrates both the importance of leak detection and the fact that a completely satisfying solution is not yet available. A basic classification of leak detection methods is, with respect to the time characteristic, intermittent off-line testing, and continuous on-line supervision. This paper discusses continuous techniques. However, a first glance at some typical intermittent leak detection methods and their general properties is helpful for comparison.

Hydrostatic pressure testing. This method is used to prove a pipeline's integrity after construction. Very small leaks can be detected, but it requires a long period of pipeline shutdown. Therefore, it can be repeated only at long intervals, i.e. every few years.
Sonic pig testing. A sonic pig is a device which is outfitted with a hydrophone to receive the ultrasonic noise created by a leak. While the pig moves through a pipeline, it can detect and locate even very small leaks. Although a pipeline shutdown is not required, it is a rather costly method which can be repeated only at lengthy intervals, i.e. once a year.
Direct inspection. This is the most commonly used technique of leak detection. Inspection is done with sensitive gas detecting sensors while walking or driving along the length of all pipelines. For this purpose, special gas collection channels or points are available along the pipelines. This technique can be performed during normal

operation and small leaks can be detected. However, the inspection intervals are again lengthy. A similar method is based on inspection by aircraft where changes in the vegetation are used as indicators for possible leaks.

The main advantage of these techniques is their capability to detect very small leaks. However, all of these methods do not permit rapid leak detection. In addition, short term leaks are discovered only by chance.

Continuous on-line leak detection techniques are required to overcome these deficiencies. To provide such means for short term pipeline integrity testing, numerous methods have been devised. They use pressure and flow measurements as provided by the SCADA system. The dynamic process of leak detection itself consists of several tasks:

- detection of the leak situation,
- estimation of the leak location,
- estimation of the leak flow rate,
- decisions, e.g. isolation of the leaking segment by closing valves.

The effectiveness of any leak detection method can be judged by the following objectives:

- the size of a detectable leak should be as small as possible,
- the time till detection should be as short as possible,
- the required duration of the leak should be short,
- and there should be no false alarms (important in practice).

The leak size is obviously of particular interest. Therefore, we classify leaks:

- pinhole leaks with a leak flow $q_L < 1\%$ of the average pipeline flow,
- small-sized leaks with $1\% < q_L < 10\%$,
- gross-sized leaks with q_L up to 100% (pipeline rupture).

In this context we are mainly interested in the detection and localization of small and gross-sized leaks. In passing we note that it is easy to detect, but difficult to locate a pipeline rupture. Let us subdivide continuous leak detection techniques into conventional and advanced methods.

Conventional leak detection methods are based on simple cause/effect models or on static process models using single measurement data or combining just a few measurement readings. Typical examples are pressure gradient methods (gradient in time and/or space) and mass balancing methods with or without steady state correction terms. These techniques were successfully applied to some fluid pipelines being operated under nearly steady state conditions. For transient gas pipelines, however, they proved to be insufficient.

Advanced leak detection methods are based on comprehensive dynamic process models being driven by all available on-line measurements or, at least a group of them. These techniques are obviously encouraging because they include essential a priori knowledge about the space-time-correlations of the measurement data. In general, this approach to instrument and process fault detection is well known in literature (e.g., Isermann, 1984; Tylee, 1986). The basic idea is rather simple. On the one hand, we have data derived from measurements (either directly measured or calculated from measurements) and, on the other hand, we have the same type of data provided by a model or estimator. We may conclude by comparison that some fault has occurred if there are any discrepancies greater than certain prespecified limits.

Before discussing these methods, it is worth considering the limits to leak detection using SCADA measurement data. Available are pressure and flow readings from discrete point locations of the network. These measurement data have a restricted accuracy due to A/D conversion and data transmission typically using an 8 bit data representation. Additionally, intermittent gross errors may occur which must be taken into account. In general, all pressure measurements as well as all flow measurements at boundary points of the network are of a medium

TABLE 1 Typical Accuracy of Measurement Data

	pressure	flow
accuracy	0.5% .. 1.0%	0.2% .. 0.5% .. 1.0%
range	10 bar .. 80 bar	dependent on maximal flow
quantization error	±0.25 bar	dependent on flow range
absolute error	±0.5 bar	dependent on flow range
bias	±1.0 bar	-

quality, whereas flow measurements at interior network points are less accurate. From this we may conclude that a leak detection scheme should be applied to the whole network and not to subsystems being artificially decoupled by interior flow measurements. Typical properties of the available measurement data are summarized in Table 1. These errors as well as the data sampling rate of typically 1 min up to 10 min obviously restrict the possible performance of any leak detection scheme being based on an evaluation of these data.

ADVANCED MODEL-BASED LEAK DETECTION METHODS

In literature we find several proposals to use on-line models of transient flow for leak detection purposes (Goldberg, 1978; Covington, 1979; Dupont and co-workers, 1980). Schmidt, Weimann and Lappus (1978) proposed a leak detection scheme for gas transport pipelines based on an observer approach. Chapman and Pritchard (1982) as well as Billmann and Isermann (1984) basically used the same approach (with some modifications). Lappus (1985) presented a leak detection scheme for gas transmission networks based on a network observer. All methods mentioned above do not include the leak in the model, i.e. they are based on a fault sensitivity approach. Digerness (1980) presented a fault model approach for leak detection in oil pipelines. He used a bank of Kalman filters, each of which models a leak at a predefined location in the pipeline. Benkherouf, Allidina and Singh (1984) presented a leak detection scheme for gas pipelines which is also based on a Kalman filter. They used a single model but included artificial leak states at predefined locations along the pipeline. The following sections compare these leak detection methods. We focus our discussion on structural similarities and basic properties. We will not go into details since these concepts are presented in the literature.

Common Functional Scheme of Model-based Leak Detection Techniques.

A first comparison of model-based leak detection methods shows that all are based on a similar hierarchical structure, see Fig. 2. On level 0, the actual measurement data are preprocessed in order to eliminate gross measurement errors. The valid data are fed into level 1 where we find either a cause/effect model, a simple static process model or a comprehensive transient model which can be a process model, an observer or a filter. Eventually, fault models are included. On this level the measurement data are merged with the a priori knowledge of the physical process and its dynamics in order to obtain complete information about the instantaneous network and/or fault state. The results of this level, i.e. cause/effect relations and/or the computed network state are input to the level 2 fault analysis system. There, data evaluation and data aggregation schemes are employed in order to extract the relevant information required for the detection and localization of a leak.

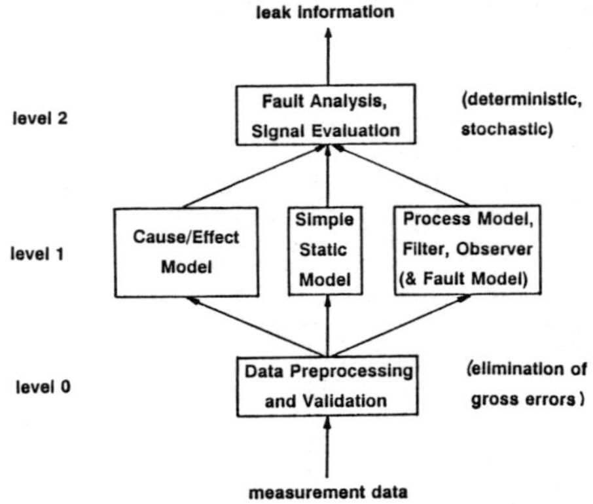

Fig. 2. Functional scheme of model-based on-line leak detection techniques.

Method (I): Parallel Model

Several leak detection techniques using an on-line model of transient gas flow were presented since about 1978, e.g. Goldberg (1978), Covington (1979), Dupont and co-workers (1980). The basic functional structure of this concept which is also offered by some companies, is shown in Fig. 3 in case of a single pipeline with pressure and flow measurements at both ends,

$$p_1(t)=p(0,t),\ p_2(t)=p(L,t),\ q_1(t)=q(0,t),\ q_2(t)=q(L,t). \tag{2}$$

We will use the same example for the presentation of all other methods. The parallel model describes normal operation of the pipeline, i.e. a non-leaking pipeline. The boundary conditions (BC) of the model are prescribed by the measured pressures $p_1(t)$ and $p_2(t)$. As the leak is not modeled, systematic modeling errors develop if a leak flow q_{leak} occurs at an arbitrary position z_L of the real pipeline. These errors cause, for instance, unexpected differences between the measured and computed flows at both ends of the pipeline,

$$e_{1q}(t)=q_1(t)-\hat{q}_1(t), \quad e_{2q}(t)=q_2(t)-\hat{q}_2(t). \tag{3}$$

By proper evaluation of these differences we can detect the occurence of a leak and estimate its position z_L and flow rate q_{leak}. This method has successfully been applied to an oil pipeline, however, in 1985 it failed in case of a gas transmission network. This insufficent performance is due to the high sensitivity of the computed flows to disturbed or faulty pressure measurements which are used as boundary conditions. It is noteworthy that, on the one hand, this model shows a differential behaviour between pressure and flow which amplifies the effects of disturbed pressure measurements. On the other hand, at least one pressure defining boundary condition is required to make the model asymptotically stable.

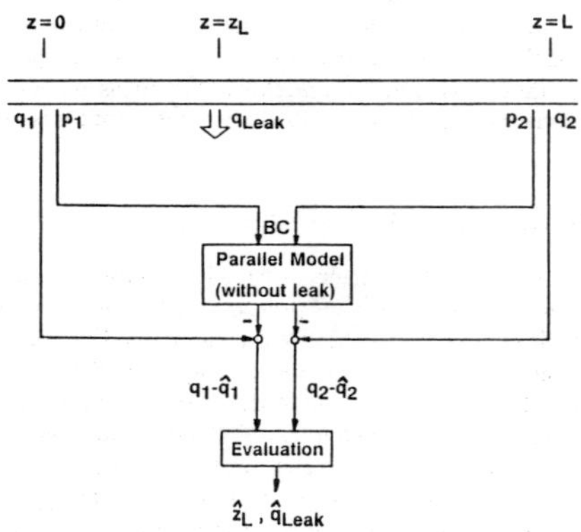

Fig. 3. Basic functional scheme of leak detection method (I).

<u>Method (II): Observer / Parallel Model</u>

Billmann and Isermann (1984) presented a leak detection scheme, see Fig. 4, which is based on a parallel model and a Luenberger-type observer. The observer is applied in the start-up phase in order to improve the convergence of the computed pipeline state to the real state. After this start-up phase, the observer feedback loop is opened and the leak detection scheme begins to work, i.e. leak detection is based again on a parallel model. The

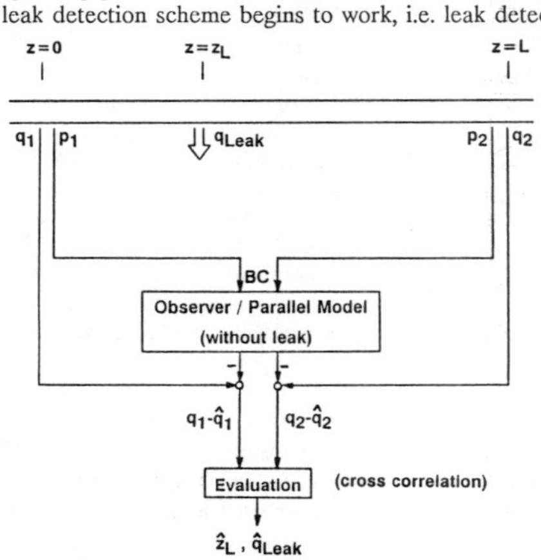

Fig. 4. Basic functional scheme of leak detection method (II).

boundary conditions of the model are prescribed by the pressure measurements $p_1(t)$ and $p_2(t)$. Leak detection and localization is based again on an evaluation of the differences e_{1q} and e_{2q}, see Eq. (3). However, this approach includes a parameter adaption scheme tracking the pipeline friction factor (part of variable a_3 in Eq. (1)). Second, the evaluation of the differences e_{1q} and e_{2q} is based on a sophisticated cross-correlation technique which reduces the inherent sensitivity to disturbed pressure measurements (see method I). Industrial on-line applications of this technique are not known to the authors.

Method (III): Luenberger-type Observer

Carmichael and Pritchard (1981) investigated leak detection based on a Luenberger-type observer. The functional scheme is shown in Fig. 5. In contrast to the methods above and similar to method (IV), flow defining boundary conditions are employed. This reduces the sensitivity to measurement disturbances. Leak detection is performed by evaluation of the observer residuals, i.e. the differences between measured and computed pressures at both ends of the pipeline,

$$e_{1p}(t)=p_1(t)-\hat{p}_1(t), \quad e_{2p}(t)=p_2(t)-\hat{p}_2(t). \tag{4}$$

Theoretically, these residuals are good indicators for a leak. However, variations caused by a leak are rather small when compared to other sources, such as faulty measurements. This small signal-to-noise ratio is certainly a difficulty in practice if leak detection is based only on an evaluation of these residuals. Industrial on-line applications of this technique are not known to the authors.

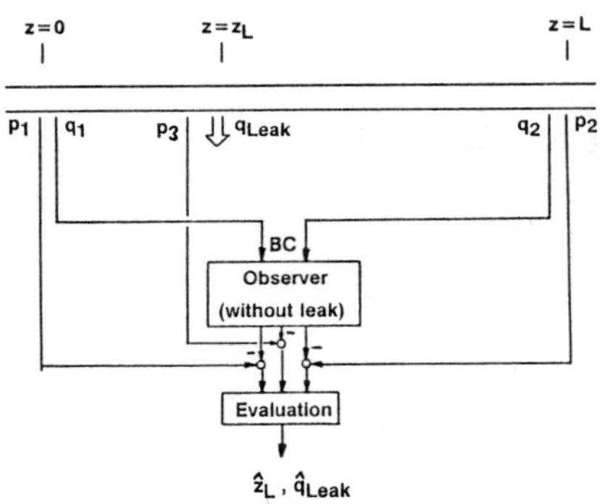

Fig. 5. Basic functional scheme of leak detection method (III).

Method (IV): Luenberger-type Observer

A Luenberger-type observer for leak detection in gas pipelines was presented first by Schmidt, Weimann and Lappus (1978). This approach was then refined and extended to gas transmission networks (Lappus, 1985). The functional scheme is shown in Fig. 6. The network observer is driven by flow defining boundary conditions. Leak detection is based on a two-stage evaluation scheme to increase reliability. Stage 1 detects the occurence of a leak, and estimates the leak flow rate. Both is done by evaluating the computed network state which provides a means for dynamic mass balancing. Once, a leak has been detected, then leak localization begins on stage 2 based on an evaluation of the observer residuals e_{1p} and e_{2p}. The drawback of this scheme, i.e. the small signal-to-noise ratio, is partially compensated by the a priori knowledge obtained at stage 1 that there is a leak. The leak detection scheme is suitable for networks and rather insensitive to measurement errors. The observer system is approved by industry, but results from industrial leak detection applications are not yet available.

Method (V): Kalman Filter Including a Leak Model

Benkherouf, Allidina and Singh (1984) describe a method for detecting and locating leaks in long pipelines based on an extended Kalman filter, see Fig. 7. Artificial leak states at predefined positions along the pipeline are included in the filter model. The filter estimates the flow process states and the leak states. The estimated leak states are then combined to one leak flow. The sensitivity of this method to real measurement errors has not been investigated. An essential drawback is the well-known computational burden involved when applying a Kalman filter. In particular, this would be a problem when considering a medium or large-sized network. Applications to industrial problems are not known to the authors.

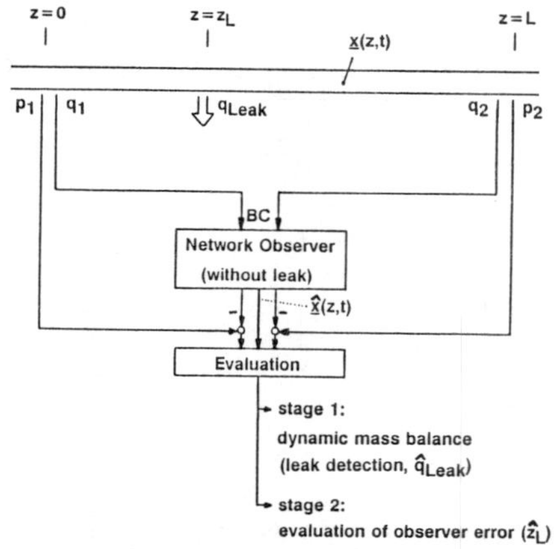

Fig. 6. Basic functional scheme of leak detection method (IV).

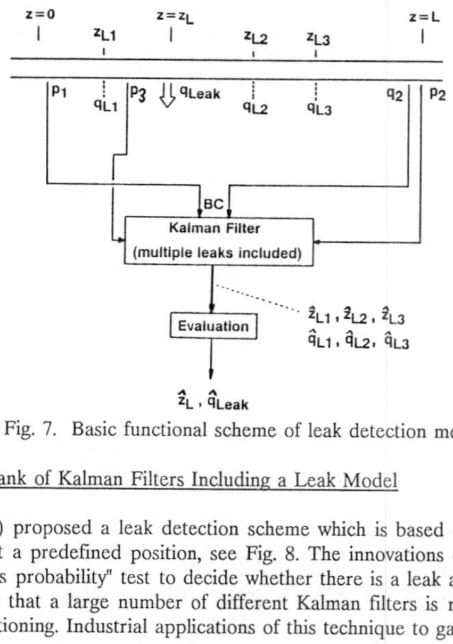

Fig. 7. Basic functional scheme of leak detection method (V).

Method (VI): Bank of Kalman Filters Including a Leak Model

Digerness (1980) proposed a leak detection scheme which is based on a bank of Kalman filters, each of which models a leak at a predefined position, see Fig. 8. The innovations of the filters are then used by a "multiple model hypothesis probability" test to decide whether there is a leak and also to locate it. The main drawback of this technique is that a large number of different Kalman filters is required to obtain a reasonable resolution on the leak positioning. Industrial applications of this technique to gas pipelines are not known to the authors.

CONCLUSIONS

The detection and localization of a suddenly occuring leak in a gas pipeline or gas transmission network is a challanging task. Difficulties are mainly due to the compressible medium and the always transient flow conditions. Conventional leak detection methods are insufficient because they do not allow rapid detection of small-sized leaks. Improvements are possible by continuously operating model-based techniques. A critical assessment of various approaches recently presented in literature shows that all of them are very similar in structure despite the differences in type and number of models employed, e.g. a simulation model, a Luenberger-type observer, one or more Kalman filters. Of course, each author has tested his leak detection scheme. The published results show that the methods have a comparable effectiveness. However, all results were obtained by simulation studies, only. As no results from industrial on-line applications are available, it is not yet possible to compare the most essential property of a leak detection scheme, i.e. its sensitivity to real disturbances in the measurement data and the number of false alarms caused by disturbed measurements.

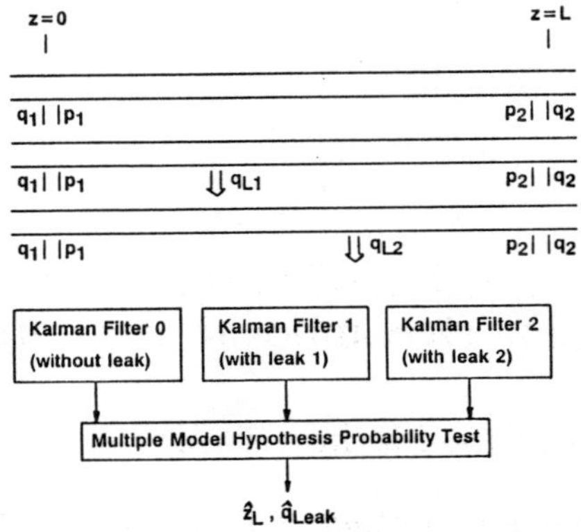

Fig. 8. Basic functional scheme of leak detection method (VI).

Further considerations show that industrial application of on-line leak detection techniques requires additional, situation-dependent expert-knowledge and global databases. These requirements can be met by a comprehensive decision support system for network monitoring and control which is composed of subsystems of two basically different types, i.e. modules providing algorithmic operations or problem solvers, and modules providing knowledge-based operations and problem solvers. A description of the structure of such a scheme is given by Lappus (1986).

REFERENCES

Benkherouf, A., Allidina, A.Y. and M.G. Singh (1984). *Detection and location of leaks in gas pipelines*. UMIST Control Systems Centre Report No. 603, Manchester.

Billmann, L. and R. Isermann (1984). *Leak detection methods for pipelines*. 9th IFAC World Congress, Budapest, Vol.3, 213-218.

Chapman, M.J., Jones, R.P. and A.J. Pritchard (1982). *State observers for monitoring gas pipelines*. 3rd IFAC/AFCET Symposium on Control of Distributed Parameter Systems, Toulouse.

Covington, M.T. (1979). Transient models permit quick leak identification. *Pipeline Industry*, August, 71-73.

Digerness, T. (1980). Real time failure detection and identification applied to supervision of oil and transport in pipelines. *Modelling, Identification and Control, 6*, 39-49.

Dupont, T., Rachford, H.H. Jr., McDonald, R.E., Gould, T.L. and H.J. Heinze (1980). *A transient remote integrity monitor for pipelines using standard scada measurements*. Interpipe Conference, 6. February 1980, Houston, Texas.

Fincham, A.E. and M.H. Goldwater (1979). Simulation models for gas transmission networks. *Transactions Measurement and Control, 1*, 3-13.

Goldberg, D.E. (1978). *On-line real time beneath the Gulf of Mexico*. Annual PSIG Meeting 1978.

Isermann, R. (1984). Process fault detection based on modeling and estimation methods. *Automatica, 20*, 387-404.

Lappus, G. (1985). *Gas dispatching by means of applied automatic control theory*. PSIG Annual Meeting, Albuquerque, New Mexico.

Lappus, G. (1986). *Knowledge-based process monitoring and control of gas transmission networks*. 1st European Workshop on Fault Diagnostics, Reliability, and Related Knowledge-Based Approaches. Aug. 31 - Sept. 3, Rhodes, Greece.

Schmidt, G.; Weimann,A. and G. Lappus (1978). Application of simulation techniques to planning, supervision and control of natural gas distribution networks. In: Carver, M.B. and M.H. Hamza (Ed.), *Simulation, Modelling and Decision in Energy Systems*. Acta Press Anaheim, 404-409.

Sedykh, A., Mirzandjanzade, A. and B.L. Kuchin (1979). *New methods of estimating and diagnosing in gas transportation systems*. World Gas Conference.

Tylee, J.L. (1986). Model-based approaches to instrument failure detection. *Intech, 33*, 59-62.

Weimann, A. and G. Schmidt (1977). Transient simulation of natural gas distribution networks by means of a medium-sized process computer. In: Van Nauta Lemke, H.R. and H.B. Verbruggen (Ed.), *Digital Computer Applications to Process Control*. Preprints of the 5TH IFAC/IFIP Conference, Hague, Netherlands, 315-320, North Holland Publ. Co., Amsterdam.

On the Sensor Fault Detection of Large Scale Systems Using the Overlapping Decomposition Approach

Spyros Tzafestas and Marinos Skolarikos

Control and Robotics Group, Computer Engineering Division
National Technical University, Zografou, Athens, Greece

ABSTRACT

The purpose of this paper is to present a method for the automatic detection of faults occuring in sensor elements of large scale systems. This method is largely based on the technique of overlapping decomposition. Specifically,using this technique decentralized observers for the overlapped subsystems are designed, which are then used to compare the values of their common state variables, as they are generated by them. Since this comparison is not sufficient to detect all possible failures, a signal is computed which in the steady-state can act as a correction to the wrongly measured variable, and produce a correct measured value for each subsystem. This corrected measured value in then used as a comparison basis for the detection of the remaining possible faults. The method was applied with success to a coupled - core nuclear reactor system with three cores, three inputs and three outputs.

KEYWORDS

Large scale systems, sensor faults, overlapping decomposition, decentralized observers, coupled-core nuclear reactors.

INTRODUCTION

Among the primary problems in the design, control and operation of multicomponent and large-scale technological systems, is the problem of effectively detecting and localizing instrument (component/ sensor) faults. All methods for solving this fault detection problem are based on some kind of **redundancy** either **hardware redundancy** or **analytical (functional) redundancy**. Hardware redundancy methods use multiple identical instruments (sensors etc) which with the aid of appropriate majority logic circuits detect the faults. Obviously this approach, due to the used back-up of instruments, leads to increased weight, volume and cost of the respective systems. Naturally, the analytical redundancy approach, which is based on some kind of mathematical model reflecting the existing inherent relationships among the measured variables of the system, has not these disadvantages. This fact together with the recent developments on control and estimation theory, as well as on distributed computing using microprocessors, has led many investigators to develop a bulk of analytical redundancy fault detection/location methods for dynamic systems.

The majority of these methods employ state (or output) estimation schemes, such as Kalman filters and Luenberger observers, and are more suitable for low-order systems.

Surveys of the entire field of fault detection/location of technological systems can be found in Willsky (1976), Iserman (1984), Tzafestas (1983,1985) and Frank (1987). The alternative approach to the system reliability/availability improvement is through repair/ maintenance (preventive or other) policies. Discussions on this approach can be found in Tzafestas (1980,1985), and Botsaris and Tzafestas (1984). This paper is devoted to the study of large-scale systems and provides a technique for the automatic detection of faults occuring in the sensor elements. This technique employs the overlapping decomposition procedure of Siljak (1982) and a set of observers designed in a decentralized way (Siljak and Vukcevic, 1976). A similar fault detection technique was firstly proposed by Singh, Hassan Chen, Li and Pan (1983). The present technique provides a refined way of detecting the faults and is also successful when more than one sensors are simultaneously faulty. This capability is achieved through the use of an appropriate signal that corrects the qrongly measured variables which are used as comparison a basis for the detection of the faults. Also the design of the observers is somehow different. The method was tested on a coupled-core nuclear reactor model with three cores, three inputs and three outputs. Some of the results derived are included. A pictorial representation of the fault detection technique developed in this paper is shown in Fig.1.

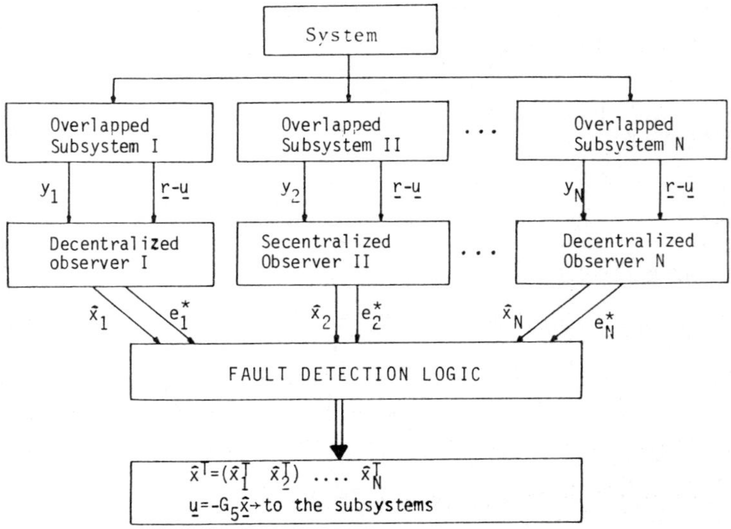

Fig.1. Block diagram of the fault detection procedure.

Here $y_1, y_2, \ldots, y_N$ are the scalar outputs of the original system, which also are outputs of the overlapped subsystems obtained after the application of the overlapping decomposition technique, $\mathbf{r}$ is the system reference input vector; $\mathbf{u}$ is the decentralized state feedback control input $\mathbf{u}=-G_s\hat{\mathbf{x}}$; $\hat{\mathbf{x}}_1, \hat{\mathbf{x}}_2, \ldots, \hat{\mathbf{x}}_N$ are the subsystems' state estimates provided by the observers and $\mathbf{e}_1^*, \mathbf{e}_2^*, \ldots, \mathbf{e}_N^*$ are correcting signals which are used in the fault detection algorithm.It is noted that to save space here we only show how to compute the matrix gain $\mathbf{G}_E$ of the observers'control law (see (18)). The procedure for computing G_s is identical.

In Section 2 we give a brief exposition of the overlapping decomposition technique, and in Section 3 we present the design of the decentralized observers for the overlapped subsystems. Section 4 describes the procedure proposed for the detection of sensor faults. Finally, section 5 provides the application of the method to a coupled-core nuclear system with three cores, three inputs and three outputs.

THE OVERLAPPING DECOMPOSITION TECHNIQUE

Consider a large-scale interconnected system S which is described by the state-space model:

$$S: \dot{\mathbf{x}} = \mathbf{A}\mathbf{x} + \mathbf{B}\mathbf{u}, \quad \mathbf{y} = \mathbf{C}\mathbf{x} \tag{1}$$

where $\mathbf{x} \varepsilon R^n$, $u \varepsilon R^m$, $y \varepsilon R^r$; $\mathbf{A}, \mathbf{B}, \mathbf{C}$ are matrices of appropriate dimensionality and $\mathbf{B}, \mathbf{C}$ are block diagonal matrices, each diagonal block of which corresponds to one of the N subsystems. Upon application of a full column rank matrix similarity transformation $\mathbf{T}$, where:

$$\tilde{\mathbf{x}} = \mathbf{T}\mathbf{x}, \quad \tilde{\mathbf{x}} \varepsilon R^{\tilde{n}} \tag{2}$$

the system (1) is transformed to the new system

$$\tilde{S}: \dot{\tilde{\mathbf{x}}} = \tilde{A}\tilde{\mathbf{x}} + \tilde{B}\mathbf{u}, \quad \mathbf{y} = \tilde{C}\tilde{\mathbf{x}} \tag{3}$$

with

$$\tilde{\mathbf{A}} = \mathbf{T}\mathbf{A}\mathbf{T}^{\dagger} + \mathbf{M}, \quad \tilde{B} = \mathbf{T}\mathbf{B} + \mathbf{N}, \quad \tilde{C} = \mathbf{C}\mathbf{T}^{\dagger} + \mathbf{L} \tag{4}$$

Here $\mathbf{T}^{\dagger}$ denotes the generalized inverse of T, i.e.

$$\mathbf{T}^{\dagger} = (\mathbf{T}^T\mathbf{T})^{-1}\mathbf{T}^T, \tag{5a}$$

which exists if $\mathbf{T}$ is of full column rank, and $\mathbf{M}, \mathbf{N}, \mathbf{L}$ are matrices satisfying the relations:

$$\mathbf{M}\mathbf{T} = 0, \quad N = 0, \quad LT = 0 \tag{5b}$$

It is clear that for a given $\mathbf{T}$ the choice of $\tilde{S}$ is not unique but depends on the choice of $\mathbf{M}, \mathbf{N}$ and $\mathbf{L}$.

The required form of the transformation $\mathbf{T}$ such as to obtain an overlapping decomposed system $\tilde{S}$ is determined as follows.

Let the state equation of the ith subsystem $S_i (i=1,2,\ldots,n)$ of S be

$$\dot{\mathbf{x}}_i = A_i\mathbf{x}_i + \sum_{\substack{j=1 \\ i \neq j}}^{N} \mathbf{A}_{ij}\mathbf{x}_j + B_i\mathbf{u}_i, \quad \mathbf{y}_i = C_i\mathbf{x}_i \tag{6}$$

where $\mathbf{x}_i \varepsilon R^{n_i}$, $\mathbf{u}_i \varepsilon R^{m_i}$ and $\mathbf{y}_i \varepsilon R^{r_i}$ with:

$$n=\sum_{i=1}^{N} n_i, \quad m=\sum_{i=1}^{N} m_i, \quad r=\sum_{i=1}^{N} r_i$$

For each subsystem i, the matrices $\mathbf{A}_{ij}$ are the parts of $\mathbf{A}$ that lie outside the diagonal and determine the interconnection of the states of the other subsystems with subsystems i. Suppose that $\mathbf{x}_i$ is decomposed as

$$\mathbf{x}_i = \begin{bmatrix} \mathbf{x}_{i1} \\ \hdashline \mathbf{x}_{i2} \end{bmatrix} \tag{7a}$$

where $x_{i1} \varepsilon \mathbf{R}^{n_{i1}}$ and $\mathbf{x}_{i2} \varepsilon \mathbf{R}^{n_{i2}}$ $(n_{i1}+n_{i2}=n_i)$. It is now obvious that to achieve overlapping decomposition of the original system we must extend the ith sybsystem state vector such as

$$\tilde{\mathbf{x}}_i = \begin{bmatrix} \mathbf{x}_{i-1,2} \\ \hdashline \mathbf{x}_{i,1} \\ \hdashline \mathbf{x}_{i,2} \end{bmatrix}, \quad i=1,2,\ldots,N \tag{7b}$$

The total state vector of the extended system $\tilde{S}$ is

$$\tilde{x}=\left[\tilde{\mathbf{x}}_1^T \mid \tilde{\mathbf{x}}_2^T \mid \ldots \mid \tilde{\mathbf{x}}_N^T\right]^T \tag{8}$$

where $\tilde{\mathbf{x}} \varepsilon \mathbf{R}^{\tilde{n}}$ and $\tilde{n}=n+\sum_{i=2}^{N} n_{i2}$.

The matrix $\mathbf{T}$ must be selected such that the new state vector $\tilde{x}$ has ths form (8). Thus if

$$\mathbf{x}=\left[\mathbf{x}_1^T \mid \mathbf{x}_2^T \mid \ldots \mid \mathbf{x}_N^T\right]^T \quad \text{or} \quad \mathbf{x}=\left[(\mathbf{x}_{11}^T \; \mathbf{x}_{12}^T) \mid \ldots \mid (\mathbf{x}_{N1}^T \; \mathbf{x}_{N2}^T)\right]^T \tag{9}$$

the relation (2) takes the form

$$\underbrace{\begin{bmatrix} \mathbf{x}_{11} \\ \mathbf{x}_{12} \\ \hdashline \vdots \\ \hdashline \mathbf{x}_{i-1,2} \\ \mathbf{x}_{i,1} \\ \mathbf{x}_{i,2} \\ \hdashline \vdots \\ \hdashline \mathbf{x}_{N-1,2} \\ \mathbf{x}_{N,1} \\ \mathbf{x}_{N,2} \end{bmatrix}}_{(\tilde{n}x1)} = \underbrace{\begin{bmatrix} \mathbf{I}_{11} & & & & & & & & & & \\ & \mathbf{I}_{12} & & & & & & & & & \\ & \mathbf{I}_{12} & & & & & & & & \bigcirc & \\ & & \mathbf{I}_{21} & & & & & & & & \\ & & & \ddots & & & & & & & \\ & & & & \mathbf{I}_{i-1,2} & & & & & & \\ & & & & \mathbf{I}_{i-1,2} & & & & & & \\ & & & & & \mathbf{I}_{i,1} & & & & & \\ & \bigcirc & & & & & \mathbf{I}_{i,2} & & & & \\ & & & & & & \mathbf{I}_{i,2} & & & & \\ & & & & & & & \ddots & & & \\ & & & & & & & & \mathbf{I}_{N,2} & & \end{bmatrix}}_{\mathbf{T}(\tilde{n}xn)} \underbrace{\begin{bmatrix} \mathbf{x}_{11} \\ \mathbf{x}_{12} \\ \hdashline \vdots \\ \hdashline \mathbf{x}_{i,1} \\ \mathbf{x}_{i,2} \\ \hdashline \vdots \\ \hdashline \mathbf{x}_{N,1} \\ \mathbf{x}_{N,2} \end{bmatrix}}_{(nx1)} \tag{10}$$

where

$$\mathbf{I}_{i1}=\begin{cases}(n_i-n_{i2})\text{-dimensional unity matrix for } i=1,N\\ (n_i-n_{i-1,2}-n_{i2})\text{-dimensional unity matrix for } i=2,3,\dots,N-1\end{cases}$$

and

$$\mathbf{I}_{i2}=\begin{cases}n_{i2}\text{-dimensional unity matrix for } i=1,2,\dots,N-1\\ (n_i-n_{i1})\text{-dimensional unity matrix for } i=N.\end{cases}$$

After the application of this transformation T, one obtains the state equations of each subsystem as

$$\tilde{S}_i:\dot{\tilde{\mathbf{x}}}_i=\tilde{\mathbf{A}}_i\tilde{\mathbf{x}}_i+\sum_{\substack{j=1\\j\neq i}}^{N}\tilde{\mathbf{A}}_{ij}\tilde{\mathbf{x}}_j+\tilde{\mathbf{B}}_i\mathbf{u}_i+\sum_{\substack{j=1\\j\neq i}}^{N}\tilde{\mathbf{B}}_{ij}\mathbf{u}_j \qquad (11a)$$

$$y_i=\mathbf{C}_i\tilde{\mathbf{x}}_i \qquad (11b)$$

It is noted that for the application of our fault detection technique it is helpful to use as overlapping state-vectors (variables) all the measured ones. This would lead to interconnections of the elements of the control vectors of the various subsystems. Obviously this affects the computations of the control law, but it does not present a real problem as long as this control law leads to an asymptotically stable system. Criteria for doing this have been developed by Siljak as described in the next section.

DECENTRALIZED OBSERVER DESIGN

The problem will be treated in the subsystem level (Siljak 1982; Singh, Hassan and Chen, Li and Pan 1983). For each one of the resulting overlapped subsystems (11a,b) a state observer must be designed of the type

$$\dot{\hat{\mathbf{x}}}_i=\tilde{\mathbf{A}}_i\hat{\mathbf{x}}_i+\sum_{\substack{j=1\\j\neq i}}^{N}\tilde{\mathbf{A}}_{ij}\hat{\mathbf{x}}_j+\tilde{\mathbf{B}}_i\mathbf{u}_i+\sum_{\substack{j=1\\j\neq i}}^{N}\tilde{\mathbf{B}}_{ij}\mathbf{u}_j+G_{iE}(\mathbf{y}_i-\tilde{C}_i\hat{\mathbf{x}}_i) \qquad (12a)$$

or

$$\dot{\hat{\mathbf{x}}}_i=\tilde{\mathbf{A}}_i\hat{\mathbf{x}}_i+\sum_{j=1}^{N}\tilde{\mathbf{A}}_{ij}\hat{\mathbf{x}}_j+\tilde{\mathbf{B}}_i\mathbf{u}_i+\sum_{\substack{j=1\\j\neq i}}^{N}\tilde{\mathbf{B}}_{ij}\mathbf{u}_j+G_{iE}(\tilde{\mathbf{C}}_{iF}\tilde{\mathbf{x}}_i-\tilde{C}_i\hat{\mathbf{x}}_i) \qquad (12b)$$

where $\mathbf{G}_{iE}$ is the control gain of each observer, and $\tilde{\mathbf{C}}_{iF}$ is the measurement matrix of the system outputs which are erroneously measured by the sensors.

To compute $\mathbf{G}_{iE}$ we shall assume that

$$\tilde{C}_{iF}=\tilde{\mathbf{C}}_i \qquad (13)$$

i.e. that no fault accurs. Using (13) the observer (12b) becomes

$$\dot{\hat{\mathbf{x}}}_i=\tilde{\mathbf{A}}_i\hat{\mathbf{x}}_i+\sum_{\substack{j=1\\j\neq i}}^{N}\tilde{\mathbf{A}}_{ij}\hat{\mathbf{x}}_j+\tilde{\mathbf{B}}_i\mathbf{u}_i+\sum_{\substack{j=1\\j\neq i}}^{N}\tilde{\mathbf{B}}_{ij}\mathbf{u}_j+G_{iE}\tilde{\mathbf{C}}_i(\tilde{\mathbf{x}}_i-\hat{\mathbf{x}}_i) \qquad (14)$$

Defining the observer state error as

$$\mathbf{e}_i=\tilde{\mathbf{x}}_i-\hat{\mathbf{x}}_i \tag{15a}$$

and subtracting (14) from (11a) yields the observer error dynamic equation:

$$\dot{\mathbf{e}}_i=(\tilde{\mathbf{A}}_i-\mathbf{G}_{iE}\tilde{\mathbf{C}}_i)\mathbf{e}_i+\sum_{\substack{j=1\\j\neq i}}^{N}\tilde{\mathbf{A}}_{ij}\mathbf{e}_i \tag{15b}$$

The gain matrix $\mathbf{G}_{iE}$ must be selected such that the error system (15b) to be asymptotically stable. One way to do this is to compute the $\mathbf{G}_{iE}^T$ of the corresponding dual error system:

$$\dot{\bar{\mathbf{e}}}_i=(\tilde{\mathbf{A}}_i^T-\tilde{\mathbf{C}}_i^T\mathbf{G}_{iE}^T)\bar{\mathbf{e}}_i+\sum_{\substack{j=1\\j\neq i}}^{N}\tilde{\mathbf{A}}_{ji}^T\bar{\mathbf{e}}_j \tag{16}$$

In this case we have the problem of minimizing the quadratic criterion

$$J=\sum_{i=1}^{N}\frac{1}{2}\int_0^{\infty}(||\bar{\mathbf{e}}_i||^2_{\bar{Q}_{iE}}+||\bar{\mathbf{u}}_i||^2_{\bar{R}_{iE}})dt \tag{17a}$$

with respect to the control function $\bar{\mathbf{u}}_i(t)$, subject to the dymanic constraint

$$\dot{\bar{\mathbf{e}}}_i=\tilde{\mathbf{A}}_i^T\bar{\mathbf{e}}_i+\tilde{\mathbf{C}}_i^T\bar{\mathbf{u}}_i \tag{17b}$$

where $\bar{Q}_{iE}$ is a symmetric semipositive definite $\tilde{n}_i x\tilde{n}_i$ matrix and $\bar{\mathbf{R}}_{iE}$ is a symmetric positive definite $r_i x r_i$ matrix.

Under the condition that the system (17b), i.e. that the matrix pair $(\tilde{\mathbf{A}}_i^T,\ \tilde{\mathbf{C}}_i^T)$ is completely controllable, the solution to this optimal control problem is known to be:

$$\bar{\mathbf{u}}_i(t)=-\mathbf{G}_{iE}^T\bar{\mathbf{e}}_i,\quad \mathbf{G}_{iE}^T=-\bar{\mathbf{R}}_{iE}^{-1}\tilde{\mathbf{C}}_i\mathbf{P}_{iE} \tag{18}$$

where the $\tilde{n}_i x\tilde{n}_i$ symmetric positive definite matrix $\mathbf{P}_{iE}$ is given by the solution of the algebraic (steady state) Riccati equation:

$$\tilde{\mathbf{A}}_i\mathbf{P}_{iE}+\mathbf{P}_{iE}\tilde{\mathbf{A}}_i^T-\mathbf{P}_{iE}\tilde{\mathbf{C}}_i^T\bar{\mathbf{R}}_{iE}^{-1}\tilde{\mathbf{C}}_i\mathbf{P}_{iE}+\bar{Q}_{iE}=0 \tag{19}$$

Moreover, if the matrix pair $(\tilde{\mathbf{A}}_i^T,\ \bar{Q}_{iE}^{1/2})$ is completely observable, then the closed-loop system obtained (using (18) in (17b)) is known to be asymptotically stable.

Clearly, the control law (18) is optimal only for the system (17b) (not for the system (16)) and leads to a closed-loop system that does not involve the interconnection term $\sum_{j=1}^{N}\tilde{\mathbf{A}}_{ji}^T\bar{\mathbf{e}}_j$ which exists in the dual error system (16). Thus we have to see how one can find certain criteria which will ensure that using the decentralized controllers found (eqs.(18) and (19)) the interconnected subsystems (16) will be asymptotically stable. The overall system corresponding to the decoupled subsystems (17b) has the state equation:

$$\dot{\bar{e}}_D=\tilde{A}_D^T\bar{e}_D+\tilde{C}_D\bar{u} \quad (20a)$$

Similarly, the overall system which also involves the existing interconnections (i.e. the terms $\Sigma A_{ji}^T\bar{e}_j$) has the state equation:

$$\dot{\bar{e}}=\tilde{A}_D^T\bar{e}+\tilde{C}_D^T\bar{u}+\tilde{A}_C^T\bar{e} \quad (20b)$$

where

$$\tilde{A}_D=\text{diag}(\tilde{A}_1,\ldots,\tilde{A}_N),\ \tilde{C}_D=\text{diag}(\tilde{C}_1,\ldots,\tilde{C}_N),\ \tilde{A}_C=(\tilde{A}_{ij}) \quad (20c)$$

The closed-loop system obtained by applying to (20b) the decentralized control (18) is

$$\dot{\bar{e}}=(\tilde{A}^T-\tilde{C}_D^T G_E^T)\bar{e} \quad (21a)$$

where

$$\tilde{A}=\tilde{A}_D+\tilde{A}_C,\quad G_E=\text{diag}(G_{1E},\ldots,G_{NE}) \quad (21b)$$

The optimal value J^o of the performance index J (in (17a)), obtained if the decentralized controller $\bar{u}=-G_E^T\bar{e}$ is applied to the decoupled system (20a) is equal to:

$$J^o=\frac{1}{2}||\bar{e}_D^o||^2_{P_E} \quad (22a)$$

where $P_E=\text{diag}(P_{1E},\ldots,P_{NE})$ and $\bar{e}_D^o$ is the resulting closed-loop (optimal) error state. The suboptimal value J* of J obtained if $\bar{u}=-G_E^T\bar{e}$ is applied to the interconnected error system (20b) is equal to:

$$J^*=\frac{1}{2}||\bar{e}^*||^2_{\tilde{H}_E} \quad (22b)$$

where $\bar{e}^*$ is the resulting closed-loop (suboptimal) error state, and

$$\tilde{H}_E=\int_0^\infty e^{\tilde{A}_{FE}^T t}\tilde{K}_E e^{\tilde{A}_{FE}^T t}dt$$

$$\tilde{A}_{FE}=\tilde{A}^T-\tilde{C}_D^T G_E^T,\quad \tilde{K}_E=\bar{Q}_E+\bar{G}_E\bar{R}_E G_E^T \quad (23)$$

$$\bar{Q}_E=\text{diag}(\bar{Q}_{1E},\ldots,\bar{Q}_{NE}),\ \bar{R}_E=\text{diag}(\bar{R}_{1E},\ldots,\bar{R}_{NE})$$

The following theorems hold (Siljak 1982):

Theorem 1

The decentralized controller (18) is suboptimal with degree

$$\tilde{\mu}_E^*=\lambda_{max}^{-1}(\tilde{H}_E P_E^{-1}) \quad (24a)$$

for the overall interconnected error system (20b) if and only if the matrix $\tilde{H}_E$ is finite.

Here, λ_{max} (A) indicates the largest eigenvalue of the matrix A. The controller (18) is said to be suboptimal with degree $\tilde{\mu}$ for the system (20b) if there exists a positive number $\tilde{\mu}$ such that

$$J^*(\bar{e})\leq\tilde{\mu}^{-1}J^o(\bar{e}) \quad (24b)$$

for all $\bar{e}$.

The matrix $\tilde{H}_E$ is finite if the matrix $\tilde{A}_{FE}$ is stable (all of its eingenvalues have negative real parts) and can be computed as the unique solution of the Lyapunov equation (Bellman 1970):

$$\tilde{A}_{FE}\tilde{H}_E+\tilde{H}_E\tilde{A}_{FE}=-\tilde{K}_E \tag{24c}$$

The above show that stability of $\tilde{A}_{FE}$ implies suboptimality. The converse is true only if some additional observability conditions hold as is shown by the following theorem.

Theorem 2

The decentralized controller (18) is a stabilizing control law of the overall interconnected error system (20b) (i.e. $\tilde{A}_{FE}$ is a stable matrix) if it is suboptimal and the matrix pair $(\tilde{A}_{FE}, \tilde{K}_E^{\frac{1}{2}})$ is observable.

According to Wonham (1968) the observability of the pair $(\tilde{A}_{FE}, \tilde{K}_E^{\frac{1}{2}})$ is implied by the observability of the pair $(\tilde{A}, \bar{Q}_E^{\frac{1}{2}})$ where $\tilde{A}=\tilde{A}_D+\tilde{A}_C$ (see (21b)). Thus, under this observability condition for the interconnected error system matrix $\tilde{A}$ and the performance index matrix $\bar{Q}_E$, the controller (18) is suboptimal if and only if it is stabilizing.

Using the procedure described in this section we compute (design) the gain matrices G_{iE} (i=1,2,...,N) which are used in the observers (12a) to obtain the state estimates $\mathbf{x}_i$ of the overlapped subsystems. Now we shall see how the sensor faults can be detected using the $\hat{\mathbf{x}}_i$'s (i=1,2,...,N).

THE SENSOR FAULT DETECTION PROCEDURE

The state vectors of the i-1, i and i+1 overlapping subsystems have the form

$$\tilde{\mathbf{x}}_{i-1}=\begin{bmatrix}\mathbf{x}_{i-2,2}\\ \mathbf{x}_{i-1,1}\\ \mathbf{x}_{i-1,2}\end{bmatrix},\quad \tilde{\mathbf{x}}_{i}=\begin{bmatrix}\mathbf{x}_{i-1,2}\\ \mathbf{x}_{i,1}\\ \mathbf{x}_{i,2}\end{bmatrix},\quad \tilde{\mathbf{x}}_{i+1}=\begin{bmatrix}\mathbf{x}_{i,2}\\ \mathbf{x}_{i+1,1}\\ \mathbf{x}_{i+1,2}\end{bmatrix} \tag{25}$$

One can see that $\tilde{\mathbf{x}}_{i-1}$ and $\tilde{\mathbf{x}}_i$ have common the state variable (or subvector) $\mathbf{x}_{i-1,2}$, and $\tilde{\mathbf{x}}_i$ and $\tilde{\mathbf{x}}_{i+1}$ have common the state variable (or subvector) $\mathbf{x}_{i,2}$.

Let $\Phi_1, \Phi_2, \ldots, \Phi_N$ denote the N local observers which have been designed for the N overlapping subsystems, and let $(\mathbf{x}_{i-1,2})_{\Phi_{i-1}}$, $(\mathbf{x}_{i-1,2})_{\Phi_i}$ be the estimate of $\mathbf{x}_{i-1,2}$ provided by the (i-1)th and ith local observer, correspondingly. Using these estimates together with the estimates of $\mathbf{x}_{i,2}$ given by Φ_i and Φ_{i+1}, we can form the following quotients:

$$\mathbf{w}_1=(\mathbf{x}_{i-1,2})_{\Phi_{i-1}}/(\mathbf{x}_{i-1,2})_{\Phi_i} \tag{26a}$$

$$\mathbf{w}_2=(\mathbf{x}_{i,2})_{\Phi_i}/(\mathbf{x}_{i,2})_{\Phi_{i+1}} \tag{26b}$$

With the help of these two quotients we can detect (isolate) faults in the sensors of the i-1, i and i+1 subsystems as follows. We assume that under normal conditions (i.e. no faults) the deviation of the joint variables cannot exceed a certain percentage, say 2%. That is if:

$0.98 < \mathbf{w}_1 < 1.02$ and $0.98 < \mathbf{w}_2 < 1.02$,

then we have no faults in the subsystems sensors. The choise of the percentage of deviation depents on the actual type of sensor, and may be different for different sensors.

Now we have the following:

c1. If $(0.98<\mathbf{w}_1<1.02)$ **AND** $(\mathbf{w}_2<0.98$ **OR** $\mathbf{w}_2>1.02)$

Then the fault occurs at the sensor of the (i+1)th subsystem.

c2. If $(\mathbf{w}_1<0.98$ **OR** $\mathbf{w}_1>1.02)$ **AND** $(0.98<\mathbf{w}_2<1.02)$.

Then the fault occurs at the sensor of the (i-1)th subsystem.

c3. If $(\mathbf{w}_1<0.98$ **OR** $\mathbf{w}_1>1.02)$ **AND** $(\mathbf{w}_2<0.98$ **OR** $\mathbf{w}_2>1.02)$

Then we cannot surely say at which sensor (or sensors) the fault occurs.

To overcome this uncertainty we will construct a corrective signal which corrects the erroneous measurements, and then use the corrected measured values, so obtained, as the basis of our comparison basis.
Going back to (11a) and (12b) we write them as:

$$\dot{\tilde{\mathbf{x}}}_i=\tilde{\mathbf{A}}_i^*\tilde{\mathbf{x}}+\tilde{\mathbf{B}}_i^*\mathbf{u} \quad (27a)$$

$$\dot{\hat{\mathbf{x}}}_i=\tilde{\mathbf{A}}_i^*\hat{\mathbf{x}}+\tilde{\mathbf{B}}_i^*\mathbf{u}+\mathbf{G}_{iE}(\tilde{\mathbf{C}}_{iF}\tilde{\mathbf{x}}_i-\tilde{\mathbf{C}}_i\hat{\mathbf{x}}_i) \quad (27b)$$

where the $\tilde{n}_i x \tilde{n}$ matrix $\tilde{\mathbf{A}}_i^*$ and the $\tilde{n}_i x\, m$ matrix $\tilde{\mathbf{B}}_i^*$ have the form:

$$\tilde{\mathbf{A}}_i^*=(\tilde{\mathbf{A}}_{i1},\ \tilde{\mathbf{A}}_{i2},\ldots,\tilde{\mathbf{A}}_{i,i-1},\ \tilde{\mathbf{A}}_i,\ \tilde{\mathbf{A}}_{i,i+1},\ldots,\tilde{\mathbf{A}}_{iN}) \quad (27c)$$

$$\tilde{\mathbf{B}}_i^*=(\tilde{\mathbf{B}}_{i1},\ \tilde{\mathbf{B}}_{i2},\ldots,\tilde{\mathbf{B}}_{i,i-1},\ \tilde{\mathbf{B}}_i,\ \tilde{\mathbf{B}}_{i,i+1},\ldots,\tilde{\mathbf{B}}_{iN}) \quad (27d)$$

Since, we are interested in faults at the steady-state operation of the system, we form the steady-state error $\mathbf{e}_i(\infty)=\tilde{\mathbf{x}}_i(\infty)-\hat{\mathbf{x}}_i(\infty)$ which is described by (see (27a,b)):

$$\dot{\mathbf{e}}_i(\infty)=\tilde{\mathbf{A}}_i^*\mathbf{e}(\infty)+\mathbf{G}_{iE}\{\tilde{\mathbf{C}}_i\hat{\mathbf{x}}_i(\infty)-\tilde{\mathbf{C}}_{iF}\tilde{\mathbf{x}}_i(\infty)\} \quad (28)$$

But in the steady-state $\dot{\mathbf{e}}_i(\infty)=0$, and if the observer has been designed correctly we have $\mathbf{e}(\infty)=0$. When a fault occurs in some sensor (which means that $\tilde{\mathbf{C}}_{iF}\neq\tilde{\mathbf{C}}_i$) then an error appears in the estimates provided by the corresponding observer. This error is due to the term

$$\mathbf{G}_{iE}\{\tilde{\mathbf{C}}_i\hat{\mathbf{x}}_i(\infty)-\tilde{\mathbf{C}}_{iF}\tilde{\mathbf{x}}_i(\infty)\}$$

of (28). Thus following (Li, Pan and Singh, 1982) we construct the signal

$$\mathbf{e}_i^*(\infty)=\mathbf{G}_{iE}\{\tilde{\mathbf{C}}_i\hat{\mathbf{x}}_i(\infty)-\tilde{\mathbf{C}}_{iF}\tilde{\mathbf{x}}_i(\infty)\} \quad (29)$$

Using this signal we can compute the corrected value $\mathbf{x}_i^*(\infty)$ of the estimate $\hat{\mathbf{x}}_i(\infty)$ as

$$\hat{\mathbf{x}}_i^*(\infty)=\hat{\mathbf{x}}_i(\infty)+\mathbf{e}_i^*(\infty) \tag{30}$$

and the corrected value of assosiated sensor measurement (output) as

$$\mathbf{y}_i^*(\infty)=\tilde{\mathbf{C}}_i\hat{\mathbf{x}}_i^*(\infty) \tag{31}$$

Recalling that the measured state variables constitute the overlapped state variables of the N subsystems, we have that

$$y_{i-1}=x_{i-1,2};\ y_i=x_{i,2};\ y_{i+1}=x_{i+1,2} \tag{32}$$

It is noted that if r variables are measured then we construct r overlapped subsystems which have r-1 overlapping (common) variables. Thus one measured variable will not be overlapping. This variable belongs to the last subsystem.

Now by using the corrected measurements (31) we can form the following quotients (in addition to $\mathbf{w}_1$ and w_2 given by (26a,b)):

$$w_3=\{\hat{\mathbf{x}}_{i-1,2}(\infty)\}_{\Phi_{i-1}}/\{\mathbf{y}_{i-1}^*(\infty)\} \tag{33a}$$

$$\mathbf{w}_4=\{\hat{\mathbf{x}}_{i-1,2}(\infty)\}_{\Phi_i}/\{\mathbf{y}_{i-1}^*(\infty)\} \tag{33b}$$

$$\mathbf{w}_5=\{\hat{\mathbf{x}}_{i,1}(\infty)\}_{\Phi_i}/\{\mathbf{y}_i^*(\infty)\} \tag{33c}$$

$$\mathbf{w}_6=\{\hat{\mathbf{x}}_{i,2}(\infty)\}_{\Phi_{i+1}}/\{\mathbf{y}_i^*(\infty)\} \tag{33d}$$

$$\mathbf{w}_7=\{\hat{\mathbf{x}}_{i+1,2}(\infty)\}_{\Phi_{i+1}}/\{\mathbf{y}_{i+1}^*(\infty)\} \tag{33e}$$

On the basis of these quotients we can proceed to detect where the failures occur in the uncertain case (condition C3)) discussed above.

c4. If

$$\left\{\begin{matrix} w_3<0.98 \\ \textbf{OR} \\ w_3>1.02 \end{matrix}\right\} \textbf{ AND } \left\{\begin{matrix} w_6<0.98 \\ \textbf{OR} \\ w_6>1.02 \end{matrix} \quad \textbf{OR} \quad \begin{matrix} \mathbf{w}_7<0.98 \\ \textbf{OR} \\ \mathbf{w}_7>1.02 \end{matrix}\right\}$$

$$\textbf{AND } \{0.98<\mathbf{w}_4<1.02\} \textbf{ AND } \{0.98<\mathbf{w}_5<1.02\}$$

Then the faults occur at the sensors of the i-1 and i+1 subsystems.

c5. If

$$\{0.98<w_3<1.02\} \textbf{ AND } \{0.98<\mathbf{w}_6<1.02\} \textbf{ AND } \{0.98<\mathbf{w}_7<1.02\} \textbf{ AND}$$

$$\left\{\begin{matrix} \mathbf{w}_4<0.98 \\ \textbf{OR} \\ \mathbf{w}_4>1.02 \end{matrix}\right\} \textbf{ AND } \left\{\begin{matrix} w_5<0.98 \\ \textbf{OR} \\ w_5>1.02 \end{matrix}\right\}$$

Then the fault occurs at the sensor of the ith subsystem.

c6. If

$$\left\{\begin{matrix} w_3<0.98 \\ \text{OR} \\ w_3>1.02 \end{matrix}\right\} \text{ AND } \left\{\begin{matrix} w_4<0.98 & & w_5<0.98 \\ \text{OR} & \text{OR} & \text{OR} \\ w_4>1.02 & & w_5>1.02 \end{matrix}\right\}$$

$$\text{AND } \{0.98<w_6<1.02\} \text{ AND } \{0.98<w_7<1.02\}$$

Then the faults occurs at the sensors of the i-1 and i subsystems.

c7. If

$$\{0.98<w_3<1.02\} \text{ AND } \left\{\begin{matrix} w_4<0.98 & & w_5<0.98 \\ \text{OR} & \text{OR} & \text{OR} \\ w_4>1.02 & & w_5>1.02 \end{matrix}\right\}$$

$$\text{AND } \left\{\begin{matrix} w_6<0.98 & & w_7<0.98 \\ \text{OR} & \text{OR} & \text{OR} \\ w_6>1.02 & & w_7>1.02 \end{matrix}\right\}$$

Then the failures occur at the sensors of the i and i+1 subsystems.

c8. If

$$\left\{\begin{matrix} w_3<0.98 \\ \text{OR} \\ w_3>1.02 \end{matrix}\right\} \text{ AND } \left\{\begin{matrix} w_4<0.98 & & w_5<0.98 \\ \text{OR} & \text{OR} & \text{OR} \\ w_4>1.02 & & w_5>1.02 \end{matrix}\right\}$$

$$\text{AND } \left\{\begin{matrix} w_6<0.98 & & w_7<0.98 \\ \text{OR} & \text{OR} & \text{OR} \\ w_6>1.02 & & w_7>1.02 \end{matrix}\right\}$$

Then all sensors have faults.

APPLICATION TO A COUPLED-CORE NUCLEAR REACTOR SYSTEM

Consider a nuclear reactor interconnected system with three coupled cores described by the following model (Tzafestas and Chrysochoides, 1977):

$$\dot{\mathbf{x}}=\mathbf{A}\mathbf{x}+\mathbf{B}\mathbf{u}, \quad \mathbf{y}=\mathbf{C}\mathbf{x}$$

where

$$A=\left[\begin{array}{cc|cc|cc} -10^{-2} & 10^{-5} & 0 & 0 & 0 & 0 \\ -10^{3} & -2 & 0 & 1 & 0 & 1 \\ \hline 0 & 0 & -10^{-2} & 10^{-5} & 0 & 0 \\ 0 & 1 & -10^{3} & -2 & 0 & 1 \\ \hline 0 & 0 & 0 & 0 & -10^{-2} & 10^{-5} \\ 0 & 1 & 0 & 1 & -10^{3} & -2 \end{array}\right] \qquad B=\left[\begin{array}{c|c|c} 0 & 0 & 0 \\ 10^{6} & 0 & 0 \\ \hline 0 & 0 & 0 \\ 0 & 10^{6} & 0 \\ \hline 0 & 0 & 0 \\ 0 & 0 & 10^{6} \end{array}\right]$$

$$C=\left[\begin{array}{cc|cc|cc} 0 & 1 & 0 & 0 & 0 & 0 \\ \hline 0 & 0 & 0 & 1 & 0 & 0 \\ \hline 0 & 0 & 0 & 0 & 0 & 1 \end{array}\right]$$

The matrix blocks indicated by the discontinuous lines correspond to the three interconnected subsystems (reactor cores).

The first step is to carry over the overlapped decomposition. Here

$$\mathbf{x}=(x_1 \quad x_2 \;\vdots\; x_3 \quad x_4 \;\vdots\; x_5 \quad x_6)^T$$
$$\tilde{\mathbf{x}}=(x_1 \quad x_2 \;\vdots\; x_2 \quad x_3 \quad x_4 \;\vdots\; x_5 \quad x_6)^T$$

and the 8x6 matrix **T** has the form

$$\mathbf{T}=\begin{bmatrix}1&0&0&0&0&0\\0&1&0&0&0&0\\0&1&0&0&0&0\\0&0&1&0&0&0\\0&0&0&1&0&0\\0&0&0&1&0&0\\0&0&0&0&1&0\\0&0&0&0&0&1\end{bmatrix}$$

This matrix is of full-column rank, and its generalized inverse $T^{\dagger}=(T^TT)^{-1}T^T$ is found to be:

$$\mathbf{T}^{\dagger}=\left[\begin{array}{cc|ccc|ccc}1&0&0&0&0&0&0&0\\0&1/2&1/2&0&0&0&0&0\\\hline 0&0&0&1&0&0&0&0\\0&0&0&0&1/2&1/2&0&0\\\hline 0&0&0&0&0&0&1&0\\0&0&0&0&0&0&0&1\end{array}\right]$$

Now using (4) together with the conditions (5b) the matrix $\tilde{\mathbf{A}}$, $\tilde{\mathbf{B}}$ and $\tilde{\mathbf{C}}$ of the overlapped decomposed system are found to be:

$$\tilde{\mathbf{A}}=\left[\begin{array}{cc|ccc|ccc}-10^{-2}&10^{-5}&0&0&0&0&0&0\\-10^{3}&-2&0&0&1&0&0&1\\\hline -10^{3}&0&-2&0&1&0&0&1\\0&0&0&-10^{-2}&10^{5}&0&0&0\\0&0&1&-10^{3}&-2&0&0&1\\\hline 0&1&0&-10^{3}&0&-2&0&1\\0&0&0&0&0&0&-10^{-2}&10^{-5}\\0&1&0&0&0&1&-10^{3}&-2\end{array}\right],\tilde{\mathbf{B}}=\left[\begin{array}{c|c|c}0&0&0\\10^{6}&0&0\\\hline 10^{6}&0&0\\0&0&0\\0&10^{6}&0\\\hline 0&10^{6}&0\\0&0&0\\0&0&10^{6}\end{array}\right]$$

$$\tilde{\mathbf{C}}=\left[\begin{array}{cc|ccc|ccc}0&1&0&0&0&0&0&0\\\hline 0&0&0&0&1&0&0&0\\\hline 0&0&0&0&0&0&0&1\end{array}\right]$$

The next step is to check whether the required controllability and observability conditions hold.
It is easy to verify that:

(I) Rank $(\tilde{\mathbf{B}}_1,\ \tilde{\mathbf{A}}_1\tilde{\mathbf{B}}_1)=\tilde{n}_1=2$

Rank $(\tilde{\mathbf{B}}_i,\ \tilde{\mathbf{A}}_i\tilde{\mathbf{B}}_i,\ \tilde{\mathbf{A}}_i\tilde{\mathbf{B}}_i)=\tilde{n}_i=3 \quad (i=2,3)$

Thus all subsystems 1,2 and 3 are completely controllable.

(II) Rank $(\tilde{Q}_1^{\frac{1}{2}})^T, \tilde{A}_1^T(\tilde{Q}_1^{\frac{1}{2}})^T = \tilde{n}_1 = 2$

for $Q_1 = \begin{matrix} 1 & 0 \\ 0 & 1 \end{matrix}$ and any $\tilde{A}_1$.

Thus $(\tilde{A}_1, \tilde{Q}_1^{\frac{1}{2}})$ is observable. Similarly choosing $Q_{iE}=Q_{is}=I$, $Q_E=Q_S=I$, where I is a unit matrix of appropriate dimension in each case, we can verify that $(\tilde{A}_i, \tilde{Q}_i^{\frac{1}{2}})$, i=2,3 are observable.

(III) All decoupled error subsystems $(\tilde{A}_i^T, \tilde{C}_i^T)$ (i=1,2,3) are controllable.

Thus the decontralized design method described in Sec.3 for the overlapped subsystems is applicable. The three observers Φ_1, Φ_2 and Φ_3 were designed and used for the estimation of $\tilde{x}_1=(x_1,x_2)$, $\tilde{x}_2=(x_2, x_3, x_4)$ and $\tilde{x}_3=(x_4, x_5, x_6)$. Fig. 2a shows the trajectories of the common state variables x_2 and x_4 under normal operation. Fig. 2b shows the trajectories of their estimates $(\hat{x}_2)_{\Phi_1}$, $(\hat{x}_2)_{\Phi_2}$; $(\hat{x}_4)_{\Phi_2}$ and $(x_4)_{\Phi_3}$ generated by the observers Φ_1, Φ_2 and Φ_3. The software package was written in **FORTRAN** and run on a **PRIME P-450** computer system under PRIMOS operating system.

Then the values of the parameters of the measurement (sensor) matrix **C** were succesively changed as follows:

i) $c_{12}=0.30$ (affects subsystem i-1=1)

ii) $c_{12}=0.30$, $c_{25}=0.32$ (affect subsystems i-1=1, i=2)

iii) $c_{25}=0.18$ (affect subsystem i=2)

iv) $c_{12}=0.15$, $c_{38}=0.64$ (affect subsystems i-1=1, i+1=3)

v) $c_{25}=0.43$, $c_{38}=0.12$ (affect subsystems i=2, i+1=3)

vi) $c_{38}=0.24$ (affects subsystem i+1=3)

vii) $c_{12}=0.28$, $c_{25}=0.74$, $c_{38}=0.56$ (affects all subsystems).

The behaviour of the observer outputs is as shown in Fig. 2c.

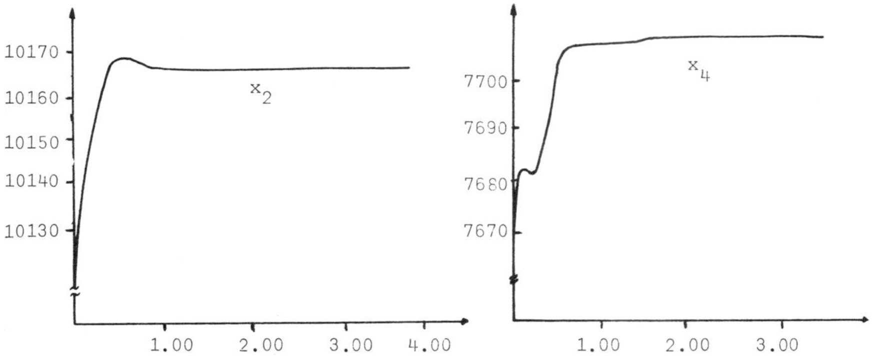

Fig.2a. Trajectories of common state variables x_2, x_4 under normal (no faulty) conditions

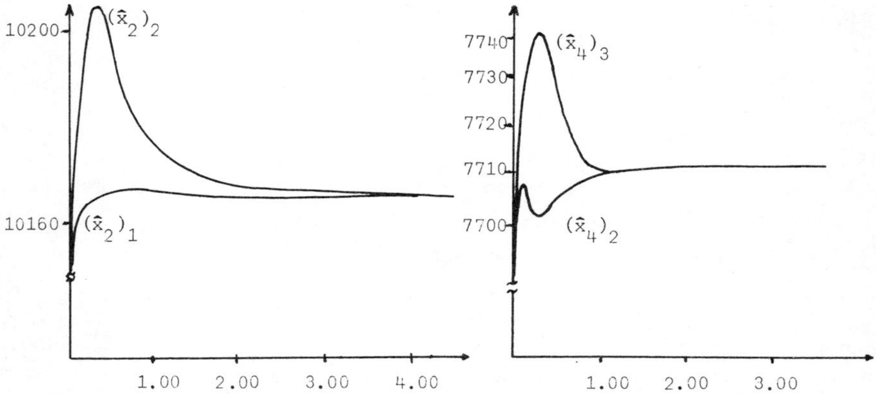

Fig.2b. Trajectories of $(\hat{\mathbf{x}}_2)_{\Phi_1}$, $(\hat{\mathbf{x}}_2)_{\Phi_2}$, $(\hat{\mathbf{x}}_4)_{\Phi_2}$ and $(\hat{\mathbf{x}}_4)_{\Phi_3}$ provided by the observers Φ_1, Φ_2 and Φ_3

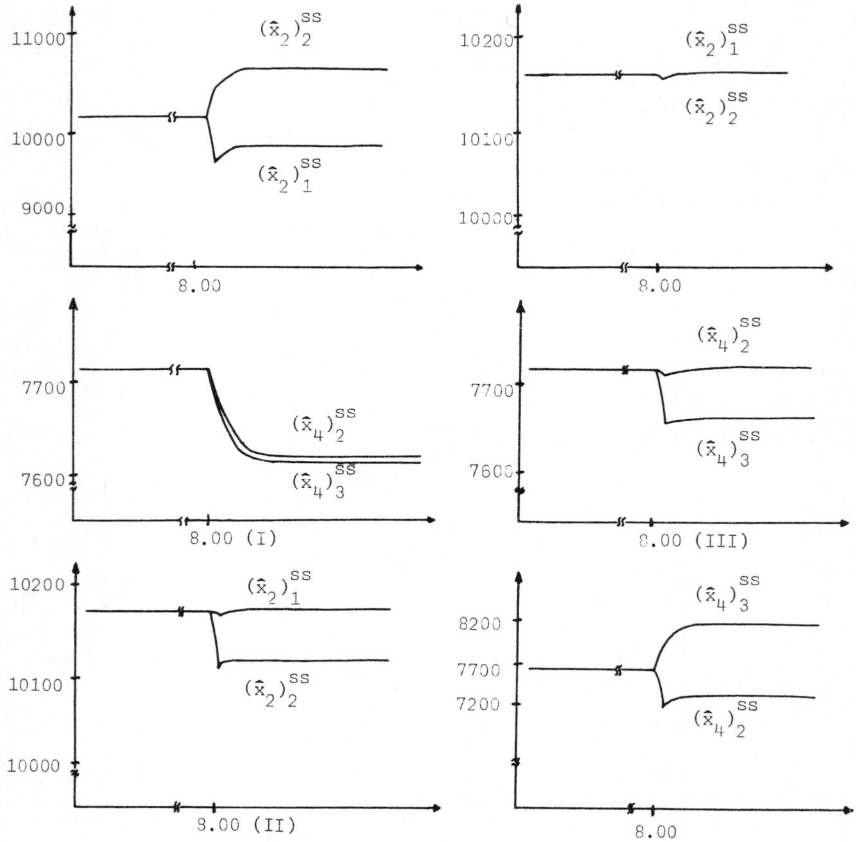

Fig.2c. Behaviour of the observer outputs under faulty conditions

Figures 3÷9 show the steady-state values of the quotients w_i $w_1, w_2, \ldots, w_7$, the corresponding conditions that satisfy, and the conclusion that is drawn in each case.

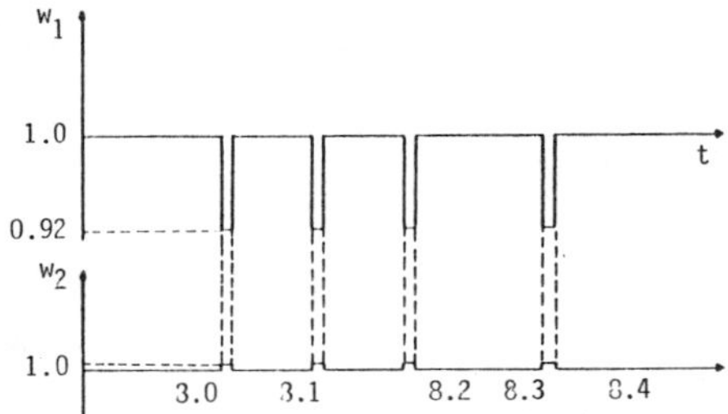

Fig.3

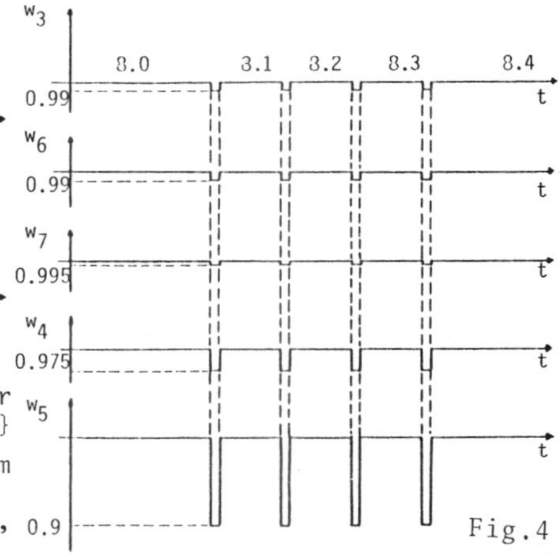

Fig.4

Fig.3. Fault in the 1st subsystem sensor $\{w_1=0.92<0.98;\ 0.98<w_2=1.005<1.02\}$

Fig.4. Faults in the 2nd (ith) subsystem sensor $\{0.98<w_3=0.99<1.02,\ w_4=0.975<0.98,\ 0.98<w_6=0.99<1.02,\ 0.98<w_7=0.995<1.02,\ w_5=0.090<0.981\}$

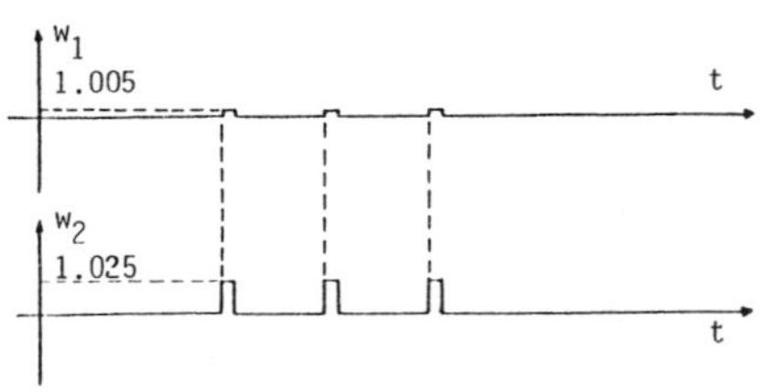

Fig.5. Fault in the 3rd {(i+1)th} subsystem sensor $\{0.98<w_1=1.005<1.02,\ w_2=1.025>1.02\}$

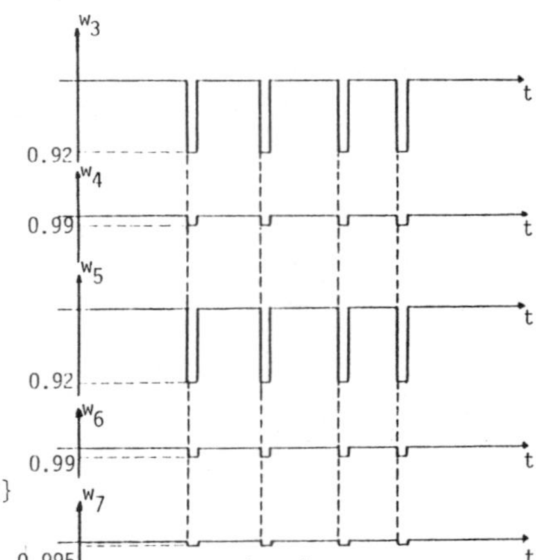

(I,II)

Fig.6. Faults in the (i-1)th and ith sensors $w_3=0.92<0.98,\ w_5=0.92<0.98,\ 0{,}98<w_6=0.99<1.02,\ 0.98<w_7=0.995<1.02\}$

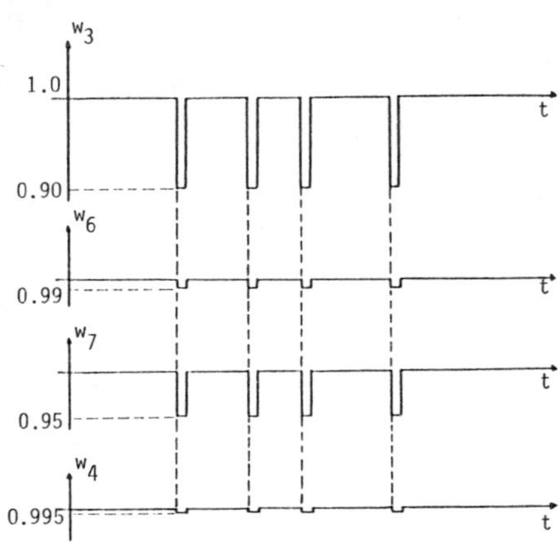

Fig.7. Faults in the (i-1)th and (i+1) sensors $\{w_3=0.90<0.98,\ w_7=0.95<0.98,\ 0.98<w_4=0.995<1.02\}$

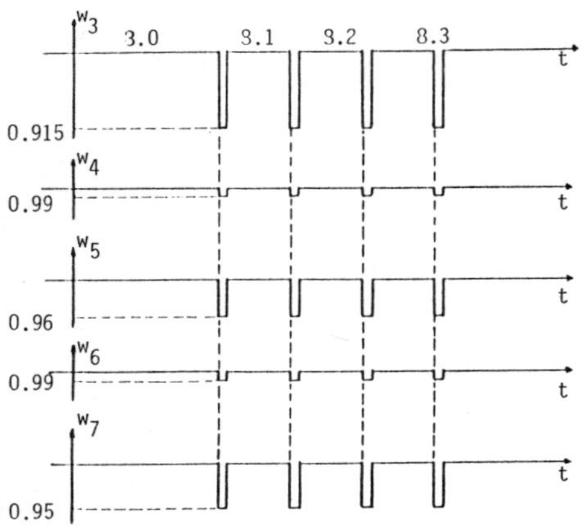

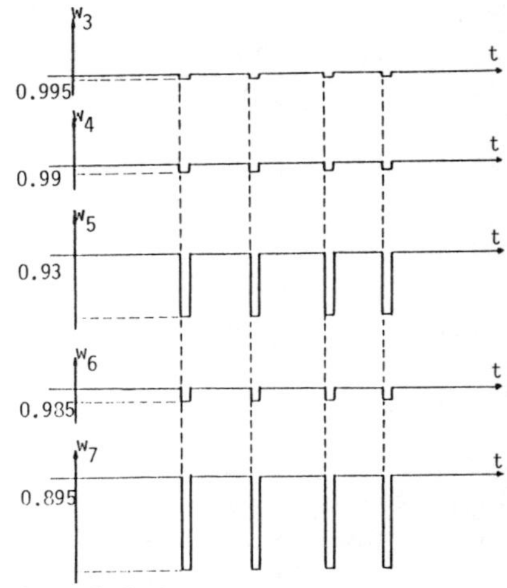

Fig.8. Faults in the sensors i-1, i and i+1
{w_3=0.915<0.98, w_5=0.96<0.98, w_7=0.95<0.98}

Fig.9. Faults in the sensors i and i+1
{0.98<w_3=0.995<1.02, w_5=0.93<0.98, w_7=0.895<0.98}

CONCLUSIONS

In this paper the overlapping decomposition approach, combined with a decentralized observer design for the resulting overlapped subsystems, and an appropriate corrective signal for the fault measurements, has been applied for treating the sensor fault detection of linear large scale systems. Since all measured variables (physical quantities) must be accomodated as common state variables of the overlapped subsystems we may need a large number of subsystems and corresponding observers, but actually their dimensionality is very much lower than the overall system dimensionality. This combined with the decentralized way of designing the observers leads to acceptable computational burden.

The method is especially appropriate for practical implementation using parallel microprocessor set-ups. This problem as well as the problem of investigating the case of nonlinear systems, including robustness aspects, are the subject of further control. Some other results on instrument and actuator fault detection can be found in Clark (1978); Clark, Fosth and Walton (1975); and Pouliezos, Stavrakakis and Tzafestas (1987).

REFERENCES

Botsaris, C.A. and S.G.Tzafestas (1984). Optimal design of a standby system with repair facilities subject to breakdown, In S.Tzafestas and M.Hamza (Eds.) **Methods and Applications of Measurement and Control**, ACTA Press, Calgary, Vol.2, pp.632-634.

Clark, R.N. (1978). Instrument fault detection **IEEE, Trans. on Aerospace and Electronic Systems**, Vol. AES-14, No.3, pp.456-465.

Clark, R.N., D.C. Fosth, V.M. Walton (1975). Detecting instrument malfunctions in Control Systems, **IEEE, Trans. on Aerospace and Electronic Systems**, Vol. AES, 11, No.4, pp. 465-473, July 1975.

Isermann, R. (1984). Process fault-Detection based on Modelling and Estimation Methods, **Automatica**, Vol.20, pp.387-404.

Li, D.S., Q.R. Pan, M.G. Singh (1982). A novel approach to the identification of failed sensor in dynamical systems, **UMIST, Control System Centre**, Report No. 532.

Pouliezos, A., S.G.Tzafestas and G.Stavrakakis (1987). Fault detection and location in MRAC robotic systems, In S.Tzafestas, M.Singh and G.Schmidt (Eds.), **System Fault Diagnostics, Reliability and Related Knowledge-Based Diagnostics**, Vol.1, D.Reidel, Dordrecht, pp.200-214.

Siljak, D.D. (1982). The Inclusion principle for Dynamic Systems, **Final Report**, The Univ. of Santa-Clara, California, April 1982.

Siljak, D.D., M.B.Vukcevic (1976). Decentralization, stabilization and Estimation of Large-Scale linear Systems, **IEEE Trans. on Automatic Control**, pp.363-366.

Singh, M.G., M.F.Hassan, Y.L.Chen, D.S.Li, Q.R.Pan (1983). New approach to failure detection in large-scale systems, **IEE Proceedings**, Vol. 13, Pt. D.No 5, pp.243-249.

Tzafestas, S.G. (1980). Optimization of systems reliability: A survey of problems and techniques, **Int. J. Systems Sci.**, Vol.11, pp.455-486.

Tzafestas, S.G. (1982). Optimal control policies in system reliability and maintanance, In S.Tzafestas (Ed.), **Optimization and Control of Dynamic Operational Research Models**, North-Holland, Amsterdam, pp.395-434.

Tzafestas, S.G. (1983). Optimal Supervision and failure detection of dynamic technological systems, **NRC "Demokritos", Report** DEMO **83/19** G.

Tzafestas, S.G. (1985). Reliability and fault detection techniques of large scale systems. In G.Schmidt et.al. (Eds.), **Real Time Control of Large Scale Systems**, Springer, Berlin, pp.35-

Willsky, A.S. (1976). A survey of Design Methods for failure detection in Dynamic Systems, **Automatica**, Vol.12, pp.601-611.

Fault Detection in an Evaporator via Parameter Estimation in Real Time

D.T. Dalle Molle and D.M. Himmelblau

Dept. Chem. Engr., University of Texas, Austin, Texas 78712, U.S.A.

ABSTRACT

Detection of faults (here meaning process degradation) can be accomplished by estimating the parameters in a process model in real time. As a specific example, a model of continuous evaporator is used to illustrate the application of (1) a state observer combined with a least squares procedure and (2) an extended Kalman filter to the model when faults were introduced. Noise was added to the model inputs and outputs to represent what might occur in the actual operation of equipment.

KEYWORDS

Fault detection; parameter estimation; evaporator; least squares estimation; Kalman filter.

INTRODUCTION

Because process degradation may lead to inefficient and even unsafe process operation, fault diagnosis has received considerable attention in the literature. A variety of methods and applications for fault detection have been developed and implemented. Watanabe and Himmelblau (1983) demonstrated fault detection strategies for nonlinear chemical reactors. Geiger (1984) used parameter estimation and pattern classification to identify faults in a motor/pump system. Shiozaki and others (1985) diagnosed faults in pipeline systems using signed directional graphs. Other techniques and applications can be found in a book by Himmelblau (1978) while techniques of fault detection via state estimation have been reviewed by Frank (1986). Here, by fault we mean degradation of some process characteristic from its acceptable range in contrast to failure. The presence of a fault does not imply that the process is inoperative, but rather that the process can only operate at a suboptimal level with respect to efficiency and safety.

Fault detection via parameter estimation relies on the principle that possible faults in the process can be associated with specific parameters as well as states of a mathematical model of the process. For example, in the chemical industries, the fouling of a heat exchanger can be related to the heat transfer coefficient in a process model, and catalyst activity can be associated with a reaction rate constant and a process temperature. Hence, to apply parameter estimation to time dependent processes, it is necessary to have a dynamic model of the process. In addition, a minimum number of independent state measurements exist that must be available to estimate various states and parameters. For linear systems, specific conditions exist on the model for state observability (Chen, 1970). Parameter estimation, however, is inherently a nonlinear procedure (when coupled with state estimation) since the model parameters multiply the states. For a nonlinear model, the criteria for observability are different than those for linear systems, and observing some parameters may be difficult and lead to erroneous estimates (Astrom and Eykhoff, 1971). Nevertheless, a variety of nonlinear estimation techniques have been established and applied. Two such techniques, (a) a nonlinear state observer combined with a least squares parameter estimation scheme and (b) extended Kalman filtering, are considered here.

GENERAL PROCESS DESCRIPTION

Let the process considered for estimation be represented by the following set of ordinary nonlinear stochastic differential equations:

$$\dot{\underline{x}} = \underline{f}(\underline{x},\underline{u},\underline{p},\underline{t}) + G(\underline{x},\underline{p},\underline{t})\underline{w}(t) \tag{1}$$

$$\underline{y} = \underline{h}(\underline{x},t) + \underline{v}(t) \tag{2}$$

where

$\underline{x}$ ≡ vector of states of the system (n x 1)
$\underline{u}$ ≡ vector of inputs to the system (d x 1)
$\underline{p}$ ≡ vector of parameters associated with faults (s x 1)
$\underline{f}$ ≡ vector of nonlinear functions of state dynamics (n x 1)
$\underline{y}$ ≡ vector of measured outputs of system (m x 1)
$\underline{h}$ ≡ vector of nonlinear functions of measurement dynamics (m x 1)
$\underline{w}$ ≡ vector of uncorrelated white input noise (n x 1)
$\underline{v}$ ≡ vector of uncorrelated white measurement noise (m x 1)
$\underline{G}$ ≡ input noise coefficient matrix (n x d)

It is assumed that the noise vectors $\underline{w}(t)$ and $\underline{v}(t)$ have the following properties:

$$\underline{w}(t)\text{: } N[0,Q(t)] \tag{3}$$

$$\underline{v}(t)\text{: } N[0,R(t)] \tag{4}$$

$$E[\underline{w}(t_1)\underline{v}(t_2)] = 0 \text{ for all } t_1 \text{ and } t_2 \tag{5}$$

where N[0,*] means that each noise vector is normally distributed with zero mean and covariance matrices Q(t) and R(t) respectively. Furthermore, $\underline{w}(t)$ and $\underline{v}(t)$ are uncorrelated in time and to each other. Two techniques for estimating the parameters $\underline{p}$ of the process will be examined.

STATE OBSERVER WITH LEAST SQUARES PARAMETER ESTIMATION

One approach to parameter estimation in nonlinear systems with a limited number of state measurements is to first obtain estimates of all of the states of the process; then, using these estimates, estimate the parameters of the model. In this approach, however, care must be taken in reconstructing the unobserved states in the presence of model parameters that may be changing in time due to process faults. If the states are not estimated correctly, parameter estimation based on improper state estimates may lead to meaningless results. Under certain circumstances, though, this approach can be applied to nonlinear systems with parameters that vary in time as discussed below.

State Estimation in the Presence of Faults

Most state estimation methods assume that the model parameters remain constant. Under certain conditions, however, the model parameters may be time-varying and the states can still be estimated correctly. Watanabe and Himmelblau (1983) proposed a special case of Eq. (1)

$$\underline{f}(\underline{x},\underline{u},\underline{p},t) = A\underline{x} + B\underline{u} + Qg(\underline{x},\underline{u},\underline{p},t) \tag{6}$$

where

$A\underline{x} + B\underline{u}$	contains the linear terms of $\underline{f}$
$Q\underline{g}(\underline{x},\underline{u},\underline{p},t)$	contains the nonlinear terms of $\underline{f}$ and the terms that contain the parameters associated with faults (1 x 1)
$g(\underline{x},\underline{u},\underline{p},t)$	is a vector of nonlinear terms and terms (linear or nonlinear) that contain parameters associated with faults (1 x 1)
Q	is a coefficient matrix (n x 1)

They further decomposed the $Qg(\underline{x},\underline{u},\underline{p},t)$ term as follows:

$$g(\underline{x},\underline{u},\underline{p},t) \equiv [g_s^T(\underline{x},\underline{u},t) \mid g_{1-s}^T(\underline{x},\underline{u},y)]^T \tag{7}$$

$$g(\underline{x},\underline{u},\underline{p},t) \equiv [p_1g_1\ p_2g_2\ \cdots\ p_sg_s \mid g_{s+1}\ g_{s+2}\ \cdots\ g_1]^T \tag{8}$$

$$Q = \begin{bmatrix} Q_1 \\ Q_2 \end{bmatrix} = \begin{bmatrix} Q_{11} & Q_{12} \\ Q_{21} & Q_{22} \end{bmatrix} \begin{matrix} (s) \\ (n-s) \end{matrix}$$
$$\qquad\qquad (s) \quad (1-s)$$

where $g_i = g_i(\underline{x},\underline{u},t)$. And by defining:

coefficient terms of the fault parameters

$$S(t) \equiv \begin{bmatrix} Q_{11} \\ \\ Q_{21} \end{bmatrix} \begin{bmatrix} g_1 & & 0 \\ & g_2 & \\ & & \cdot \\ 0 & & g_s \end{bmatrix}$$

other nonlinear terms

$$\underline{r}(t) \equiv \begin{bmatrix} Q_{12} & g_{s+1} \\ & \vdots \\ Q_{22} & g_1 \end{bmatrix}$$

Then: $Qg(\underline{x},\underline{u},\underline{p},t) = S(t)\underline{p} + \underline{r}(t)$ (9)

and a special form of a Luenberger observer can be designed to give asymptotically unbiased estimates of the remaining (n - m) unmeasured states.

Parameter Estimation from Full State Availability

Once estimates are available for all of the states, either through direct measurement or through an observer, estimating the model parameters is quite straightforward. By using Eqs. (6) and (9), Eq. (1) can be written as

$$\dot{\underline{x}} = A\underline{x} + B\underline{u} + S(t)\underline{p} + \underline{r}(t) + G\underline{w}(t) \tag{10}$$

The derivatives, $\dot{\underline{x}}$, can be approximated as

$$\dot{\underline{x}} = [\underline{x}(t) - \underline{x}(t - \tau)]/\tau$$

where τ is a time interval that is small relative to the shortest time constant of the process and the highest frequency of the observation noise. By letting $t = k\tau$, equation (10) can be written as

$$[\underline{x}(k) - \underline{x}(k-1)]/\tau = A\underline{x}(k) + B\underline{u}(k) + S(k)\underline{p} + \underline{r}(k) + G\underline{w}(k) \tag{11}$$

For estimation purposes, equation (11) can be written conveniently as

$$S(k)\underline{p} = \underline{b}(k) + \varepsilon(k) \tag{12}$$

where

$$\underline{b}(k) = [\underline{x}(k) - \underline{x}(k-1)]/\tau - A\underline{x}(k) - B\underline{u}(k) - \underline{r}(k) \tag{13}$$

$\varepsilon \equiv$ unknown error vector due to input and measurement noise

It is desired to find $\underline{p}$ that minimizes the square of the error, $[\varepsilon(k)]^2$.

Recursive least squares algorithms can be found frequently in the literature and in textbooks (Gelb, 1974; Sorenson, 1980; Young, 1969), but most of these algorithms provide estimates that converge to an average value as the number of measurements increases. Beyond a certain point, new measurements no longer provide any significant information to the estimate. Hence, if a fault were to develop after the algorithm had been in operation for some time, the estimate of the fault parameter would not show any change as the algorithm would be insensitive to new measurements. This problem could be resolved in several ways. One method is to restart the algorithm when the gains have fallen below a prespecified value by adding large values to the covariance matrix (thus increasing the gains). Another common technique used for keeping least squares algorithms sensitive to new measurements is to incorporate a "forgetting factor" the purpose of which is to weight the new estimate from current measurements so that the new estimate becomes a weighted average of the new and old estimates. We used the forgetting factor method and the following recursive least squares estimation scheme for the fault parameters $\underline{p}$:

$$\hat{\underline{p}}(k+1) = I_s - U(k)V(k)S(k+1)][\hat{\underline{p}}(k) + U(k)\underline{b}(k+1)] \tag{14}$$

where

$$U(k) \equiv (1/\lambda)R(k)^T R(k)]^{-1}S(k+1)^T \tag{15}$$

$$V(k) \equiv [I_n + S(k+1)U(k)]^{-1} \tag{16}$$

$$[R(k+1)^T R(k+1)]^{-1} \equiv (1/\lambda)]R(k)^T R(k)]^{-1} - U(k)V(k)U(k)^T \tag{17}$$

Here, the forgetting factor λ ranges from 0 to 1. The initial values for the algorithm are given by

$$\hat{\underline{p}}(0) = [R(0)^T R(0)^{-1}\underline{b}(0)] \qquad [R(0)^T R(0)]^{-1} = S^T(0)S(0)]^{-1} \tag{18}$$

THE EXTENDED KALMAN FILTER

The Extended Kalman Filter is so well known both for discrete and continuous measurements that the equations for updating the states, the error covariance matrix, and the gain matrix do not have to be repeated here (see Gelb, 1974 or Jazwinski, 1970 for the various formula). Recall that parameter estimation can be regarded as a special case of state estimation. If the parameters of a model that are to be estimated are constants of the system dynamics, then the parameters can be modelled by the differential equation.

$$\dot{\underline{p}}(t) = [0]\underline{p}(t) + \underline{w}_p(t) \tag{19}$$

where $\underline{w}_p(t)$ is normally distributed white noise with covariance $Q_p(t)$. Thus, the parameters have no dynamics but are random variables with Gaussian noise. Even for linear systems, parameter estimation is a nonlinear problem because $\underline{p}$ and $\underline{x}$ appear in a nonlinear fashion in the linear state equations

$$\dot{\underline{x}}(t) = F(\underline{p})\underline{x}(t) + B\underline{u}(t) \tag{20}$$

Hence, for linear or nonlinear systems, the extended Kalman filter can be applied to parameter estimation by treating the parameters as additional states augmented to the original state vector. The augmented system dynamics are modelled by

$$\dot{\underline{x}}_a = \begin{bmatrix} \dot{\underline{x}} \\ \dot{\underline{p}} \end{bmatrix} = \begin{bmatrix} f(\underline{x},\underline{u},\underline{p},t) \\ 0 \end{bmatrix} + \begin{bmatrix} G(\underline{x},\underline{p},t) & 0 \\ 0 & I_p \end{bmatrix} \begin{bmatrix} \underline{w}(t) \\ \underline{w}_p(t) \end{bmatrix} \tag{21}$$

THE EVAPORATOR MODEL

Evaporation is a common process in chemical plants. Fig. 1 shows a flow diagram. Some of the complexities of a real evaporator have been simplified here so that the model reduces to the same one found in the design literature (Andre and Ritter, 1968; Burdett and Holland, 1971), namely

$$\frac{dx_1}{dt} = F - (\omega x_1 + E_c) - V \tag{22}$$

$$\frac{dx_2}{dt} = [\beta F x_F + (V - F)(x_2 - T_B)]/x_1 \tag{23}$$

where

$$[x_1 \quad x_2]^T = [W \quad T]^T$$
$$V = [UA(T_S - T) - FC_p(T - T_F) - Q_L]/\Delta H_v$$

and UA is the (heat transfer coefficient) x (area of heat transfer), T_S is the temperature of steam in the steam chest, T_B is the normal boiling point of the solvent, C_p is the heat capacity of the solution, T_F is the temperature of the feed stream, Q_L is the rate of heat loss to the surroundings, ΔH_v is the heat of vaporization of the solvent, Q_s is the total rate of heat transfer from the steam, t_i is the mass fraction of solute in the stream, ω is a constant (0.6), β is the boiling point elevation per mass fraction of solute, and E_c is a constant (0.1). Figure 1 shows the rest of the notation. Hence, the states of the model are the holdup and the temperature (or equivalently composition or enthalpy). The specific values of the coefficients and inputs for the model can be found in Dalle Malle (1985).

To carry out parameter estimation by the method of Watanabe and Himmelblau for the evaporator, for every nonlinear term and linear or nonlinear term that contains a fault parameter in the state equations, an independent state measurement must be made. From the state equations of the model, it can be seen immediately that the expression for x_2, Eq. (23), is entirely nonlinear because x_1 appears in the denominator of each term. Hence, measuring the temperature could be sufficient for estimating the holdup through an observer, but no parameters could be estimated because the model has no more states to measure. Alternatively, if x_1 and x_2 were both measured, then two parameters could be estimated and there would be no need for a state observer. The latter mode was selected to determine whether or not faults associated with model parameters exist. Noise was also added to the process measurements to represent randomness, and was also introduced into the inputs. For the simulations in this work, all process parameters were assumed to remain constant (except for the fault parameters). The standard deviations of the noise factors are listed in Dalle Molle (1985).

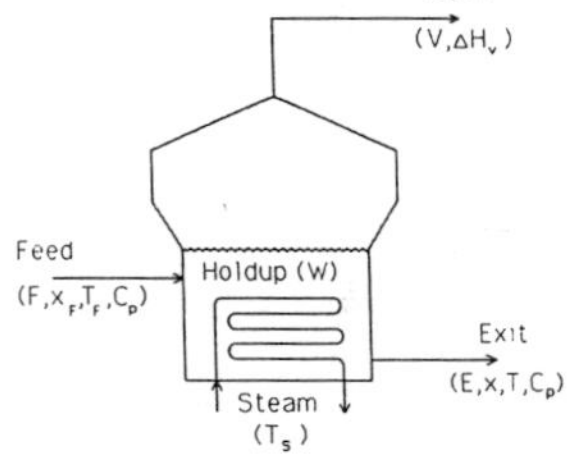

Fig. 1. Evaporator configuration and flow chart.

RESULTS

One process parameter of interest associated with process degradation or faulty operation is the heat transfer coefficient, U (actually, UA here), which cannot be measured directly by any means. As the heat transfer surface becomes fouled or scaled, the heat transfer rate is decreased and the efficiency of the process is reduced. Another parameter of interest for fault detection is the composition of the feed, x_F. Depending on the component, a composition analysis might be difficult to make because of poor instrument sensitivity for the component or other problems such as toxicity. If the feed to the evaporator is the product of some other unit operation such as a reactor, then knowing the composition at the input of the evaporator could be useful in determining if the previous unit was operating properly.

To illustrate the types of trajectories that occur for the two parameters, the following faults were introduced by way of example:

	UA	x_F
% change in value	-10.0 ramp	-20.0 square
Starting time of change (min)	75.0	165.0
Stopping time of change (min)	375.0	285.0

Figure 2 shows the estimate of UA, and Fig. 3 shows the estimate of x_F when $\lambda = 0.95$.

These results demonstrate that the least squares estimation scheme is valid even when two faults occur simultaneously. However, it can be seen in Fig. 3 that, for abrupt changes such as that applied to x_F, the estimate responds slowly and convergence to the new value takes nearly 60 minutes. The rate of convergence of the estimate could be increased by decreasing the weighting factor, but the variance of the faster response would be larger. The effect of making process measurements at discrete intervals rather than continuously is a slower response in the estimates. A larger sampling interval reduces the effect of measurement noise on the approximation of the derivative of the states because the approximations are averaged over a longer time. Hence, the variance of the estimates does not necessarily increase with the slower response due to sampling. The speed of the response for the discrete case can also be increased by decreasing λ, but only at the expense of the variance at a constant sampling interval.

The same example was used with the extended Kalman filter. Because the extended Kalman filter involves the solution of differential equations for both the state

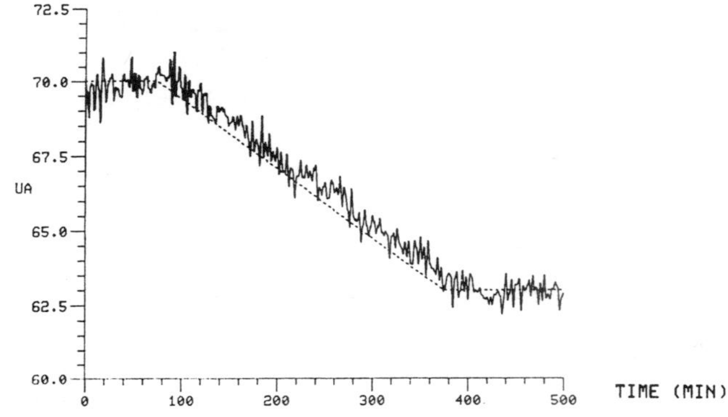

Fig. 2. Estimate of UA for λ = 0.95; (——) Estimate; (-----) Measured Value

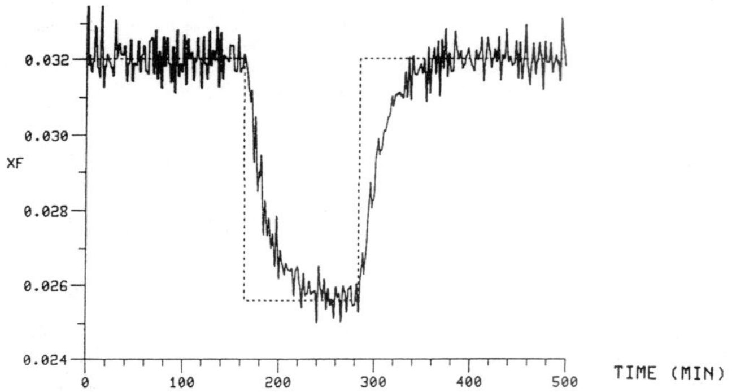

Fig. 3. Estimate of x_F for λ = 0.95; (——) Estimate; (-----) Measured Value

and error covariance propagation, initial conditions must be supplied for the states, parameters, and the error covariance matrix of the augmented state vector. For the states and parameters, the design or normal operating values can be used as the initial conditions. The initial error covariance matrix is usually assumed to be diagonal with large values of the elements to express uncertainty in the initial values of the states and parameters. Simulations (not shown here) demonstrate that the initial values of the state vector and the error covariance matrix had no effect on the operation of the algorithm. Even with poor initial guesses, the estimates converged to unbiased values immediately. In addition to the initial values, the noise covariance matrices R(t) and Q(t) must also be supplied. Normally, the measurement noise covariance matrix, R(t), is assumed to be diagonal. The variances of each measurement can be guessed or estimated from sample output values. The input noise covariance matrix, Q(t), is also assumed to be diagonal. Although the values of the elements in R(t) and Q(t) might be obtained from process measurements, the filter is usually "tuned" to dynamics of the process so that the response for the parameter trajectories is reasonably fast and their covariance matrix elements are of reasonable size. Figures 4a and 4b illustrate the trajectories of x_F and UA for the Kalman filter in which confidence limits (P = 0.95) have been put in place based on the period of normal operation.

The extended Kalman filter (EKF) with continuous measurements requires 15 times more computation time (on a Cyber computer) than the least squares method for a sampling interval of 1 minute. If a constant gain matrix is used, however, only 4 times more CPU time is needed. For a sampling interval of 8 minutes, the least squares method runs 6 times faster than it did for a 1 minute sampling interval.

Examples not described above because of lack of space demonstrated that there was a need for some heuristics in the analysis of the estimates to avoid misdiagnosing nonexistent faults when more than one fault occurs at a time. Factors such as decision rules, confidence coefficients, and so on must be selected based on the filter parameters selected and the dynamics of the process, the latter of which may not be known exactly.

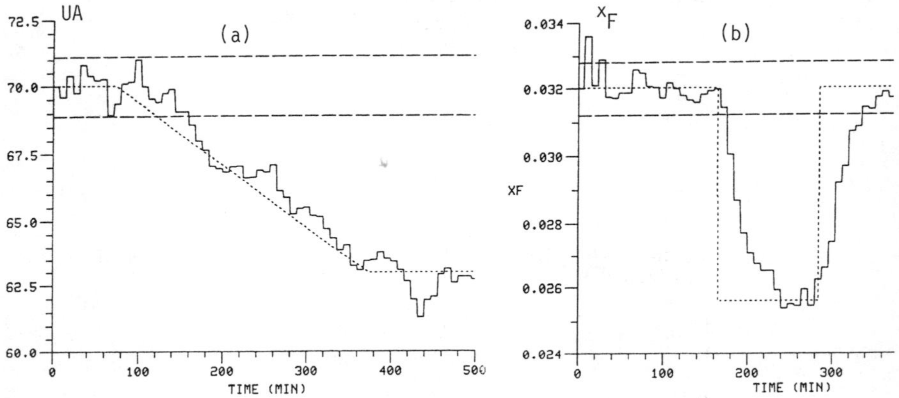

Fig. 4. EKF estimate of UA and x_F with confidence intervals

(———) Estimate; (-----) True Value; (— —) Confidence Limits

REFERENCES

Andre, H. and R. A. Ritter (1968). The Can. Chem. Engr. 46, 259-264.
Astrom, K. L. and P. Eykhoff (1971). Automatica 7, 123-162.
Burdett, J. W., and C. D. Holland (1971). A.I.Ch.E. Jl. 17, 1080-1086.
Chen, C. T. (1970). Introduction to Linear System Theory. Holt, Rinehart, and Winston, New York.
Dalle Molle, D. T. (1985). Fault Detection via Parameter Estimation in a Single Effect Evaporator. MS Thesis, University of Texas, Austin, Texas.
Frank, P. M. (1986). Fault diagnosis in dynamic systems via state estimation - a survey. Paper presented at the First European Workshop on Fault Diagnosis and Related Knowledge-Based Systems, Rhodos, Greece, 1-3 Sept., 1986.
Geiger, G. (1984). Fault Identification of a Motor-Pump System Using Parameter Estimation and Pattern Classification. Proc. IFAC-Congress 7/84.
Gelb, A. (1984). Applied Optimal Estimation, The MIT Press, Cambridge, MA.
Himmelblau, D. M. (1978). Fault Detection and Diagnosis in Chemical and Petrochemical Processes, Elsevier, Amsterdam.
Jazwinski, A. H. (1970). Stochastic Processes and Filtering Theory, Academic Press, New York.
Shiozaki, J., H. Matsuyama, K. Tano, and E. O'shima (1985). Int. Chem. Engr. 25, 561-659.
Sorenson, H. W. (1980). Parameter Estimation, Marcel Dekker, New York.
Watanabe, K. and D. M. Himmelblau (1983). A.I.Ch.E. Jl. 29, 243-249.
Watanabe, K. and D. M. Himmelblau (1982). Int. J. Syst. Sci. 13, 137-158.
Young, P. C. (1969). Control Engng. 119-125.

Filtering and Estimation with Antagonist Hypothesis for Failure Detection

D. Sauter, M. Aubrun, T. Cecchin, J. Merckle

Laboratoire d'Automatique et de Recherche Appliquée
C.R.A.N., Université de NANCY I
B.P. 239, 54506 Vandoeuvre Cedex

ABSTRACT :

This paper is concerned with the problem of a failure detection for one dimensional systems evolving in a steady state.

The detected failures are characterized by a second order model. To defect a failure we compute the innovation sequences generated by two filters elaborated with the two hypothesis :

- H_0 normal mode
- H_1 failure mode

Under H_0 hypothesis we suppose that process is corrupted by a white gaussian noise. The model of the process is known and we calculate the innovation sequence generated by a kalman filter.

Under H_1 hypothesis we suppose that a failure exist and we estimate the parameters of the failure model in order to compute the innovation sequence.

To detect the presence of a failure, a test on the two innovation sequences is proposed.

A pratical application concerning the detection of surface defect of steel products illustrates the proposed method.

KEYWORDS :

Failure detection, Kalman filter, Antagonist hypothesis, Iron working units.

1. INTRODUCTION

In modern ironworking units wich use continuous casters, it is very important to control, in real time, the quality of steel slab before rolling.

Detection of surface defects (craks) allows to only roll products without failures, in order to avoid important failures in steel sheet.

For failure detection, an eddy current sensor is used. A simple method wich was restricted to checking sensor output for upward or downward transgression of fixed limits is not sufficient. Indeed, oscillation ridges generated by the process can have the same magnitude as failures but are not pejudicial to rolling.

To resolve this problem of craks detection, we use a method based on the knowledge of the approached signal model under normal mode (H_0 hypothesis) and of the signal model under failure mode (H_1 hypothesis). The shape and duration too of the failures are known; only the magnitude and apparition time of failures are unknown.

The proposed approach for failure detection consists of the following steps :

- A first model describing the behavior of the system under normal operating conditions is defined. It also specifies the statistics of the measurement noise in the output variable.
A CHI-square test based on the innovation process generated by a Kalman Filter allows the detection of failures.

- A second model describing the evolution of the system under failure conditions is proposed too. The possible presence of failures detected by the precedent test is introduce in an adaptative model. A test on the innovation sequence generated by a Kalman filter confirms if a failure is present.

In this paper, to generalize the proposed method, we suppose :

- the process evolves in a steady state
- a failure produce an additive effect on process observation ; the jump model, wich characterize the failure, is an autoregressive model.

The application of the general case to failure detection on steel slab is presented in part 6.

2. PROBLEM STATEMENT

Under steady state conditions, the normal mode H_0 evolution of the process is described by the following equations :

$$\begin{cases} x(k+1) = x(k) + v(k) & (2.1) \\ y(k) = x(k) + w(k) & (2.2) \end{cases}$$

Where $x(k)$: is the state variable
$y(k)$: represents the observation of the process

$\{v(k)\}$ and $\{w(k)\}$ are zero mean, independant, white gaussian sequences, with covariances defined by :

$$q(k) = E\,[v^2(k)] \qquad (2.3)$$

$$r(k) = E\,[w^2(k)] \qquad (2.4)$$

Under H_0 hypothesis we suppose that model parameters and noise statiscal properties are known.

Now let use introduce the model of the process under H_1 conditions. If we suppose that failures present an additif effect on the precedent model, we obtain :

$$\begin{cases} x(k+1) = x(k) + v(k) & (2.5) \\ \mu(k+1) = q(k)\ \mu(k) + v'(k) & (2.6) \\ y(k) = x(k) + d(k)\ \mu(k) + w(k) & (2.7) \end{cases}$$

Where :

- q(k) cararacterize the presence of a failure during the duration $[K_1\ K_2]$

$$q(k) = 1 \qquad k \in [K_1\ K_2] \tag{2.8}$$

$$q(k) = 0 \qquad k \notin [K_1\ K_2] \tag{2.9}$$

- d(k) is assumed to be the effect of the failure on the observation
- $\mu(k)$ represents the magnitude of the failure

3. STATE VARIABLE ESTIMATION

3.1 Estimation under normal condition :

With the equations (2.1) and (2.2) an optimal estimation of the state variable is given by the following recurrences :

$$\hat{x}(k+1/k+1) = \hat{x}(k+1/k) + K(k+1)\ [g_0(k+1)] \tag{3.1}$$

$$\hat{x}(k+1/k) = \hat{x}(k/k) \tag{3.2}$$

$$K(k+1) = P(k+1/k+1)\ /\ r(k+1) \tag{3.3}$$

$$P(k+1/k+1) = \left[\frac{1}{p(k+1/k)} + \frac{1}{r(k+1)}\right]^{-1} \tag{3.4}$$

$$P(k+1/k) = P(k/k) + q(k) \tag{3.5}$$

$$g_0(k+1) = y(k+1) - \hat{x}(k+1/k)$$

Where $g_0(k+1)$ is the innovation sequence. Under H_0 conditions it represents a white gaussian random sequence with variance :

$$V_0 = p(k/k-1) + r(k)$$

3.2 Failure sensitive filter

A state variable representation :

$$\begin{cases} X(k+1) = AX(k) + V(k) & (3.6) \\ Y(k) = C(k)\ X(k) + W(k) & (3.7) \end{cases}$$

Can now be given with following definitions :

$$X(k) = \begin{bmatrix} x(k) \\ \mu(k) \end{bmatrix} \qquad A = \begin{bmatrix} 1 & 0 \\ 0 & q(k) \end{bmatrix} \tag{3.8}$$

$$Y(k) = (x(k)) \qquad C(k) = (1 \ d(k)) \tag{3.9}$$

$$V(k) = \begin{bmatrix} v(k) \\ v'(k) \end{bmatrix} \qquad W(k) = w(k) \tag{3.10}$$

Where W(k) is the process noise with known statistics :

$$E[V(k)] = 0 \quad , \quad E[V(k)\ V^T(k)] = Q(k) \tag{3.11}$$

If the effect of the failure on the observation variable is supposed been defined with a second order recurrence then :

$$d(k) = a.d(k-1) + b.d(k-2) \tag{3.12}$$

The problem is now to estimate the state variables x(k) and $\mu(k)$.

Under H_1 hypothesis the optimal state estimation is given by the discrete Kalman filter equations :

$$\hat{x}(k/k-1) = A\hat{X}(k-1/k-1) \tag{3.13}$$
$$\hat{X}(k/k) = \hat{X}(k/k-1) + K(k)\ g_1(k) \tag{3.14}$$
$$g_1(k) = y(k) - C(k)\ \hat{X}(k/k-1) \tag{3.15}$$

Where $g_1(k)$ is the zero-mean, Gaussian innovation process, and the gain K(k) is calculated from the equations :

$$P(k+1/k) = P(k/k) + Q(k) \tag{3.16}$$

$$V(k) = C(k)\ P(k/k-1)\ C^T(k) + r(k) \tag{3.17}$$

$$K(k) = P(k/k-1)\ C^T(k)\ V^{-1}(k) \tag{3.18}$$

$$P(k/k) = P(k/k-1) - K(k)\ C(k)\ P(k/k-1) \tag{3.19}$$

4. DETECTION

To take into account the presence of a failure different methodes based on the observation of the innovation sequence are used.

- A whiteness and a chi-squared test of the residual of the normal Kalman filter (Mehra and Peschon (1971).

- A sequential probability ratio test (Newbold (1968)) where H_0 hypothesis is inprove against H_1 hypothesis.

-A generalized likelihood ratio test wich results in a correlation of the observed residuals with the precomputed filter responses due to certain faults (Willsky and Jones (1974).

A recent description of these different, methods is given by M. Basseville and A. Benveniste (1983), A. Willsky (1976) and R. Isermann (1984).

Let'us examine in more details, the CHI-square test proposed by Mehra and Peschon (1971) and applied by Willsky, Dest and Crawford (1974).

Let $g_i(k)$ been the innovation for a filter F_i elaborated with model H_i. If the system is operating normally, according to model H_i, the innovation is zero mean and white with known covariance $V_i(k)$; the following quantity :

$$I_i(k) = \sum_{j=k-N}^{k} \frac{g_i^2(j)}{V_i(j)} \tag{3.20}$$

is a CHI-squared random variable with N+1 degrees of freedom. If a change in model H_i occurs, the statistical properties of g_i change. To take into account an abnormallity one can compare $I_i(k)$ to a predefinite threshold :

$I_i(k) > A_i$ ---> Failure

Where A_i can be computed with the aid of chi-squared tables in order to reduce the probability of false alarm or missed alarm.

In our case, with the two antagonist models H_0 and H_1 and the two associated filters defined in 3.1 and 3.2 one obtain when a failure is present :

$I_0(k) > A_0$

$I_1(k) < A_1$

Therefore to detect failures caracterized by a 2nd order model, one can consider a detection rule of the forme :

$(I_0(k) > A_0)$ and $(I_1(k) < A_1)$ ---> 2nd order failure
$(I_0(k) < A_0)$ ---> no failure
$(I_0(k) > A_0)$ and $(I_1(k) > A_1)$ ---> undefined failure

Concerning F_1 filter defined with H_1 model a problem is still to be solved : the apparition time of a failure (represented by q(k) = 1) is supposed to be known :

In that sense an estimation $\hat{q}(k)$ of parameter q(k) is given by F_0 filter as follows :

$I_0(k) > A_0$ ---> $\hat{q}(k) = 1$

$I_0(k) < A_0$ ---> $\hat{q}(k) = 0$

The configuration for the proposed detection and estimation mechanis m is shown in Fig.1

5. APPLICATION : detection of surface defects on steel slab

The state equations of the process are the same as in (2.1) and (2.2) :

$$\begin{cases} x(k+1) = x(k) + v(k) & (5.1) \\ y(k) = x(k) + w(k) & (5.2) \end{cases}$$

x(k) represente the thickness of the slab "see" by the sensor at time k. The sensor is in a fixed plan, the slab is moving with a constant speed.

v(k) is an additive noise, representative of surface state.

w(k) is the measurement noise of eddy current sensor.

The failure, model can be simulated by a second-order model (see Fig 2) :

$$d(k) = a.d(k-1) + b.d(k-2) \qquad (5.3)$$

For pratical implementation, we must define :

- the variance measurement and process noises
- the size of the sliding window

The variance of the noise measurement r(k) is approximated by:

$$\hat{r}(k) = \sum_{k-N}^{k} (y(j) - x(j/j-1))^2 \qquad (5.4)$$

In this way, $\hat{r}(k)$ is an adaptative weighted factor in equation (3.1) :

- under H_0 hypothesis, $\hat{r}(k)$ is minimal and the filter is oblivious to old measurements,
- under H_1 hypothesis, $\hat{r}(k)$ is maximal and the filter is oblivious to new measurements.

The primary criterion in the choice of $\hat{r}(k)$ is not that it provides a good estimate of the state x(k), but rather that the effects of perturbations are accentuated in the filter residual.

Variance q(k) for filter F_0 and covariance matrix Q(k) for filter F_1 are choosen by learning on well known signals. The size of the sliding window is computed according to probable length of the failures. In our case, only the mangitude of the failure is varying.

The numerical results shown in Fig. 2-6 indicate that the two antagonist filters technique proposed here is more improved than a one filter technique elaborated with H_0 hypothesis.

6. CONCLUSION

This paper has proposed a technique of the failure detection system for discrete-time linear systems evolving in a steady state. The failure detection system developed consists of two parts :

- a normal mode filter
- a failure mode filter.

When anomaly state is detected by normal mode filter it gives an alarm. The failure mode filter takes it into account and confirms the effective presence of a failure characterized by a second order model.

Numerical results obtain from a pratical application show that the method proposed here is effective for the detection and estimation for failure parameter.

References :

M. BASSEVILLE ; A. BENVENISTE (1983) :
Design and comparative study of some sequentiel jump detection algorithms for digital signals ; IEEE transactions on acoustics, speech, and signal processing ; vol. ASSP-31, n°3 ; pp 521-535.

R. ISERMANN (1984) :
Process fault detection based on modeling and estimation methods. A survey ; Automatica ; vol. 20, n°4, pp 387-404.

R.K. MEHRA ; J. PESCHON (1971) :
An innovations approach to fault detection and diagnosis in dynamic systems ; Automatica, vol. 7, pp 637-640.

P.M. NEWBOLD ; Y.C. HO (1968) :
Detection of changes in the characteristics of a gauss-markov process ; IEEE transactions on Aerospace and Electronic systems : vol. AES-4, n°5, pp 707-718.

A. S. WILLSKY ; J.J. DEYST ; B.S. CRAWFORD (1974)
Adaptative filtering and self-test methods for failure detection and compensation. Proc. of the 1974 JACC. Austin, Texas.

A.S. WILLSKY ; H.L. JONES (1976)
A generalized likelihood ratio approach to the detection and estimation of jumps in linear systems ; IEEE transactions on automatic control AC-21 ; pp 108-112.

A. S. WILLSKY (1976) :
A survey of design methods for failure detection in dynamic systems ;
Automatica ; vol. 12, pp 601-611.

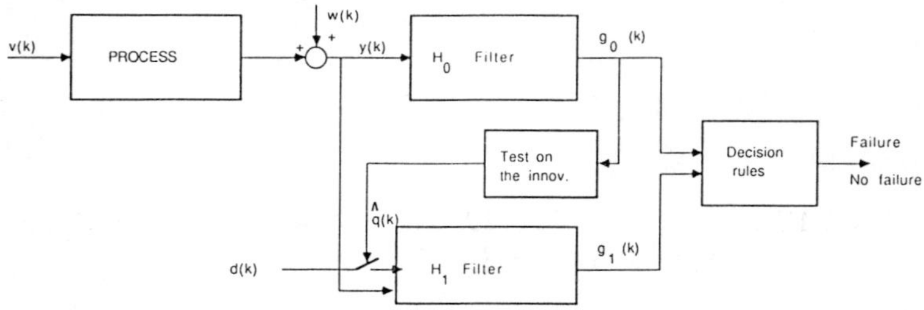

Fig. 1: Failure detection mechanism

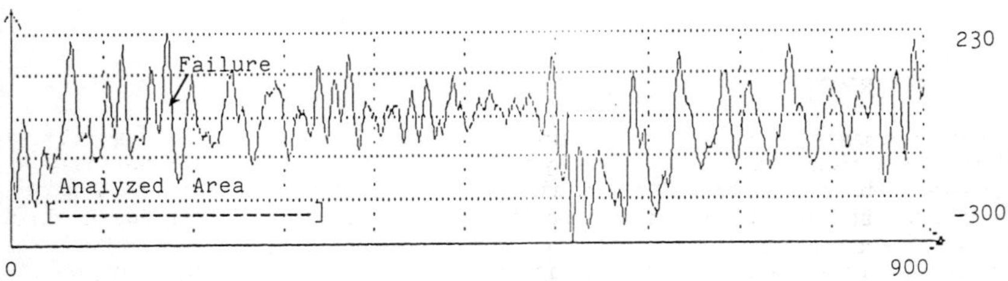

Fig 2: Measured signal

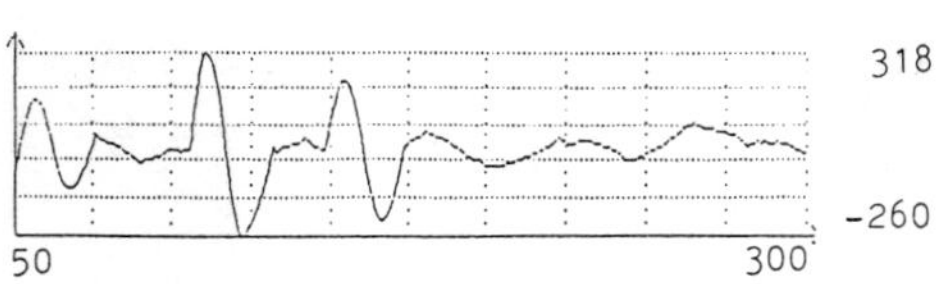

3 (a) Filtered signal

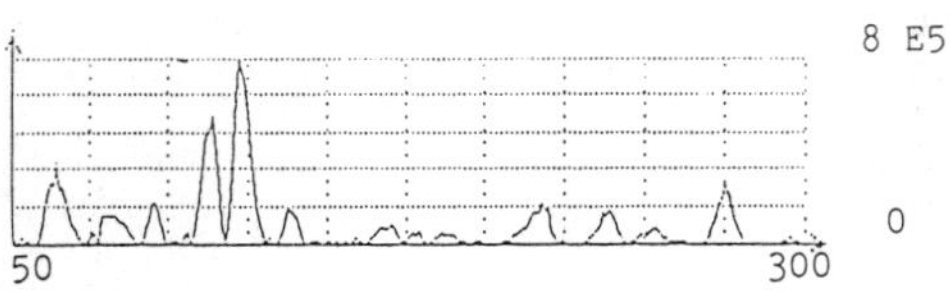

3 (b) Innovation

Fig .3: Filtering with H_1 model

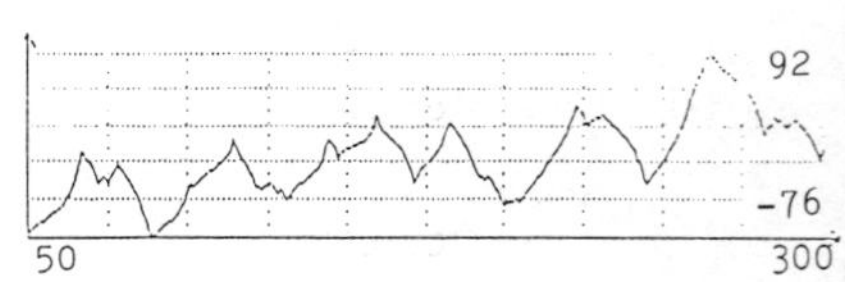

4 (a) Filtered signal

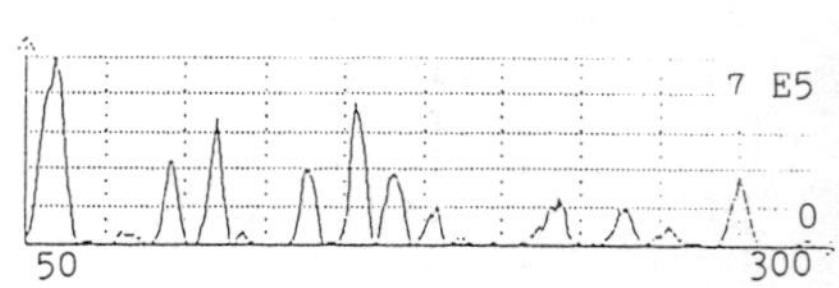

4 (b) Innovation

Fig. 4: Filtering with H_0 model

Component Failure Detection Using Local Second-Order Observers

D. Hengy, P. M. Frank

University of Duisburg,
Department of Electrical Engineering, Duisburg, FRG

ABSTRACT

This paper presents a component failure detection scheme based on analytical redundancy. The redundant signals are generated by state estimation using local second-order observers that are designed for each component of a given nonlinear process. The scheme may be considered as a generalization for a class of nonlinear systems of the method proposed by Janßen and Frank (1984, 1986) for linear or linearized systems. The introduction of the nonlinearities of the process into the structure of the observers contributes to improve the efficiency of the failure location scheme. Some results of a digital simulation illustrate the method in the context of an example.

KEYWORDS

Complex systems - Local nonlinear state estimation - Analytical redundancy - Failure detection and isolation.

INTRODUCTION

With the increasing complexity of technical dynamical systems the notions of security and supervision are of increasing importance. The recent progress in digital computing make possible the development of efficient failure detection and isolation methods using analytical redundancy. Methods using state observers or Kalman filters have been already widely studied in the past. A recent survey on the state of the art has been given by Frank (1986).
Janßen and Frank (1984, 1986) proposed a method of component failure detection (CFD) based on state estimation by means of local linear observers. The aim of this approach is to detect and to locate failures in physical subsystems (component) of a process. In this way two groups of components are considered. There are the so-called ASCC's (Available State Coupled Components) coupled each other with measurable states for which the observer scheme ASCOS (Available State Coupled Observer Scheme) is applied. On the other hand, there are the so-called USCC's (Unknown State Coupled Components) coupled by unknown states for which the observer scheme ESCOS (Estimated State Coupled Observer Scheme) is applied.
This paper discusses a new approach by using local nonlinear observers in the

scheme. Since most of the processes encountered are nonlinear, imbedding these nonlinearities in the structure of the observer may improve the efficiency of the scheme in order to avoid discrepancies between the system model and the linear estimator.

SECOND-ORDER NONLINEAR OBSERVER FOR ASCC's

Let us consider a component of a deterministic nonlinear system in the following state-space form:

$$\dot{x} = f(x, \bar{x}, u) \tag{1a}$$

$$y = c(x, u) \tag{1b}$$

$$\bar{y} = \bar{c}(\bar{x}, u) \tag{1c}$$

$$x(o) = x_o \quad , \quad \bar{x}(o) = \bar{x}_o \tag{1d}$$

with x the $(n \cdot 1)$ state vector, $\bar{x}$ the $(\bar{n} \cdot 1)$ coupling vector, u the $(p \cdot 1)$ input vector, y the $(m \cdot 1)$ output vector, $\bar{y}$ the $(\bar{m} \cdot 1)$ output vector of the coupled components, f, c and $\bar{c}$ assumed vector valued functions at least twice differentiable with respect to their arguments.
In general, to get an efficient failure isolation, the following output decoupling conditions are required:

$$\frac{\partial c}{\partial \bar{x}} \equiv 0 \quad \text{and} \quad \frac{\partial \bar{c}}{\partial x} \equiv 0 \ . \tag{2}$$

According to the definition of the ASCC's, the coupled state vector $\bar{x}$, is measurable, e.g.

$$\bar{y} = \bar{x} \tag{3}$$

and is, therefore, considered as an additional input vector. With respect to the failure isolation conditions (2), a coupling between the output of the subsystem under consideration and the output of the coupled subsystems is allowed in this case, e.g.,

$$y = c(x, \bar{x}, u) \tag{4}$$

The structure of a second-order nonlinear observer proposed for an ASCC (1a-b) and (3) is then given by

$$\dot{\hat{x}} = f\,(\hat{x}, \bar{x}, u) + K(y - \hat{y}) + \frac{1}{2} \sum_{j=1}^{n} \alpha_j \,(y - \hat{y})^T\, G_j(y - \hat{y}) \tag{5a}$$

$$\hat{y} = c\,(\hat{x}, \bar{x}, u) \tag{5b}$$

$$\hat{x}(o) = \hat{x}_o \tag{5c}$$

$$\text{with } \alpha_j = [0,\ldots,0,\overset{(j)}{1},0,\ldots,0]^T \tag{6}$$

the natural $(n \cdot 1)$ basis vector, where superscript T denotes transpose, $\hat{x}$ the estimated state vector, K the $(n \cdot m)$ and G_j the $(m \cdot m)$ matrices, which have to be specified. In this way let us introduce the estimation error

$$e = x - \hat{x} \tag{7}$$

and expand the nonlinear vector function (1a) and the nonlinear measurement vector function (4) in a Taylor series with respect to the estimated vector $\hat{x}$ neglecting third and higher order terms

$$\dot{x} = f(x, \bar{x}, u) = f(\hat{x}, \bar{x}, u) + f_x e + \frac{1}{2} \sum_{j=1}^{n} \alpha_j \, e^T f_{jxx} e + \ldots \tag{8}$$

$$y = c(x, \bar{x}, u) = c(\hat{x}, \bar{x}, u) + c_x e + \frac{1}{2} \sum_{j=1}^{n} \alpha_j \, e^T c_{jxx} e + \ldots \tag{9}$$

with f_x the Jacobian matrix

$$f_x = \frac{\partial f}{\partial x} \Big|_{\hat{x},\bar{x},u} = \left[\frac{\partial f_j}{\partial x_k}\right]_{\hat{x},\bar{x},u} \tag{10}$$

f_{jxx}, $j=1,\ldots,n$, the Hessian matrices

$$f_{jxx} = \frac{\partial^2 f_j}{\partial x^2} \Big|_{\hat{x},\bar{x},u} = \left[\frac{\partial^2 f_j}{\partial x_k \partial x_l}\right]_{\hat{x},\bar{x},u} \quad . \tag{11}$$

The same definitions hold for c_x and c_{jxx}.

According to (5a-b) and (7-9) the error estimation e is governed by the differential equation

$$\dot{e} = (f_x - Kc_x)e + \frac{1}{2} \sum_{j=1}^{n} \alpha_j e^T \left(f_{jxx} - \sum_{l=1}^{m} K_{jl} c_{lxx} - c_x^T G_j c_x\right) e + \ldots \tag{12}$$

where again third and higher order terms have been neglected. In order to achieve asymptotic stability for (12) one has to obtain stable eigenvalues for

$$F = (f_x - Kc_x) \tag{13}$$

by an appropriate choice of the matrix K. With the method proposed in (Zeitz, 1979), one has to assign constant stable eigenvalues for F after compensation of its time-variant elements. But this is in general not applicable for higher order systems ($n > 2$). Seidel (1980) proposed to determine a constant matrix K which minimizes the performance index

$$J = \int_0^\infty e^T Q \, e \, dt \quad . \tag{14}$$

In this case K is fixed for a certain operating point and is the solution of the Ljapunov equation

$$(A - KC)^T P + P(A - KC) + Q = 0 \tag{15}$$

with P, Q some positive definite matrices and A, C the constant matrices

$$A = f_x \quad \text{and} \quad C = c_x \tag{16}$$

defined at the operating point $(x_s, \bar{x}_s, u_s)$ instead of $(\hat{x}, \bar{x}, u)$. The stability of the observer scheme has then to be tested for other operating points.

The quadratic form in (12) is minimized choosing G_j as

$$G_j = c_x^{+T} \left(f_{jxx} - \sum_{l=1}^{m} K_{jl} c_{lxx}\right) c_x^{+} \; , \; j=1,\ldots,n \tag{17}$$

where the sign + denotes generalized inverse. Further, the initial conditions $e_o = e(o)$ have to be small to avoid that the neglected terms in (12) including the minimized quadratic form influence the stability of the observer scheme. This means that the initial conditions x_o have to be known with a sufficient precision.

Defining a component failure z(t) as an additional unknown function to $f(x,\bar{x},u)$ in (1a), the error estimation equation becomes

$$\dot{e} = F\, e + z(t) \tag{18}$$

Therefore failures occuring in this component affect the output estimation error of the failed component only, according to

$$y - \hat{y} = c_x e + \frac{1}{2}\sum_{j=1}^{n} \alpha_j e^T c_{jxx}\, e + \ldots \tag{19}$$

In other words, failures in a given component are detected and located by the corresponding local observer. This allows us to consider the scheme as a generalization of ASCOS for this class of nonlinear systems.

SECOND-ORDER NONLINEAR OBSERVER FOR USCC's

For such components the entire coupling vector $\bar{x}$ is not measurable but some information is available by means of the output vector of the coupled components (1c). In this case the failure isolation condition (2) holds strictly. As the coupling vector $\bar{x}$ is not available, its corresponding estimated $\hat{\bar{x}}$ is used in the observer equation. The structure of a second-order nonlinear observer is proposed for the subsystem (1a-e) as follows

$$\begin{aligned}\dot{\hat{x}} = {} & f(\hat{x}, \hat{\bar{x}}, u) + K(y - \hat{y}) + \frac{1}{2}\sum_{j=1}^{n} \alpha_j (y - \hat{y})^T G_j(y - \hat{y}) \\ & + \bar{K}(\bar{y} - \hat{\bar{y}}) + \frac{1}{2}\sum_{j=1}^{n} \alpha_j (\bar{y} - \hat{\bar{y}}) \bar{G}_j(\bar{y} - \hat{\bar{y}}) \\ & + \frac{1}{2}\sum_{j=1}^{n} \alpha_j [(y - \hat{y})^T H_{1j}(\bar{y} - \hat{\bar{y}}) + (\bar{y} - \hat{\bar{y}})^T H_{2j}(y - \hat{y})]\end{aligned} \tag{20a}$$

$$\hat{y} = c(\hat{x},u) \tag{20b}$$

$$\hat{\bar{y}} = \bar{c}(\hat{\bar{x}},u) \tag{20c}$$

$$\hat{x}(o) = \hat{x}_o,\ \hat{\bar{x}}(o) = \hat{\bar{x}}_o \tag{20d}$$

K, G_j are the above defined matrices and $\bar{K}$ $(n\cdot\bar{m})$-, $\bar{G}_j$ $(\bar{m}\cdot\bar{m})$-, H_{1j} $(m\cdot\bar{m})$-, and H_{2j} $(\bar{m}\cdot m)$-matrices, that have to be determined.

Introducing the error estimation coupling vector

$$\bar{e} = \bar{x} - \hat{\bar{x}} \tag{21}$$

and expanding $f(x, \bar{x}, u)$ in a Taylor series with respect to the estimated vectors $\hat{x}$ and $\hat{\bar{x}}$ and neglecting third and higher order terms yields

$$\begin{aligned}\dot{x} = f(x, \bar{x}, u) = {} & f(\hat{x}, \hat{\bar{x}}, u) + f_x e + f_{\bar{x}}\bar{e} \\ & + \frac{1}{2}\sum_{j=1}^{n}\alpha_j(e^T f_{jxx}e + \bar{e}^T f_{j\bar{x}\bar{x}}\bar{e} + \bar{e}^T f_{j\bar{x}x}e + e^T f_{jx\bar{x}}\bar{e})\end{aligned} \tag{22}$$

$c(x, u)$ in (1b), resp. $\bar{c}(\bar{x}, u)$ in (2b) is expanded with respect to $\hat{x}$, resp. $\hat{\bar{x}}$ in the same way as in (9). The error estimation equation becomes

$$\begin{aligned}\dot{e} = {} & (f_x - Kc_x)e + (f_{\bar{x}} - \bar{K}\bar{c}_{\bar{x}})\bar{e} + \\ & + \frac{1}{2}\sum_{j=1}^{n} \alpha_j [e^T(f_{jxx} - \sum_{k=1}^{m} K_{jk}c_{kxx} - c_x^T G_j c_x)e + \bar{e}^T(f_{j\bar{x}\bar{x}} - \sum_{k=1}^{\bar{m}} \bar{K}_{jk}\bar{c}_{k\bar{x}\bar{x}} \\ & - \bar{c}_{\bar{x}}^T\bar{G}_j\bar{c}_{\bar{x}})\bar{e}] + \frac{1}{2}\sum_{j=1}^{n} \alpha_j[e^T(f_{jx\bar{x}} - c_x^T H_{1j}\bar{c}_{\bar{x}})\bar{e} + \bar{e}^T(f_{j\bar{x}x} - \bar{c}_{\bar{x}}^T H_{2j}c_x)e]\end{aligned} \tag{23}$$

For the definitions of the matrices $f_{j\bar{x}x}$, $f_{jx\bar{x}}$, $f_{j\bar{x}\bar{x}}$ and $\bar{c}_{\bar{x}}$, $c_{j\bar{x}\bar{x}}$ we refer analogously to (10) and (11). All derivatives of f are taken at the point $(\hat{x},\hat{\bar{x}},u)$ instead of $(\hat{x},\bar{x},u)$, the derivatives of c at $(\hat{x},\bar{u})$ and the ones of $\bar{c}$ at $(\hat{\bar{x}},u)$.

The matrices K and G_j are determined in the same manner as in (15) and (17). In order to minimize the different quadratic forms one choose

$$\bar{G}_j = \bar{c}_{\bar{x}}^{+T} (f_{jxx} - \sum_{k=1}^{\bar{m}} \bar{K}_{jk}\bar{c}_{k\bar{x}\bar{x}})\bar{c}_{\bar{x}}^{+} \tag{24}$$

$$H_{1j} = c_x^{+T} f_{jx\bar{x}} \bar{c}_{\bar{x}}^{+} \tag{25}$$

$$H_{2j} = \bar{c}_{\bar{x}}^{+T} f_{j\bar{x}x} c_x^{+} \tag{26}$$

Assuming now that the above quadratic forms can be neglected, (23) is reduced to

$$\dot{e} = F e + \bar{F}\bar{e} \tag{27}$$

$$\text{with} \quad \bar{F} = f_{\bar{x}} - \bar{K}\bar{c}_{\bar{x}} \tag{28}$$

$\bar{K}$ is chosen in order to minimize the influence of the coupling term $\bar{F}\bar{e}$, eg., as

$$\bar{K} = f_{\bar{x}} \bar{c}_{\bar{x}}^{\pm} \tag{29}$$

This time-variant $\bar{K}$ matrix, though minimizing $\bar{F}$ may sometimes unstabilize the global observer scheme. A compromise between stability and efficiency of the failure isolation method has to be found - a task that is for nonlinear systems more difficult to solve than for time invariant linear systems.
The convergence of the observer states to the system states depends on the good knowledge of the initial conditions of the system. Thus, in order to avoid a possible influence of the neglected terms in (23), the error estimation vectors e_o and $\bar{e}_o$ have to be small.

In this way the nonlinear observer scheme depicted above may be considered as an extension of ESCOS for a class of nonlinear systems.

Note that equations (5a) and (20a) can be found using a more general formulation for the observer equations, e.g.

$$\dot{\hat{x}} = \hat{f}(\hat{x}, \bar{x}, u, y) \text{ or } \hat{f}(\hat{x}, \hat{\bar{x}}, u, y, \bar{y}) \tag{30}$$

in the case of ASCC's or USCC's. Expanding the error estimation differential equations in a Taylor series up to the second order with respect to $\hat{x}$ or $\hat{x}$ and $\hat{\bar{x}}$, the appropriate forms can be deduced for the function $\hat{f}$. This approach has been developed in (Hengy, Frank, 1986) in order to obtain the equations of local first-order nonlinear observers and is omitted here.

EXAMPLE OF APPLICATION

The above described CFD scheme is applied to the three-tank system sketched on Fig. 1. The water levels h_1,h_2,h_3 are governed for $h_1>h_2>h_3>H>o$ by the following differential equations:

$$\dot{h}_1 = -b(h_1 - h_2)^{1/2} + \frac{1}{Q} u_1 \tag{31}$$

$$\dot{h}_2 = b(h_1 - h_2)^{1/2} - b(h_2 - h_3)^{1/2} \tag{32}$$

$$\dot{h}_3 = b(h_2 - h_3)^{1/2} - b(h_3 - H)^{1/2} \tag{33}$$

with the outputs $y_i=h_i$, $i=1,2,3$, $b=(q/Q)(2g)^{1/2}$, g the gravity constant and u_1 a constant input. The three tanks represent the components of this system. Since all the states h_1,h_2,h_3 are measurable, nonlinear ASCOS is applied. For each component a nonlinear observer is designed according to (5a-b) with n=1, m=1. The detailed observer equation for the first component (31) is given as follows

$$\dot{\hat{h}}_1 = -b(\hat{h}_1 - h_2)^{1/2} + \frac{1}{Q} u_1 + K_1(h_1 - \hat{h}_1) + \frac{1}{2} G_1(h_1 - \hat{h}_1)^2 \tag{34}$$

The corresponding error estimation equation yields

$$\dot{e}_1 = [- \frac{b}{2}(\hat{h}_1 - h_2)^{-1/2} - K_1] e_1 + \frac{1}{2} [\frac{b}{4} (\hat{h}_1 - h_2)^{-3/2} - G_1] e_1^2 + \ldots \tag{35}$$

with $e_1=h_1-\hat{h}_1$. The quadratic form in (35) disappears by choosing

$$G_1 = \frac{b}{4} (\hat{h}_1 - h_2)^{-3/2} \tag{36}$$

If one assign a constant eigenvalue λ_1 to the observer, the factor K_1 in (34) becomes time-variant

$$K_1 = - \frac{b}{2} (\hat{h}_1 - h_2)^{-1/2} - \lambda_1 \quad . \tag{37}$$

If K_1 is set to a constant value, for example $K_1=1.5b$, the observer eigenvalue becomes time-variant but remains stable. The observer equations for the two last components are established similarly.

Three different component failures are considered:

1. Leak in the base of tank 2 with a cross section 0.25 q
2. Clog producing a reduction to 0.25 q in the flow between tank 1 and 2
3. Clog producing a reduction to 0.25 q in the outlet of tank 3.

The system is driven under the operating point corresponding to the constant scalar input $u_1=1.222\ m^3s^{-1}$ and the stationary state vector $h_{s1}=[9m\ 7m\ 5m]^T$. The simulations of the process and the three observers are made on a digital computer HP 1000 for a period of 3000 sec. The failures occur at 50 % of the simulation time. The simulation is started by a step of the input from u_1 to $u_2=1.5\ u_1$, e.g. a deviation of +50 % from the preceeding operating point. The initial conditions for the observers are chosen as the same as those of the system, since all the states are measurable. The three component failures under consideration have been separately simulated. The results are shown on Fig. 2, 3 and 4. The failed components are detected and isolated as the system states h_i deviate from their corresponding estimate $\hat{h}_i$. For example, on Fig. 2 a leak in tank 2 implies a deviation of the output h_2 from its estimate $\hat{h}_2$. Though the other outputs h_1 and h_3 are also affected by this failure, their estimates behave in the same manner and therefore do not indicate a failure in the corresponding components. This proofs the great strength of local observers which guarantee the failure location.

Moreover, for this example, no difference has been noted by means of local first-order (G_j=o) or second-order observers. This is not the case by means of local linear observers, using a linearized model of the system. Here some false location has occured. Consider Fig. 5 which shows the behaviour of the different absolute error estimations in consequence of a leak in tank 2. The scalar $e_i=h_i-\hat{h}_i$ denotes the difference between the system and the nonlinear observer output, the scalar $e_{iL}=h_i-\hat{h}_{iL}$ the difference between the system and the linear observer output. These error estimations have been divided by the sum $(h_1 + h_2 + h_3)_{s1}$ of the elements of the stationary state vector h_{s1} and indicated in %. As it can be seen, the estimation errors e_2 of the nonlinear observer and the one e_{2L} of the linear observer of tank 2 are reacting to the failure as expected. Additionally, the output error estimation e_{1L} of the linear observer of tank 1 is also reacting. It may be caused by the too great variation about the operating point, so that the linearized model used in the observers is not valid any more.

CONCLUSION

An on-line method of component failure detection in complex dynamic systems making use of analytical redundancy via state estimation is outlined in this paper. Local second-order nonlinear state observers are used in the scheme. The design structure may be considered as an extension of the hierarchical failure detection scheme proposed by Janßen and Frank (1984, 1986) to a more general class of nonlinear systems.

Applying the scheme in the context of a simple example, the three-tank system, simulation results have presented satisfying results in view of the ability to detect and locate component failures. Moreover, the scheme is made robust against some variations around the operating point by including the nonlinearities of the system into the observer structures.

The next steps will be the verification of the described scheme to more complex systems, especially by applying it to real processes. Further, the problem of the robustness of the observer scheme with respect to unknown parameter variations has to be theoretically studied.

REFERENCES

Frank, P. M. (1986). Fault Diagnosis in Dynamic Systems via State Estimation Methods - A Survey. First European Workshop on Fault Diagnostics, Reliability and Related Knowledge - Based Approaches, Rhodos, Greece.

Hengy, D., and P. M. Frank (1986). Component Failure Detection via Nonlinear State Observers. Proc. of the IFAC Workshop on Fault Detection and Safety in Chemical Plants, 153 - 157. Kyoto, Japan.

Janßen, K., and P. M. Frank (1984). Component Failure Detection via State Estimation. Proc. of the 9th IFAC World Congress, 147 - 152. Budapest, Hungary.

Janßen, K., and P. M. Frank (1986). Entdeckung von Komponentenfehlern in dynamischen Systemen durch hierarchische Zustandsschätzung. Automatisierungstechnik, 1, 23 - 31.

Seidel, H.(1980). Nichtlineare Beobachter zur Schätzung von Zustandsgrößen der Längsbewegung eines Flugzeuges. Doctor's dissertation, TU Braunschweig.

Zeitz, M. (1979). Nichtlineare Beobachter. Regelungstechnik, 8, 241 - 272.

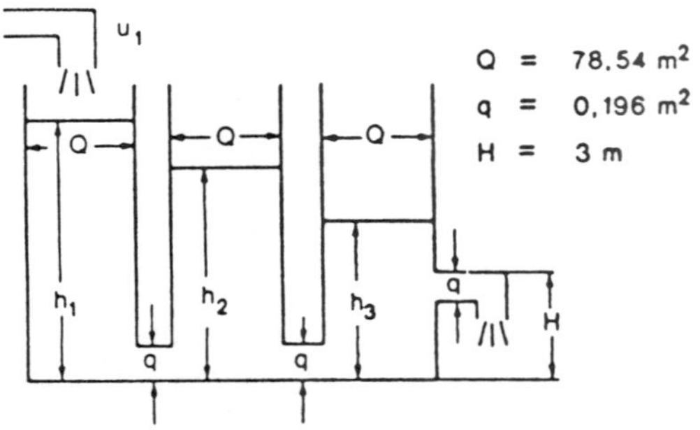

Fig. 1: Scheme of the three-tank system under consideration

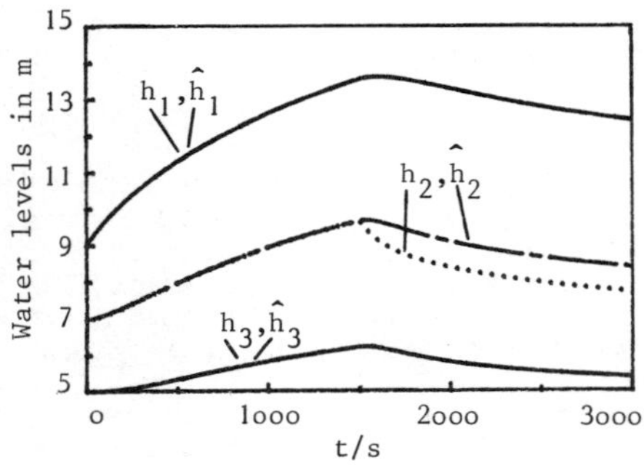

Fig.2 Evolution of the 3 water levels and their estimates due to a leak in tank 2.

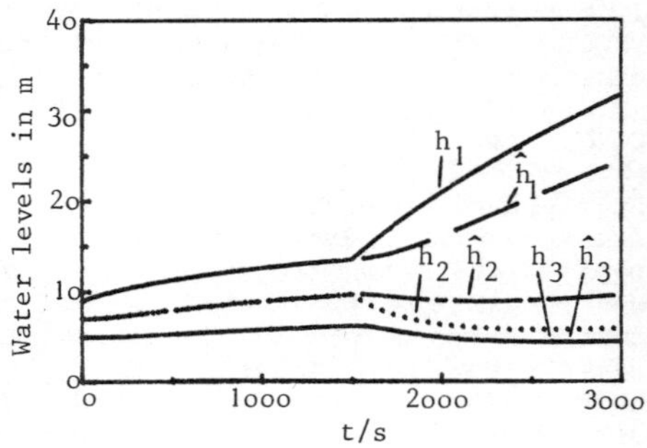

Fig.3 Evolution of the three water levels and their estimates due to a clog between tank 1 and 2.

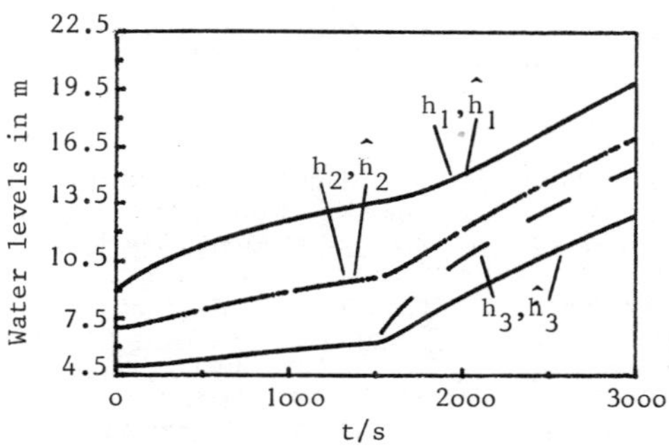

Fig.4 Evolution of the three water levels and their estimates due to a clog in the outlet of tank 3.

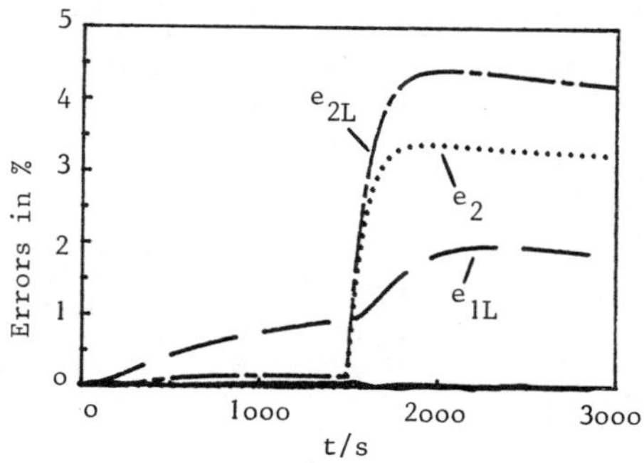

Fig.5 Evolution of the error estimations e_i, resp. e_{iL} from the nonlinear, resp. linear observers due to a leak in tank 2.

A Robust Method for Fault Diagnosis Using Parity Space Eigenstructure Assignment

R.J. Patton and S.W. Willcox

Department of Electronics, University of York

Keywords Fault detection, Analytical redundancy, Eigenstructure assignment, Observers.

Abstract This paper describes new results based on the design and application of an analytical redundancy method for instrument fault detection (I.F.D.) using an observer approach. The fault detection monitor described uses a priori information concerning the frequency domain characteristics of parameter variations and disturbance inputs acting on the system. It is assumed that the system is nth order and that the instrument vector is an mth order subset of the states and that the observers used can be designed using eigenstructure assignment of the parity (operational) space.

1. Introduction

Many systems for detecting instrument faults in dynamic systems have been introduced in recent years[1,2]. Fault tolerance in many systems especially those in which faults could result in a catastrophic event (such as in an aircraft flight control system) is usually achieved through hardware redundancy. To enable normal operation to continue in the presence of a fault, critical hardware, such as sensors and computers, are repeated three or four times in a triplex or quadruplex redundancy scheme. The repeated elements are usually distributed spatially around the system to provide protection against localised damage. This type of scheme operates by comparing the outputs from the repeated hardware in a straightforward majority voting scheme so that software and hardware faults may be detected and isolated. Hardware redundancy provides an accurate and robust method of protection against sensor faults. The major problems encountered with hardware redundancy, and those problems which have led to investigations into other forms of fault detection, are the extra weight the repeated hardware adds to the system payload, and the space required to accomodate this, which could be used for more mission orientated equipment.

In an attempt to overcome the above problems, and to try and improve the overall system reliability and performance with the high integrity control systems (now required in many aerospace applications), recent work has concentrated on the use of analytically redundant relationships to enable the detection of sensor faults. These analytical redundancy techniques use knowledge of the dynamic or analytical relationships between dissimilar outputs of the same system. The majority of analytical redundancy methods are based on the use of state estimation.

In all of the analytical techniques the general process on which the fault/no fault decision takes place is based on the comparison of the redundant data. The difference between the redundant data (usually termed the residual) in an ideal situation is a measure of the performance of the sensor outputs under consideration. In all cases the decision process applied to the one or more residuals is, in general, a threshold criterion in which a fault is signalled when the threshold is exceeded by one or more of the residuals.

The factor which is common to all state comparison techniques is the use of some form of state estimator to produce estimates of part or all of the system state[1]. The state estimates may then be compared, either against available system states or against estimates generated from different measurement subsets.

The main problem associated with all of the above I.F.D. schemes is in the certainty required in the modelling of the physical system, both in the design and implementation stages. The above techniques make certain assumptions about the physical system and/or the failure modes the may not be accurate, and may degrade the performance of the I.F.D. scheme from that which was predicted theoretically. These limitations cause linear based design approaches to be inadequate fo many real engineering applications. Sensitivity to input-induced parameter variations cause uncertain errors between redundant state estimate vectors and, in an I.F.D. scheme these errors would cause false signalling of an instrument fault. The solution proposed by previous investigators is to widen the threshold band for the fault detect signal and so reduce false alarms. This also has the adverse effect of increasing the time taken to detect a fault. This increased time for detection may prove to be unacceptable in many fault-critical systems, such as those found in aerospace applications.

Clearly any available knowledge about the uncertain effects acting on the system should be used in the design stage in an attempt to improve the performance of the I.F.D scheme. For many applications frequency domain information is available, or may be ascertained from the desired control characteristics, about the parameter variations and disturbance inputs acting on the system. The use of frequency domain sensitivity information in this way enables a new, robust approach to observer design, in the I.F.D. field, to be made. It can be shown that the error vector signals of an observer contain all the infomation relating to the parameter variations and disturbance inputs of the process being monitored[3,4]. From the consideration of the non-linear processes to be found in many real systems, a scheme has been developed which uses the above information in the design stage of a fault monitoring observer based on the 'output zeroing' of scalar combinations of elements of the observer's error vector to produce an effective solution to the problem of fault detection and isolation in many systems in which parameter variations and uncertainties prove to have an effect that may not be considered to be negligible.

2. Problem Specification

The major assumption in the alternative I.F.D. techniques is that the system to which the I.F.D. analysis is applied can be well represented by a linear model, a poor assumption for most real applications. It is instructive to consider a representation of a real system which is subject to parameter variations, and stochastic disturbances applied to an observer in which some system states have been neglected, ie. The system contains effects not modelled by the observer. eg. the coupling between the longitudinal and lateral motions. Let the system be represented by:

$$\dot{\underline{x}} = \begin{bmatrix} \underline{x}_0 \\ \hline \underline{x}_u \end{bmatrix} = \left[\begin{array}{c|c} A_0 + \Delta A & A_c \\ \hline A_{u1} & A_{u2} \end{array}\right] \begin{bmatrix} \underline{x}_0 \\ \hline \underline{x}_u \end{bmatrix} + \begin{bmatrix} B_0 + \Delta B \\ \hline B_u \end{bmatrix} \underline{u}$$

$$+ \begin{bmatrix} D_0 \\ --- \\ D_u \end{bmatrix} \underline{v} \qquad 2.1$$

$$\underline{y} = [C_0 \vdots C_u][\underline{x}_0 \vdots \underline{x}_u]^T \qquad \text{where:}$$

A_0 is the linearised model of the plant to be modelled in the observer. ΔA, A_c, A_{u1}, A_{u2}, ΔB, and B_u are the non-linear functions of the state and input vectors which describe the time varying dynamics of the plant, the subscript u denoting unmodelled effects. D_0 and D_u are the systems disturbance distribution matrices. $\underline{v}$ is a vector of disturbances acting on the plant. This may contain non-Gaussian and non-stationary noise components. C_0 is the output matrix of the linearised plant modelled in the observer.

The observer estimate of the partial state vector $\underline{x}_0$ is given by the following standard observer equations:

$$\dot{\tilde{\underline{x}}}_0 = A_0\tilde{\underline{x}}_0 + B_0\underline{u} + K(\underline{y} - C_0\tilde{\underline{x}}_0) \qquad 2.2$$

If this observer is applied to the nominal system denoted by the 0 subscripts then the state estimate error is dependent only upon the initial error between the actual state and the observer estimate. Therefore, assuming that the observer error dynamics are stable, the initial error vector will die away in some finite time. However, when this linear observer is applied to the actual plant described by the system equations 2.1 effects other than the initial error will determine the history of the estimation error. When driven by the output from the actual plant, the estimation error dynamics are as follows:

$$\dot{\underline{e}} = (A_0 - KC_0)\underline{e} + \underbrace{[\Delta A \vdots \Delta B \vdots A_c - KC_u \vdots D_0]}_{E}\underbrace{[\underline{x}_0 \vdots \underline{u} \vdots \underline{x}_u \vdots \underline{v}]^T}_{\underline{d}} \qquad 2.3$$

where $\underline{e} = \tilde{\underline{x}}_0 - \underline{x}_0$

ie. The estimation error equation now has an extra input vector which will have the effect of driving the error away from it's zero steady-state during disturbances and plant parameter variations.

The observer error dynamics may now be completely described by:

$$\dot{\underline{e}} = (A_0 - KC_0)\underline{e} + E\underline{d} \qquad 2.4$$

Where $\underline{d}$ is an unknown disturbance input vector and, E is it's associated **unknown** distribution matrix.

It should be noted that E as well as $\underline{d}$ will be time-varying and dependent upon the state and input of the system. Furthermore, in order to overcome the problems caused by the unknown inputs acting on the system it is important to use frequency domain information available about the disturbance inputs at the design stage in an attempt to make the estimation error vector insensitive to the effects of the unmodelled dynamics.

3. Non-Linear Simulation

To test the effectiveness of I.F.D. schemes it is necessary to have use realistic models to which the schemes may be applied. As stated earlier, previous studies only considered the performance of the techniques developed applied to linear models upon which the designs were based. They did not, in general, consider the application of the schemes in a real environment, where effects such as non-lineararities, stochastic disturbances and unmodelled states could degrade the performance of the fault detection process in terms of speed of response, false alarm rate, etc.

In this present work the operation of the I.F.D. scheme has been tested in as realistic an environment as possible without initially using the physical system as a test bed. The system was implemented in the form of a fully non-linear simulation.

The model used is that of a small, unmanned aircraft[5] designed for control system evaluation and performance tests. The simulation model describes the non-linear dynamic behaviour of the aircraft in all it's motions. The control of the aircraft is achieved by means of a full state feedback auto-pilot system based on modern control theory. Straight and level flight is not a particuarly severe test for any I.F.D. scheme, as false alarms are most likely to occur whilst the aircraft is performing some manoeuvre (inducing parameter variations). Therefore, the reference demands to the controller may be changed during the flight to simulate a varied range of pilotic inputs. For this particular application study it was decided to limit the I.F.D. scheme to only the lateral motion of the aircraft since longitudinal motion application has already been investigated[3]. The lateral motion state-space matrices for zero rudder and aileron deflection trim setting with an airspeed of $33 ms^{-1}$ are:

$$A_{la} = \begin{bmatrix} -0.2770 & 0.0 & -32.9 & 9.81 & 0.0 \\ -0.1033 & -8.525 & 3.75 & 0.0 & 0.0 \\ 0.3649 & 0.0 & -0.639 & 0.0 & 0.0 \\ 0.0 & 1.0 & 0.0 & 0.0 & 0.0 \\ 0.0 & 0.0 & 1.0 & 0.0 & 0.0 \end{bmatrix}$$

$$B_{la} = \begin{bmatrix} -5.432 & 0.0 \\ 0.0 & -28.64 \\ -9.49 & 0.0 \\ 0.0 & 0.0 \\ 0.0 & 0.0 \end{bmatrix}$$

where: $\dot{\underline{x}} = A_{la}\underline{x} + B_{la}\underline{u}$,

$\underline{x}$ = [v(side-slip vel.),p(roll rate),r(yaw rate),Φ(roll angle), Ψ(yaw angle)]^T, and $\underline{u}$ = [(aileron angle), (rudder angle)]^T.

The parameters of the aircraft vary from the nominal model when it deviates from the specified steady-state trim conditions. Any control system or I.F.D. scheme (even the most robust) will have it's performance degraded from the theoretical projection when implemented on the non-linear system.

The evaluation of an I.F.D. scheme requires that other effects, apart from the aircraft dynamics, should be modelled realisticaly. For this case, atmospheric effects and the type of instrument fault can have a significant effect on the performance of the I.F.D. scheme. Consequently realistic gust and fault models have been used in the simulation[6]

4. Robust Fault Detection

Section 2 has shown how unmodelled states and disturbances effect the error vector of an observer. It was shown that the estimation error vector contains information about these disturbances, ie. the error dynamics of an observer are, in the s-domain:

$$\underline{e}(s) = M(sI - J)M^{-1}[E\underline{d} + K\underline{f}] \qquad 4.1$$

Where M is the generalised eigenvector matrix forming a basis set for the Jordan form J.

The time response of the error vector is affected to various degrees, by the initial estimation error, the disturbance inputs, and by the fault vector. The component each mode contributes to the overall error output is dependent upon the full eigenstructure of the observer and on the frequency domain characteristics of the disturbance input vectors.

To design a robust I.F.D. scheme it is necessary to assign an eigenstructure to the observer to minimise the effects of disturbances and maximise the effect of any sensor fault. With aerospace and many other applications the parameter variations are strongly related to the states and inputs of the system, and therefore tend to vary with the same modal dynamics as that of the system. Assuming that the control of the system is performing well and has been designed to minimise the sensitivity of the plant eigenvalues to the control inputs, the frequency domain characteristics of the disturbances should be band-limited. This allows another dimension in the design of a robust I.F.D. scheme to be considered. This information about the frequency domain characteristics can be used to build in frequency discriminating structures into the observer's error dynamics and so provide an effective disturbance filter.

The technique described here is an attempt to design the structure of the error dynamics, to fulfil the criteria specified above. The technique attempts to maintain the error vector of an observer in a constant direction in the error space during normal operation. The observer is designed such that the error vector remains close to this 'operational' direction until such time as a sensor fault occurs. The fault has the effect of pulling the error vector away from this operational direction. This alteration of error vector direction allows the detection of an instrument fault. In contrast to earlier work the observer is now attempting to maintain a constant direction the requirements on the input to error vector component functions have been reduced and may be easily understood by consideration of the observer eigenstructure.

In the design of the robust I.F.D. filter it is necessary to use all the design freedom available in the observer eigenstructure to produce an output that is insensitive to the disturbance inputs. From the design specification described earlier this may be achieved by ensuring that the final fault detect signal dynamics only contain components from modes of the observer that are uncontrollable from the disturbance inputs, ie. the remaining modes are unobserverable in the fault detect signal.

In this technique the fault detect signal is a function of the 'closeness' of the error vector from the designed operational direction. Let $\underline{w}_1$ be a vector orthogonal to the required operational (null) space. In a no-fault situation the fault detect signal is required to satisfy the followng criterion:

$$\underline{w}_1^T C_0 \underline{e}(s) = 0; \text{ for all } s. \qquad 4.2$$

Left Eigenvectors:

0.0245	0.0221	0.0044	0.0013+0.0002j	0.0013-0.0002j
0.0018	0.0066	0.0147	0.0389-0.0104j	0.0389+0.0104j
-0.1249	-0.1418	-0.1041	-0.0574-0.0058j	-0.0574+0.0058j
-0.0532	-0.0502	-0.0215	-0.0008+0.0052j	-0.0008-0.0052j
0.9904	0.9884	0.9942	0.9975	0.9975

In this case modes 2,3,4,5 have been made unobservable in the estimation error space and the corresponding eigenvectors are in the null-space projection which is orthogonal to $\underline{w}_1^T C_0$. Therefore, the resulting disturbance filter characteristic is dependent on the dominant (range space) mode of s = -2.14. This was chosen to provide filtering of the spectrally higher disturbance effects, and so provide increased disturbance rejection. It should be noted that the space on R^n is partitioned by using the orthogonality property between the left and right eigenvectors of distinct eigenvalues.

5. Simulation Results

Figures 1 and 2 show the results of applying the above scheme to the non-linear aircraft model. In each case the model is simulating an initial offset correction applied to the roll rate whilst changing height following a stepped reference demand. Figure 1 demonstrates the system in a no-fault situation. Figure 1(a) shows the roll rate output, (b) the weighted estimation errors, ie. The components in the operational space, and (c) the fault detect signal. It can be seen in (c) that in this no-fault case the fault detect signal remains close to the desired operational space. In figure 2 the same signals now show the response to a fault. In this case a slow drift offset developing on the roll rate sensor output. It can be seen that, for this type of fault, the fault detect signal exhibits a rapid response allowing fast detection to be achieved.

6. Conclusion

It has been shown that, from a knowledge of frequency domain parameter sensitivity information, a robust fault monitor can be designed. The observer used has an eigenstructure which is assigned in a robust manner such that motions in the estimation error space become insensitive to disturbances and uncertain dynamic effects. In the no-fault case a parity signal which is a weighting of the measured estimation error components gives rise to a zero sensitivity measure for the nominal model. The bandwidth restriction imposed by the design reduces the sensitivity of this parity signal to disturbance effects such that a low threshold for detection can be set. During a fault the faster modes become observable giving rise to rapid deviations and resulting in efficient fault detection. The paper has shown that the method can be used for more than two measurement signals. An example has been given based on the lateral motion of a multivariable non-linear aircraft model comprising three measurements, five states, and additional uncertainties arising from two control inputs.

7. Acknowledgements

The authors would wish to acknowledge MoD contract support from the Royal Aircraft Establishment.

ie. during the no-fault operation the disturbances drive the error vector asymptotically towards one direction in the error space. To achieve this, each mode of the observer should either be unobservable in the estimation error space or it's associated projection C_0M should be orthogonal to $\underline{w}_1^T$, ie.

$$C_0M = 0; \text{ or } \underline{w}_1^T C_0 M = 0 \qquad 4.3$$

In practice, the number of modes which are designed to fit a particular criterion depends on the number of measurements available, the observer order, and the characteristics of the disturbance filter required. By leaving one or more of the observer modes in the range space of the estimation error vector, (ie. those modes which are to fulfill $\underline{w}_1^T C_0 M = 0$) the response of the disturbance filter may be tailored to that which will provide the greatest rejection of the disturbances, increasing further the robustness of the I.F.D. filter.

To fit the observer eigenstructure an optimisation algorithm is used which attempts to minimise a cost function which is itself a measure of the closeness of the actual observer eigenstructure to that of the required eigenstructure. For this particular application an alternating variable, direct search technique was used[7]. The cost function is of the form:

$$J = \sum_{i=1}^{n} \underbrace{\left[p_i (\lambda_{r_i} - \lambda_{a_i})^2 \right]}_{(i)} + \underbrace{q \underline{w}_1^T C_0 \underline{m}_1}_{(ii)} \qquad 4.4$$

Where λ_r and λ_a are the required, and actual closed-loop observer eigenvalues, respectively, with weighting p, and $\underline{m}_i$ is the eigenvector associated with λ_i. The terms in the cost function serve the following purposes: (i) Cost associated with the difference between the desired and fitted eigenvalues. (ii) Cost required to minimise the projected scalar product of the left and right eigenvectors of the (single) range space mode. It should be noted that $\underline{w}_1^T$ is the left eigenvector corresponding to the (single) range space mode. The procedure is an alternative to well known (exact but restrictive) eigenstructure assignment methods[8,9,10].

The actual design for the lateral motion aircraft system described in section 2, is shown below. For this application the measurements were assumed to be p, Φ, and Ψ. This gives the following output matrix:

$$C_0 = \begin{bmatrix} 0.0 & 1.0 & 0.0 & 0.0 & 0.0 \\ 0.0 & 0.0 & 0.0 & 1.0 & 0.0 \\ 0.0 & 0.0 & 0.0 & 0.0 & 1.0 \end{bmatrix}$$

The final assigned eigenstructure is:

Eigenvalues:

$$\lambda_1 = -2.14$$
$$\lambda_2 = -2.65$$
$$\lambda_3 = -9.34$$
$$\lambda_{4,5} = -19.57 \pm 2.72j$$

Right Eigenvectors:

range space			null space	
-0.0154	0.6117	0.9351	0.7509	0.7509
0.5964	0.4793	0.1907	-0.1596+0.2359j	-0.1596-0.2359j
-0.5601	-0.4142	-0.0247	-0.0527+0.0451j	-0.0527-0.0451j
0.5721	0.4718	0.3115	0.5781-0.1261j	0.5781+0.1261j
-0.0549	-0.0429	-0.0098	0.0061-0.0015j	0.0061+0.0015j

8. References

1. Frank P.M. 'Fault Diagnosis in Dynamic Systems via State Estimation - A Survey', Proceedings of the 1st European Workshop on Fault Diagnosis, Reliability and Related Knowledge-Based Approaches, 2-4 Sept. 1986, Rhodes, Greece.

2. Himmelblau D.M. 'Fault Detection and Diagnosis - Today and Tomorrow', Proceedings of the IFAC workshop on Fault Detection and Safety in Chemical Plants', 28th Sept. - 1st Oct. 1986, Kyoto, Japan, pp95-105.

3. Patton R.J. and Willcox S.W. 'Fault Diagnisos in Dynamic Systems Using a Robust Output Zeroing Design Method', Proceedings of the 1st European Workshop on Fault Diagnosis, Reliability and Related Knowledge-Based Approaches, 2-4 Sept. 1986, Rhodes, Greece.

4. Patton R.J., Willcox S.W. and Winter J.S. ' A Parameter Insensitive Technique for Aircraft Sensor Fault Analysis', AIAA Journal of Guidance Control and Dynamics, May-June 1987.

5. Aslin P.P. 'Aircraft Simulation and Robust Flight Control System Design', D.Phil Thesis, University of York, England, July 1985.

6. Gault J.D. and Gunth D.E. 'Atmospheric Turbulance Considerations for Future Aircraft Designed to Operate at Low Altitudes' (Porgram LO-LOCAT) J. Aircraft 5, (2), 1968.

7. Kuester J.L. and Mize J.H. 'Optimisation Techniques with Fortran', McGraw-Hill, 1973.

8. Moore B.C. 'On Flexibility Offered by State Feedback in Multivariable Systems Beyond Closed-Loop Eigenvalue Assignment', IEEE Trans. AC-22, pp883-885, Oct. 1977.

9. Andry A, Shapiro E. and Chung J. 'Eigenstructure Assignment for Linear Systems', IEEE Trans. AES-19, No. 5, pp711-729, Sept. 1985.

10. Mudge S.K. and Patton R.J. 'A Comparison of Two Techniques of Eigenstructure Assignment', Proceedings of the 2nd European Workshop on Fault Diagnosis, Reliability and Related Knowledge-Based Approaches, April 6-7 1987, UMIST, Manchester, UK.

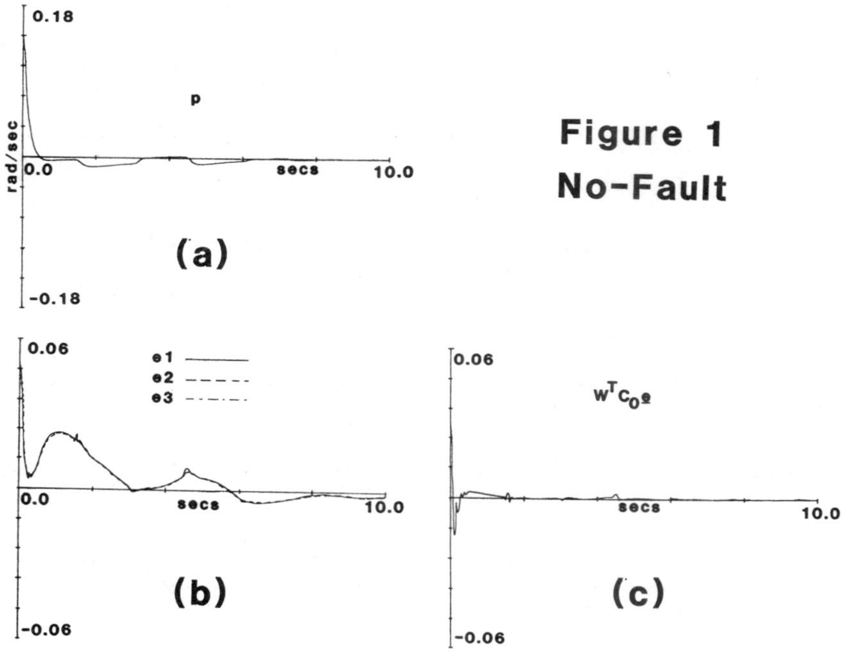

Figure 1
No-Fault

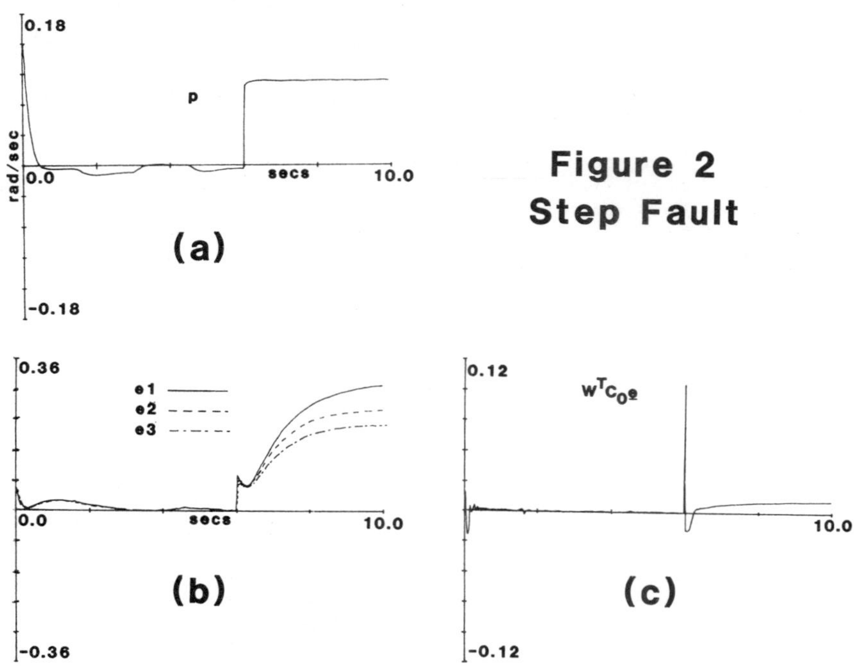

Figure 2
Step Fault

Fault Detection Mechanisms and Approaches to Autonomous Operation of Flexible Assembly Cells

O. v. Dungern and Günther K. Schmidt

Lehrstuhl für Steuerungs- und Regelungstechnik
Technische Universität München, West Germany

ABSTRACT

With the increasing interdependency of automation equipment, reliability improvement of the single unit becomes a major design goal. This paper describes an approach to reliability improvement of flexible small parts assembly through efficient fault detection and autonomous recovery. Fault detection is based on cyclical decentral comparison of actual and expected sensor values. In case of a fault, regular operation is interrupted and a recovery process is invoked. This fault detection mechanism is embedded in a general sensor/control model with several levels of abstraction. The decentral structure of the sensor data processing architecture allows for the creation of powerful subsystems.
Finally, a cell is described to illustrate these concepts which is now being implemented in our laboratory and which is designed for the assembly of an electromechanical component family. A fast serial bus is employed as a general interface for intelligent subsystems, such as the proportional gripper.

KEY WORDS

Small Parts Assembly; Reliability Improvement; Autonomous Systems; Fault Detection; Sensor Bus; Intelligent Robot Gripper.

RELIABILITY IMPROVEMENT IN MANUFACTURING SYSTEMS

In the past, manufacturing organization was such that different production stages were completed in independent islands which produced for large buffers. It was sufficient to employ *fault avoidance techniques* when designing automatic machines, since, due to loose coupling, a delay in one production island had little effect on others.

Modern trends, however, try to reduce the minimum lot size and the time a certain product requires for completion by decreasing buffer sizes, thus increasing the interdependency of the production islands. Also, Lotter (1985) reports that the most often used part feeders have an MTBF of less than 3 minutes. With severe economical consequences resulting from production delays or persistent misproduction, the improvement of reliability for the single unit is a major design goal.

Reliability or availability improvement of manufacturing systems is closely related to its autonomy. Here, autonomy is characterized by the ability of the system to run fully automatically, i.e. without human supervision or intervention. The minimum requirement for autonomous operation is *fault detection* and prevention of damage in the system through suitable *fault containment measures*, such as interrupting robot motion and calling the operator. Higher degrees of autonomy may be attained through more or less sophisticated recovery schemes using sensory information. However, no matter which degree of autonomy is desired, efficient fault detection is a prerequisite. Hence, this aspect is emphasized in this paper. It is our conviction that fault diagnosis and consequently fault handling is considerably simplified, if faults in the system are detected early.

Basically, we have to consider three types of faults:
* unsuccessful operation due to a module misfunction ("primary failure");
* defective parts after an undetected primary failure ("secondary failure");
* program and operator faults.

The first type is internally induced in the system and can be detected either by consistency checking of subsystem interaction (e.g. time-outs, protocol violations, etc.) or through unexpected sensor readings. The former method must be "built in" through careful design of the control system hierarchy, while the latter is dependent on the application given, so that the system should provide suitable mechanisms for the application programmer. The second type of faults is induced externally and can be identified through visual and functional testing of intermediate and final products or again through unexpected sensor readings. The last type applies to all programmable systems. Adverse effects can only be avoided or minimized through nested semantic, syntax and consistency checks. In this paper, we will concentrate on fault detection mechanisms which are application dependent. The main question is, therefore, how "unexpected sensor readings" can be discovered and whether or not there are generally applicable methods for making system support possible.

AN APPLICATION EXAMPLE: FLEXIBLE SMALL PARTS ASSEMBLY

In order to investigate the concepts of efficient sensor data processing and fault detection, we are developing a flexible assembly cell for a family of electromechanical relays. Except for the insertion of the contact blades into the main body, all five parts are to be assembled in this cell, which imposes that a common set of tools and sensors be used for all assembly steps.

The cell is equipped at the moment with a 6-axis robot, an electrically powered proportional gripper and 2 pneumatic valves to press a pair of yokes into the solenoid. The sensors include 3-axis force measurement in each gripper finger, 1 reflective light barrier in each finger tip as well as part detectors in each of the 3 feeders and the assembly base. This is believed to be the minimum sensory equipment required to operate the system autonomously. However, a camera for scene supervision is considered to be useful in terms of an additional "security margin".

Such an assembly cell can be regarded as a "discrete event dynamical system" (DEDS). A suitable control scheme having particular properties must be developed. One of the problems is how to transform continuous sensor data, e.g. stemming from force sensors, into the domain of discrete events. As will later be seen, this is achieved through sensor data interpretation or, more concretely, through range checks with programmable limits.

In this case, like many other applications, the mechanical construction of the relays is such that there only exists 1 sensible assembly sequence which can consequently be established off-line. With this assembly sequence we have clear cut expectations of what should be happening and what should not when assembly is in progress, making it possible to define a set of "normal operational states" corresponding to the assembly sequence. In this context, fault detection is a general term for a mechanism deciding whether or not a given system state (consisting of the internal and external state as perceived through the sensors) is normal. As soon as a fault is discovered, an alarm is issued and robot motion is interrupted. Higher degrees of autonomy would initiate fault diagnosis and decide on an appropriate control strategy to reestablish the normal operational state ("fault recovery"). Note that this hybrid control strategy - off-line preparation of the normal control sequence and on-line planning of the recovery from exception states - is both fast and highly flexible, needing comparatively little computing power. Provided that exception states "rarely" occur, a small delay when planning a recovery action does not considerably affect system performance.

BASIC FAULT DETECTION TECHNIQUES

Elkind (1982) gives a general overview of fault detection techniques in data processing. They are all based on some sort of redundancy, because the "input items" (i.e. data) either cannot or only vaguely be described beforehand and "output items", (i.e. results), being immaterial, can be discarded without loss. Obviously, the situation is completely different in manufacturing ("parts processing") systems. First of all, the operation is performed on identical input items which, in turn, are supposed to yield identical output items, each representing a certain material value.

Since well defined expectations exist along the operation, fault detection need not be based on redundant operation. Instead external "real world" states and events can be compared with suitable internal models. It is clear that, in this context, sensors and efficient sensor data processing play a key role in fault detection. Each sensor makes potentially additional failure modes "visible".

Due to the existence of only one well defined normal state, there is no need to implement a comprehensive on-line model ("observer") for the detection of false sensor data, which is investigated in other fields, such as process engineering. Instead, we can memorize expected sensor data directly, which can either be generated off-line or "recorded" by the system. This will be discussed later.

A practical solution is to divide the assembly sequence in phases in which no major changes of sensor data are anticipated and to define permissible ranges for all sensors in each phase. This reduces the amount of data and allows for easy synchronization with the control sequence. Also, this represents a perfectly decentral scheme for fault detection. Thus, if we consider all sensors as coordinates in a multisensor space, we define a permissible sensor subspace for each assembly phase. Taking the decentral approach, these subspaces are restricted to hypercubes, which is considered acceptable for this class of applications.

By using the proposed scheme, all *sensor failures* in a given sensor coordinate are also detected, as long as at least two assembly phases exist with non-overlapping permissible ranges for this coordinate. Also, most sensor coordinates are used as well for control purposes (such as touch detection for the gripper controller through force measurements in the fingers), so that faulty sensors sooner or later produce "inconsistent system behavior".

The notion of fault coverage must be addressed by selecting the appropriate sensor coordinates. However, in a given sensor configuration, the fault detection threshold can be adjusted by narrowing and broadening the permissible sensor subspaces to achieve a compromise between sufficient fault coverage and an acceptable rate of false alarms.

SENSOR DATA PROCESSING CONCEPTS

We want to distinguish normal situations and exceptions by *continuously* comparing actual and expected sensor values. This imposes two requirements. First, this fault detection mechanism should respond quickly in order to interrupt robot motion before a collision occurs. Second, supervision must be done in *parallel* to the control sequence. Otherwise, it is not continuous and the control program becomes overcharged with sensor data processing. IBM (1981), for example, has included a similar mechanism in its AML robot control language, setting up background sensor supervision functions ("MONITOR < sensor_id, range, sampling time >").

When searching for the optimal and most flexible solution, a centralized architecture is desirable. On the other hand, this approach leads to highly specialized systems which are difficult to modify if additional sensors need to be installed etc. Also there is little potential to define modules that can be transferred to similar applications. Therefore, following the guideline "configuring is easier than programming", we adopted a decentral solution consisting of self-confined modules or components with a standardized interface called intelligent subsystems.

The same argument holds for control functions as for sensor supervision functions. A more intelligent control is theoretically possible if all actuators were to be controlled based on all available sensor data. However, there is no need stressing the fact that, in practical systems, it is fairly easy to define intelligent subsystems by implementing a certain set of functions based on the relevant sensor data.

In our application, for example, the gripper controller requires no more than the position measurement as well as the force measurement in the degree of freedom of its fingers. For this reason, the gripper controller is combined with all gripper sensors into one intelligent subsystem. Since the force measurement bears more information about the progress in assembly than is needed for gripper control alone, it is also made accessible for other control functions through a sensor command interface.

A SENSOR/CONTROL MODEL FOR ASSEMBLY OPERATION

By analyzing the particular needs of assembly operation, Malcolm and Ambler (1986) have described three basic control levels using sensor data. Similarly, we have identified the following (sensor based) control functions (Fig. 1):

On the highest control level (5) an assembly strategy has to be generated or selected. In our case, this strategy is partially prepared off-line ("normal states") and partially on-line, based on the present situation ("exception states"). Without considering the manner in which it is deduced, we can conclude that we need aggregate *sensor information* at this level.

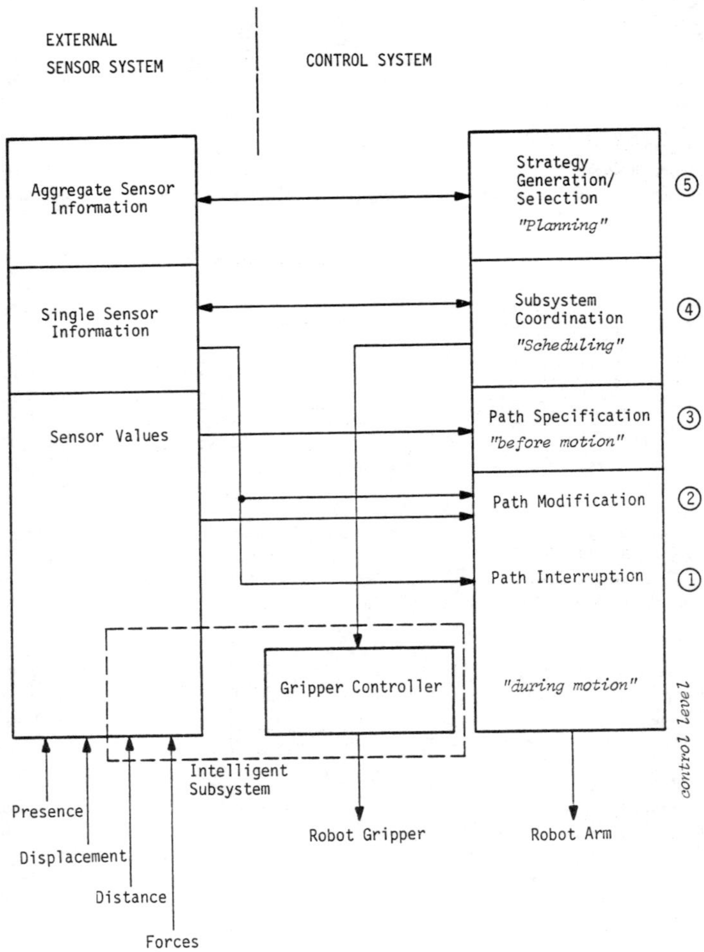

Fig. 1. A sensor/control model for flexible assembly cells

On the next level (4) we need to schedule the next operation within a given strategy, i.e. synchronize the actions of the decentral control and sensor supervision functions. Before a motion begins, we must specify the target position (3). This may be fixed or determined by sensors, e.g. a vision system providing proportional values when an object is to be picked from a conveyor belt.

A path must sometimes be interrupted during motion due to impending collision ("emergency halt"), or modified in the case of inaccurate spatial specification ("contour tracking"). These functions correspond to the lowest control levels (1 and 2).

Let us examine next the levels of sensor data processing (Fig. 2), which provide the information needed in the control system:

Transducers, such as strain gages, phototransistors or CCD-Arrays provide continuous electrical "raw" sensor signals. Preprocessing modules filter or scale the signals, thus providing binary or proportional *sensor values*. In the case of a vision system, such preprocessing modules are algorithms identifying an object or determining its displacement and rotation. According to Fig. 1, these sensor values are used by the lower control levels, such as path modification or specification.

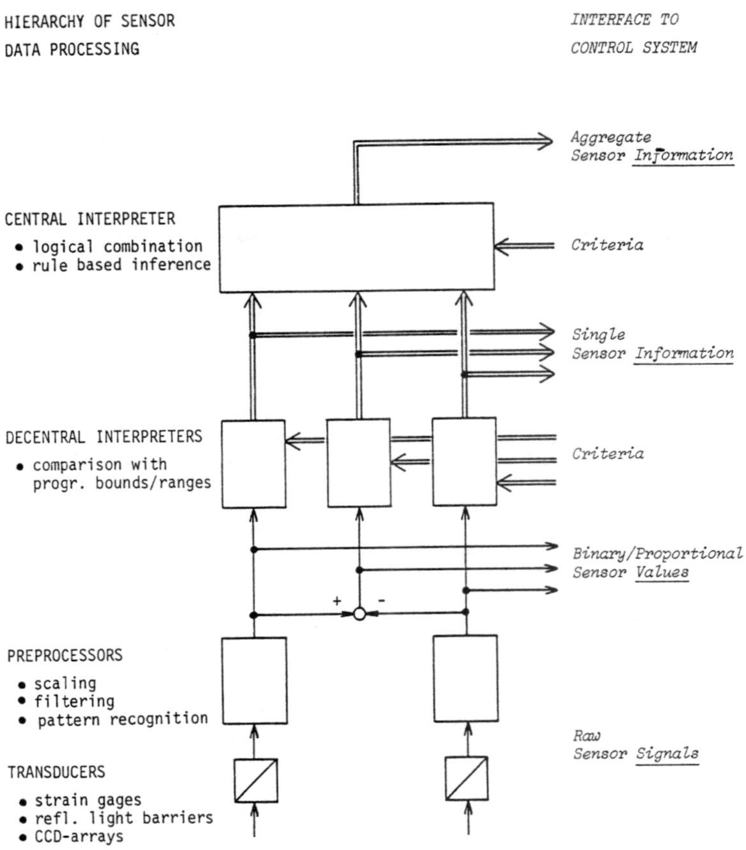

Fig. 2. The sensor data processing architecture

For decentral interpretation, these values lend themselves only for purposes of comparison with programmable bounds and ranges, more generally designated "criteria". The output of these processing blocks is *sensor information*, indicating whether a certain sensor coordinate is within a specified range or above a specified bound. If these criteria are suitably assigned to each sensor coordinate and updated from one assembly phase to the next, this level is very efficiently employed to supervise the sensor values and detect faulty external system states. Binary sensor coordinates are treated in a similar manner, except that the expectation can be expressed by the values "true", "false" or "don't care".

Another application of single sensor information is in statements of higher robot control languages, such as "MOVE UNTIL RZ_force > 4N". It is characteristic for this level of sensor data abstraction that not the value itself is of interest, but whether or not it exceeds a certain bound imposed by the application.

New sensor coordinates can be created through the linear combination of sensor values. For example, it is very useful to observe the *differences* between the left and right finger force components, since they correspond closely to the reaction force components when inserting a gripped object. They are therefore much more significant than the finger force components themselves.

Finally, when combining the single sensor information and internal system state, aggregate sensor information can be centrally derived. This organization has three main characteristics: First, sensor data is decentrally processed up to a fairly high level. Thus, the creation of powerful subsystems is encouraged. Second, a method exists in decentral range checks to detect unexpected sensor values at a fairly low and hence fast system level. Third, it provides the basic functions for easy implementation of advanced robot control languages.

AN IMPLEMENTATION EXAMPLE

After establishing basic concepts, we will now describe the structure of our laboratory assembly cell (Fig. 3). The *cell controller* selects appropriate assembly actions (1) according to aggregate sensor information (2). The *robot controller* carries out requested motions (control levels 1...4) and coordinates the actions of the subsystems (3). It uses sensor values and information. Finally, the *external sensor system* provides a common interface for all sensory and peripheral subsystems. In order to simplify the configuration, intelligent subsystems are connected through a fast serial bus, except for the vision system, which will be integrated directly for reasons of computing power.

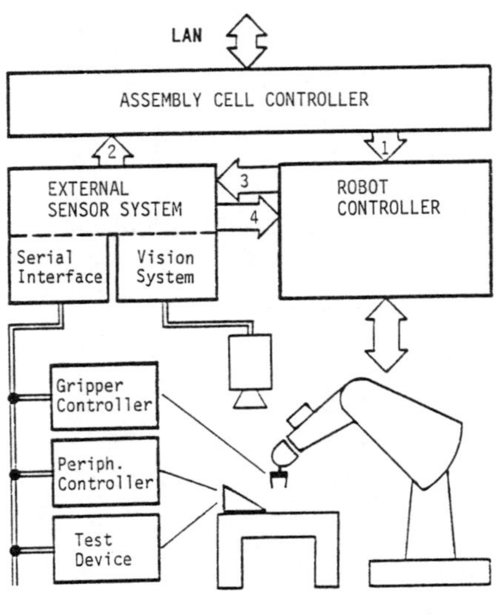

Fig. 3. Proposed architecture for a flexible assembly cell

TABLE 1. Command Interface of the Gripper Control Subsystem

Gripper Command Interface:
- **move fingers to reference position**
- **move fingers < width >**
- **grip an object < width, force >**
- **read finger position/gripper status**
- **set internal parameters**

Sensor Command Interface:
- **determine zero offsets**
- **read zero offsets**
- **read actual sensor values**
- **set sensor value limits**
- **read sensor status**
- **read and reset max/min sensor values**
- **signal < condition >**

A multidrop serial bus reduces interfacing costs considerably and allows local distribution of the subsystems as well as easy system expansion. We have chosen the Intel Bitbus™ because it is the only microcontroller network offering a fully implemented software protocol. Thus, intelligent subsystems are self-confined components based on their own microcontroller, which can be connected to the system using a standard hardware and software interface.

For example, in case of the gripper controller the interface to higher system levels is realized through two independent asynchronous command interpreters; one for the standard sensor functions and the other for the particular gripper functions. Table 1 lists the implemented commands, which are always checked for correct syntax and semantics before execution. The gripper command interpreter configures the control algorithms according to the particular command, while the sensor command interpreter basically transfers values between command messages and internal variables. The "max/min sensor values" are used to record sensor value ranges in an assembly phase by an optional teach-in method to define permissible sensor ranges. The "signal condition" command allows for (regular) termination of a robot motion, e.g. when a certain force is reached.

CONCLUSION

An efficient sensor data processing architecture has been proposed for flexible assembly cells. It is marked by the following characteristics:

* higher *reliability* of the assembly cell is achieved through comprehensive fault detection and autonomous recovery;

* external faults are efficiently and quickly detected using *decentral range checks* on all sensor coordinates;

* the fault detection mechanism itself is embedded in a more general *sensor/control model*, encouraging the employment of powerful subsystems;

* in the implementation example discussed, a fast *serial bus* is used to interface these subsystems, thus,

  - creating *inherent parallelism* with little communication and task switching overhead;

  - allowing simple system *configuration* and *expansion*;

  - off-loading low level repetitive chores from the host computer, reserving it for *information ("event-") processing*;

  - creating self-confined *components* which can be used without modification in similar applications.

Based on the proposed sensor data processing concepts, in the future we will concentrate on control in exceptional situations or, in other words, on autonomous recovery. Perez and Rouchy (1986) argue that a control system with comprehensive exception handling leads to intricate control programs, of which only 10 percent is executed in normal operation. They propose a rule-based approach to simplify modification and testing of control programs. Other artificial intelligence techniques having been successfully employed under real-time conditions are discussed comprehensively by Tsafestas (1986). Even though the prospects for assembly control systems with new properties are promising, many issues remain to be solved before all requirements of factory floor equipment is met.

REFERENCES

Lotter, B. (1985). Aufbau und Einsatz flexibler Montagelinien in der Feinwerktechnik. *Proceedings of the Kolloquium "Automatische Produktionssysteme"*, 14. - 15.2.85, Institut für Werkzeugmaschinen und Betriebswissenschaften, TU München. pp. 338-412.

Elkind, S.A. (1982). In Siewiorek, D.P. and Swarz, R.S. (Eds.), *The Theory and Practice of Reliable System Design*. Digital Press, Bedford Mass. pp. 63-182.

IBM (1981). *Robot System/1, AML Reference Manual*. IBM Information Development, Boca Raton, Fla.

Malcolm, C.A. and Ambler, A.P. (1986). Some Architectural Implications of the Use of Sensors. *Preprints of the Intelligent Autonomous Systems Conference*. Elsevier Science Publishers B.V., Amsterdam. pp. 71-78.

Perez, J.C. and Rouchy, I. (1986). Increasing Autonomy of Assembly Robots with Artificial Intelligence. *Preprints of the Intelligent Autonomous Systems Conference*. Elsevier Science Publishers B.V., Amsterdam. pp. 490-501.

Tsafestas, S.G. (1986). A Look at the Knowledge-based Approach to System Fault Diagnosis and Supervisory Control. In Tsafestas, S.G., Singh, M. and Schmidt G. (Eds.), *System Fault Diagnostics, Reliability and Related Knowledge Based Approaches, Vol. 2*, pp. 421 - 433.

Hierarchical and Fast Recursive State Estimation with Robust Bad Data Pre-Cleaning for Large-Scale Power Systems

A. E. Mansour, H. A. Nour Eldin

Group of Autom. Control and Tech. Cybern., University of Wuppertal, 5600 Wuppertal, West Germany

ABSTRACT

This paper presents a new hierarchical static state estimation method for large-scale power system.For each subsystem and its tie-lines, a robust local bad-data precleaning (BDPC) (detection, identification and correction) is achieved before applying the new fast recursive static state estimator (RSSE). The BDPC is based on non-linear transformation by which four bad-data indicators are defined. The fast RSSE is based on the direct use of the measurements redundancy together with the network flow equations to achieve recursive estimate that verifies the network conditions at stationary opertation. The proposed algorithm is applied on the AEP 30-bus test system.The results are displayed throughout this paper.

KEYWORDS

Power system monitoring; large-scale power system; hierarchical state estimation; bad-data identification and correction; recursive estimation.

INTRODUCTION

On-line state estimation is an important function in modern control center (c.c.) of electric power system. Reliable state estimation generates consistant data base to be used for power systems monitoring and control. The known estimation methods developed can be devided into centralized methods e.g. (schweppe, 1970; Dopazo, 1973; Mansour, 1987) and decentralized or hierarchical methods e.g. (Kurzyun, 1983; Van Cutsem, 1983). The hierarchical methods are more suitable in case of large-scale power systems, since the estimation problem of the overall system is decomposed into a number of subsystems estimations which are carried out locally as a low-level of estimation. They will be coordinated in the main c.c. as the high-level estimation. The decomposition of the whole estimation as well as the parallelism feature appearing in performing subsystems estimations lead to speeding up the estimation procedure of the overall system. Almost all the known hierarchical methods use the conventional least square formulation for estimation and bad-data treatment . This is releable method only when the measured data are bad data free. The existance of multiple bad-data may couse convergence or observability problems . Also, the bad measurements in neighbourhood to the boundary buses may not be detected. These are the reasons why problem of bad-data handling at tie-lines is still challinging.

This paper presents a new hierarchical method in which local (i.e. on subsystem) BDPC technique and a new algorithm for fast (RSSE) are used. Through the use of a special form of the network power flow equations for nonlinear transformation, four robust indicators (RV,RΘ',RQ,RQI) are defined to detect, identify and to correct the multiple bad data in all avialable measurements (voltage magnitudes, active and reactive powers).The complete handling of bad data in measurements of overall system (subsystems and their tie-lines) is performed locally (i.e. on subsystems c.c.s) before starting the state estimation process.After the successful handling of bad data, the network flow equations are used for the locally recursive subsystem state estimation. This state estimation does not use "WLS" formulation. As the bad-data cleaning is achived locally before the corresponding local state estimation ,the known problems of observability or convergence of the state estimator are avoided.Through exploiting the data exchange between adjacent subsystems, the bad measurements at tie-lines can be also locally pre-cleaned.

MODEL EQUATIONS OF (RSSE) AND (BDPC) TECHNIQUES

The equations on which the new technique is developed are partationed into two basic sets of equations A and B. The basic set of equations A (1-7) was derived in (Nour Eldin, 1978; Nour Eldin, 1981; Sakr, 1983). This is a node oriented set of power equations at each node describing the stationary conditions of an n-node electric network:

$$e_j = \sqrt{.5\ (1+2Q_{jj})\ (1\pm\sqrt{1-4\,\Delta_{jj}})} \quad ; \quad \Psi_j = \tan^{-1}(\,P_{jj}/(e_j^2 - Q_{jj})) \tag{1,2}$$

$$\Delta_{jj} = (P_{jj}^2+Q_{jj}^2)/(1+2Q_{jj})^2 \quad ; \quad 0 < \Delta_{jj} < (1/4) \tag{3}$$

$$\begin{bmatrix} P_{jj} \\ Q_{jj} \end{bmatrix} = 1/(Y_{jj}E_j^2) \begin{bmatrix} \sin\Phi_{jj} & \cos\Phi_{jj} \\ \cos\Phi_{jj} & -\sin\Phi_{jj} \end{bmatrix} \begin{bmatrix} P_j \\ Q_j \end{bmatrix} \tag{4}$$

$$\underline{E}_j = E_j\, e^{J\Phi_j} = \sum_{\substack{i=1 \\ i\neq j}}^{n} (\underline{Y}_{ji}/\underline{Y}_{jj}) \cdot V_i\, e^{J\Theta_i} \quad ; \quad V_j'' = E_j\, e_j \quad ; \quad \Theta_j'' = \Phi_j - \Psi_j \tag{5,6,7}$$

The 2nd set of equations B (8-11) are derived in Bonanomi (1982). They are based on the relation between the power flow and the complex voltage drop on the lines.

$$\underline{E}_{ji} = ((P_{ji} - JQ_{ji})/V_j) - \underline{Y}_{ji0}\, V_j)/\,\underline{Y}_{ji} \quad ; \quad V_i' = |\,V_j - \underline{E}_{ji}\,| \tag{8,9}$$

$$V_{ji} = V_j - V_i' \quad ; \quad \Theta_{ij} = \Theta_i - \Theta_j = \tan^{-1}[\,IM(V_j - \underline{E}_{ji}) \,/\, RE(V_j - \underline{E}_{ji})] \tag{10,11}$$

where:

V_j, Θ_j : voltage magnitude and angle at bus j.

P_j, Q_j : active and reactive injec.n at bus j.

P_{jj},Q_{jj} : oriented injection power at bus j.

P_{ji},Q_{ji} : active and reactive flow on line ji.

E_j, Φ_j : adjacent voltage and its angle of bus j.

Y_{ji}, Φ_{ji} : admittance magnitude and its angle of line j-i.

Y_{jj}, Φ_{jj} : amplitude and angle of the sum of all admittances connected to node j.

$\underline{Y}_{jio}$: line charging admittance of line j-i.

e_j : relative voltage w.r.t. the adjacent voltage.

Ψ_j : relative angle of adjacent voltage w.r.t. voltage angle at bus j.

Δ_{jj} : nodal factor is defined in equation (3).

$\underline{E}_{ji}$: complex voltage drop in line j-i.

V_{ji} : voltage magnitudes difference at buses j,i.

Θ_{ij} : voltage angles difference at buses i,j.

n : total number of buses.

For BDPC technique, equation (5) is replaced with equation (5)' (Sakr, 1983), and a new equation (12) (Mansour, 1987) will be included to the second set B.

$$\underline{E}_j = E_j\, e^{J\tilde{\Psi}_j} = \sum_{\substack{i=1\\ i\neq j}}^{n} (\underline{Y}_{ji}/\underline{Y}_{jj}) .\ V_i\, e^{J\Theta_{ij}} \tag{5)'$$

$$Q'_j = Y_{jj}(\ E'_j V'_j \sin(\Phi_{jj}+\Psi'_j) - V_j^2 \sin\Phi_{jj}) \tag{12}$$

The RSSE technique uses the following set of filtering equations (13-15)

$$V'_i = \frac{1}{K}\sum_{b=1}^{K}(V'_i)_b \quad ; \quad \Theta'_i = \frac{1}{K}\sum_{b=1}^{K}(\Theta_j - \Theta_{ij})_b \quad ; \quad V'_i = \frac{1}{K+1}\,(V_i^m + \sum_{b=1}^{K}(V'_j)_b$$

$$;\ (V'_i)_b = (|V_j - \underline{E}_{ji}|)_b \tag{13,14,15}$$

(Mansour, 1987) to estimate the voltage magnitude and angle at bus i (V'_i, Θ'_i) as an arithmatical mean taking into consideration the avialable power flow measurements in lines end with bus i. $\underline{E}_{ji}$ and Θ_{ij} are defined in equations (8) and (11) , V_i^m is the measured voltage at bus i and K is the number of lines connected to bus i and share in calculating V'_i and Θ'_i.

BAD DATA PRE CLEANING (BDPC) TECHNIQUE

The BDPC technique is based mainly on considering each node together with its neighbours and investigating whether the measurements obtained from this group of neighbours match together with the basic network equations (A,B). This matching is judged by four sensitive indices (RV,RΘ',RQ,RQI) which do not depend on the operating point. Both sets of equations (A,B) process the available measurements to compute the transformed variables (V''_j, Ψ_j, V'_j, $\tilde{\Psi}_j$, Q'_j) at each bus. These computed (transformed) variables together with the physical measurements are exploited to define a set of stationary conditions which must be satisfied in statistical sence whenever the physical measurements are free from bad data.In the presence of bad data, these stationary conditions will be statistically violated.The following bad data indices (RV,RΘ',RQ and RQI) are introduced to check the degree of this violation.

$$RV_j = \frac{|V''_j - V_j|}{\bar{\sigma} v_j} \quad ; \quad R\Theta'_j = \frac{|\Psi_j - \tilde{\Psi}_j|}{\bar{\sigma}\Theta_j} \quad ; \quad RQI_j = \frac{|Q'_j - Q_j|}{\bar{\sigma} q_j} \tag{16,17,18}$$

$$RQ_{ji} = \begin{cases} \dfrac{|\Psi_{ji} - \tilde{\Psi}_{ji}|}{\bar{\sigma}\Psi_{ji}} & \text{for lines with } (X/R) \leq 5 \\[2ex] \dfrac{|V''_j - V'_j|}{\bar{\sigma} q_{ji}} & \text{for lines with } (X/R) > 5 \end{cases} \tag{19}$$

where:

$\Psi_{ji} = \Psi_j - \Psi_i$; $\tilde{\Psi}_{ji} = \tilde{\Psi}_j - \tilde{\Psi}_i$; X,R the series reactance and resistance of line ji.

$\bar{\sigma}V_j, \bar{\sigma}\Theta_j, \bar{\sigma}\Psi_{ji}, \bar{\sigma}q_{ji}$ and $\bar{\sigma}q_j$ are the maximum accepted differences ($|V''_j - V_j|, |\Psi_j - \tilde{\Psi}_j|$, $|\Psi_{ij} - \tilde{\Psi}_{ji}|$, $|V''_j - V'_j|$ and $|Q'_j - Q_j|$. They are calculated under the assumption of additive gaussian randam noise associated with the measurements. In the absence of measurment noise, and for arbitrary stationary operating point, all indices will equal to zero. When bad-data is present in the real-time measurements, some of the indices values will be almostly greater than one, otherwise they will be less than one. The relation between the indices (RV,RΘ',RQ and RQI) and the node measurements are derived in (Nour Eldin, 1986). Exploiting the linear relation between the avialble measurements and the indicators, the BDPC technique identifies and corrects the multiple bad data according to the scheme shown in Fig. 1.

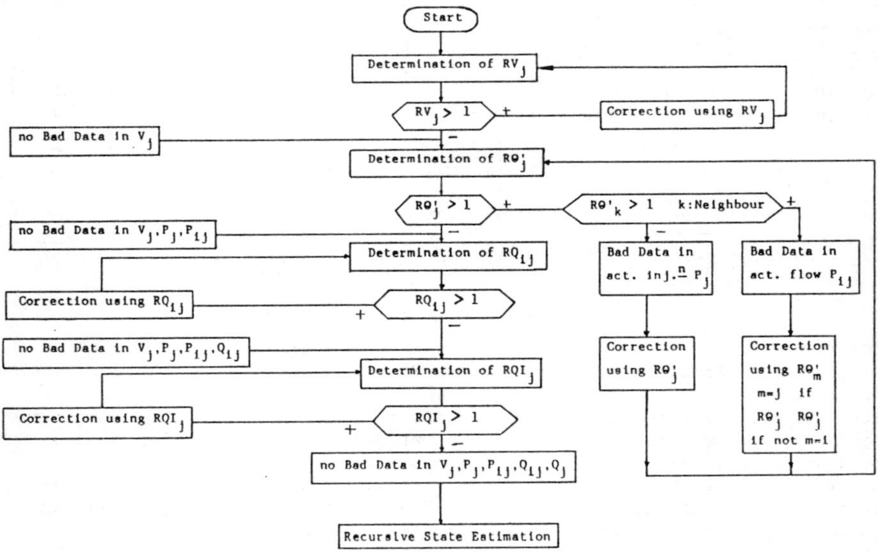

Fig. 1. Bad-data identification and correction

Using this algorithm Fig. 1, detection, identification and correction of bad measurements in voltage, active and reactive power (flow and injection) are achieved. The BDPC technique has the following features which overcome the known disadvantages of bad data detection in the other methods (Handschin, 1975; Dopazo, 1973; Sakr, 1983,1986).

- Bad data pre-cleaning for all avialable measurements is achieved before performing the state estimation and not after it.
- The accuracy and robustness of bad data detection could reach $5\sigma_V$ for voltage and $8\sigma_p$ for power measurements.
- The identification of the bad data type together with its localization is possible even for adjacent multiple bad data so long as the measurements are different in type (active and reactive).
- High capability and accuracy of correction of bad measurements.

THE RECURSIVE STATIC STATE ESTIMATION (RSSE)

The consistant set of measurements (i.e. after application of the BDPC technique) will be processed in a recursive manner using the network flow equations A, B to estimate the network states (bus voltage magnitude and angles) that must satisfy the network stationary conditions. There are some facts regarding the recursive verification used in the RSSE algorithm:

- In applying equation (9), the voltage error which can be associated with V_i, does not propagate into the calculated voltage V'_i without attenuation. This fact was exploited in (Bonanomi, 1983) for bad data validation algorithm, and it was called Stabilizing Effect.
- The standered deviation of the transformed voltage V''_j (equation (6)) is less than the standered deviation of the measured voltage V_j (Sakr, 1983).
- Using equations (13-15), introduce a filtering effect (FE) which helps to

accelerate the convergence of the recursive estimation.
In the RSSE technique Fig.2 , the set of equations B and the filtering equations (13-15) process mainly the power flow measurements under the assumption that the voltage at reference node is measured with high accuracy and its angle is equal zero. This first process (I) yields accurate calculated bus voltages V_i', Θ_i' at each node. The output of process (I) will be used as the input of the subsequent process (II) which uses the set of equations A to calculate new estimates of bus states (V_j'', Θ_j'') for all buses. At this point, the basic cycle is completed resulting in quite accurate estimation of the complex voltages. To achieve more accuracy, the second process could be repeated one or two times. More than two iterations is not necessary as the main gain of accuracy can be achieved after two iterations.

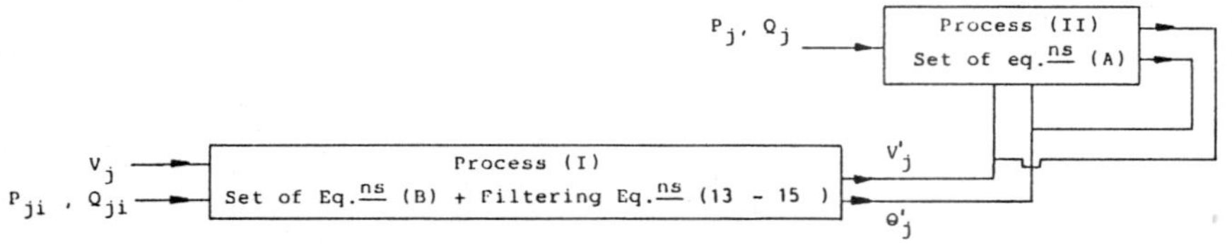

Fig. 2. The recursive estimation algorithm

DECENTRALIZED FORMULATION OF THE (BDPC) AND THE (RSSE)

Let the overall system containing N-nodes be composed of K non-overlapping subsystems. The adjacent subsystems are interconnected by means of ties (Tie-Lines or Transformers) which end at boundary buses belonging to the subsystems. Let N_i denote the number of buses of the i th subsystem, and N_{ib} denote the number of boundary buses of the i th subsystem. The remaining $N_{ir}=N_i-N_{ib}$ are termed the internal buses.; s denotes the number of ties in each interconnection.

Decomposition of Measurements System

The set of measurements of whole system are decomposed into K subsets of measurements. The local control center (c.c.) of the i th subsystem will recieve directly all avilable measurements which belongs to the i th subsystem comprising the ties measurements (power flow) which are recorded at the i th boundary buses N_{ib}. Power flow measurements at both ends of ties are not requiered. The measured voltages at boundary buses will be reinformed to the concerned c.c.s of the adjacent subsystems. It is assumed that the local reference voltages of subsystems are measured with high accuracy.

Decentralized formulation of BDPC technique

The BDPC problem of the whole system is decomposed into K subproblems which will be solved at subsystems c.c.s . In each subsystem, the pre-cleaning of bad measurements comprising the bad measurements of ties will be performed locally according to the BDPC algorithm shown in Fig. 1. The main procedure steps are :

- The measurements and the other informations belonging to the i th subsystem will be processed at the i th c.c. resulting in the transformed variables (V'', Ψ , $\hat{\Psi}$, V', Q') at all N_i nodes of the ith subsystem.
- The bad data indices (RV, $R\Theta'$, RQ, RQI) pertaining the avilable measurements in the i th subsystem are obtained using equations (16-19).
- The indices of values greater than one at the boundary nodes N_{ib} will be informed to the concerned adjacent c.c.s .

The above mentioned steps will be performed in parallel at all subsystem c.c.$\underline{s}$. Each subsystem c.c. has the required tools to detect, identify and to correct the bad data in its subsystem measurements including ties measurements before starting the estimation process.

- During performing the BDPC at subsystem c.c.$\underline{s}$, the main c.c. can regulate and coordinates (or supervise) the required data exchange between the local c.c.$\underline{s}$. The important features of this decentralized BDPC technique, which overcome the known disadvantges of bad data detection in other HSE, can be summarized as:

- The bad data in the subsystems measurements including the ties measurements are treated locally at the c.c.$\underline{s}$ of subsytems.
- The whole bad data problem is decomposed into K subproblems to be solved locally (seperatly and independently) at the local c.c.$\underline{s}$.
- Only a reasonable amount of data exchange between the adjacent subsystems to overcome the bad data problem of ties at the 1$\underline{st}$ level is necessary.
- The power flow measurements at both ends of ties are not required.

Application of the RSSE in the Hierarchical State Estimation (HSE)

Directly, after the successful application of the BDPC algorithm in each subsystem, the RSSE estimator (described in Fig. 2) will be applied, proccessing its own consistant (bad data free) measurements to estimate the subsystem states and also to estimate the voltage angle differences of its own ties at the same level (i.e. locally). At the high level, it is required only to coordinate the angles estimates utilizing the subsystems angle estimates and the angle difference estimates of all ties.This coordination procedure will be performed in two main steps:

- In the first step, the angle estimates of a subsystem neighbour to the base subsystem 1 (e.g. subsystem 2) will be reestimated in relation to the rference node. For the s ties coupling these two subsystems 1 and 2 , a coordinated angle $\Delta^{2/1}$ is estimated as:

$$\Delta^{2/1} = (1/s) \sum_{i=1}^{s} (\Theta_i^1 + \Theta_i^{2/1} - \Theta_i^2) \tag{20}$$

where i refers to tie-line number; 2,1 refer to the subsystems numbers; Θ_i^1, Θ_i^2 are the estimated angles (at low-level) of the boundary buses of tie-line number i; $\Theta_i^{2/1}$ is the estimated angle difference of that tie-line. Then the corrected voltage angle Θ^j at each bus j of subsystem 2 is calculated by using:

$$\Theta^j = \Delta^{2/1} + \Theta^j \quad ; \text{(j refers to bus number)} \tag{21}$$

After the correction of all bus voltage angles of subsystem 2, the new enlarged subsystem (1,2) adopts the name "base subsystem" and is referred afterwards as subsystem (1). In the second step , the procedures, according to the frst step is repeated in turns for the remaining K-2 subsystems. The above two coordination steps follow the coordination steps B, C in (Kurzyun, 1983; page 2057), but it is required there (in step A) to estimate the angle differences of all tie-lines at the same high level using the WLS formulation.

RESULTS OF THE DECENTRALIZED (BDPC) AND (RSSE) TECHNIQUES

Both techniques (BDPC and RSSE) are illustrated here through the application on the AEP 30-bus test system. The operating point was calculated using a load flow program based on the node oriented load flow formualtion (Nour Eldin, 1981). The random errors used have the standared deviations:
for voltage magnitude $\sigma_V = 0.5\%$
for power flow or injection $\sigma_P = 2\%+0.35\%$ full scale
The additive random noise sources used are uncorreleted with normal distribution and zero mean value. The test system is decomposed into 3 subsystems (I,II,III) as shown in APP.$\underline{x}$. The following bad data locations and their values relative to the actual values are assumed:

bad data	bad data value	bad data	bad data value
voltage V_{22}	$1.2\ V_{22\ act.}$	voltage V_{26}	$1.2\ V_{26\ act.}$
active flow P_{28-27}	$1.5\ P_{28-27\ act.}$	reactive flow Q_{17-16}	$1.5\ Q_{17-16\ act.}$
active inj.$\underline{n}$ P_1	$0.5\ P_{1\ act.}$	reactive inj.$\underline{n}$ Q_5	$0.5\ Q_{5\ act.}$

Using the decentralized BDPC algorithm, detection, identification and correction of the above assumed multiple bad data are achieved. In subsystem I, the bad data in P_1 and Q_5 are detected and corrected to new accurate values shown in Table 1. In subsystem II, the bad data in V_{22} and Q_{17-16} are detected and corrected to accurate values (Table 1). Also, in subsystem III, both bad data of voltage V_{26} and P_{28-27} are detected and corrected into an accurate ones as shown in Table 1.

Table 1 Detected bad measurements with thier measured (mea.), corrected (cor.) and actual (act.) values (in p.u.)

detected bad-data	mea. value	cor. value	act. value	detected bad-data	mea. value	cor. value	act. value	subsystem number
P_1	1.0795	2.1409	2.1589	Q_5	-0.0300	-0.0582	-0.0599	I
V_{22}	1.2234	1.0216	1.0195	Q_{17-16}	-0.0700	-0.0443	-0.0467	II
V_{26}	1.1973	0.9983	0.9977	P_{28-27}	0.2863	0.1905	0.1908	III

Direct after correcting the local bad data, the RSSE is applied locally at each subsystem. The subsystem estimation results are shown in Tables 2 , 3 and 4 for subsystems I, II, III respectively.

Table 2 Subsystem estimation for subsystem I

j	V_j^{mea}	V_j^{est}	V_j^{act}	θ_j^{est}	θ_j^{act}
2	1.0435	1.0451	1.0450	-4.38	-4.39
3	1.0410	1.0410	1.0409	-7.10	-7.12
4	1.0340	1.0362	1.0361	-8.54	-8.57
5	1.0107	1.0101	1.0100	-10.48	-10.44
7	1.0119	1.0107	1.0106	-10.66	-10.62
12	1.0584	1.0586	1.0573	-14.71	-14.65
13	1.0679	1.0723	1.0710	-14.71	-14.65
14	1.0421	1.0417	1.0404	-15.49	-15.48
15	1.0338	1.0356	1.0342	-15.52	-15.46

Table 3 Subsystem estimation for subsystem II

j	V_j^{mea}	V_j^{est}	V_j^{act}	θ_j^{est}	θ_j^{act}
17	1.0306	1.0283	1.0283	-.265	-15.05
16	1.0382	1.0391	1.0391	-.195	-14.97
20	1.0221	1.0191	1.0192	-.983	-15.76
19	1.0146	1.0162	1.0162	-1.23	-16.01
18	1.0211	1.0209	1.0209	-1.15	-15.93
21	1.0212	1.0190	1.0189	-.433	-15.23
22	1.2234	1.0196	1.0195	-.427	-15.21
24	1.0118	1.0091	1.0090	-.810	-15.58
23	1.0184	1.0198	1.0198	-.890	-15.66

Table 4 Subsystem estimation for subsystem III

j	V_j^{mea}	V_j^{est}	V_j^{act}	θ_j^{est}	θ_j^{act}
8	1.0067	1.0100	1.0100	-.497	-10.35
9	1.0497	1.0479	1.0478	-3.26	-13.11
11	1.0810	1.0818	1.0820	-3.26	-13.10
28	1.0239	1.0239	1.0239	-.681	-10.53
27	1.0260	1.0279	1.0280	-4.92	-14.78
25	1.0161	1.0152	1.0154	-5.42	-15.25
26	1.1973	.9974	.9977	-5.80	-15.67
29	1.0096	1.0082	1.0083	-6.12	-16.00
30	1.0001	.9968	.9969	-7.01	-16.87

Table 5 Coordination of subsystem II

j	θ_j^{est}	θ_j^{act}
10	-14.832	-14.776
17	-15.096	-15.045
16	-15.026	-14.974
20	-15.815	-15.756
19	-16.066	-16.007
18	-15.985	-15.925
21	-15.253	-15.227
22	-15.253	-15.212
24	-15.641	-15.583
23	-15.721	-15.664

Table 6 Coordination of subsystem III

j	θ_j^{est}	θ_j^{act}
6	-9.863	-9.839
8	-10.353	-10.347
9	-13.127	-13.101
11	-13.126	-13.101
28	-10.544	-10.524
27	-14.771	-14.778
25	-15.279	-15.252
26	-15.637	-15.673
29	-15.973	-15.996
30	-16.860	-15.871

The results of coordination steps, which are applied to coordinate the voltage angle estimates of both subsystems II and subsystem III in relation to the common reference (node 1) are recorded in tables 5 and 6.

Direct Detection and Identification of Topological Errors in EPS Data Base

M.A.M. Hassan, A. Mansour, H. Nour Eldin

Group of Autom. Control and Tech. Cybern., Wuppertal University, 5600 Wuppertal 1, West Germany

ABSTRACT

A powerfull detection of topological errors in Electric Power Systems (EPS) is presented and implemented. The method is based on a new node oriented load flow formulation which allows the non-linear transformation of the measurements. The main advantage of these transformation is the capability to detect the bad data and topological error before performing the estimation process. This powerfull detection can detect and identify all types of topological errors. These advantages will support the reliability of transformed measurements technique to be used for on-line PSSSE.

KEYWORDS

Static state estimation, power systems, bad data analysis, topological errors, network modelling, transformed measurements.

INTRODUCTION

Power System Static State Estimation (PSSSE) together with Bad-data detection, identification and correction provide a real-time data base for control and dispatch in the power system. Generally, the on-line state estimation is a main objective in control centers. On the other hand, off-line applications of the state-estimation in analysis and planning of information as well as metering system design are essential. However, a number of practical problems have to be solved in relation to the state estimation for the real power system application. Beside the Network modelling, the problem of Bad data treatment, Topological Error Analysis (TEA), the Observability problem, and the Measurements placement problems should be adequatly studied. The above mentioned problems could be treated in different ways according to the technique used in solving the PSSSE. This could be either by using the direct measurements with linearized WLS (classical PSSSE) or by the transformed measurements with linear WLS or recursive estimation (Mansour, 1987). The mathematical formulation of the static state estimation according to the above mentioned techniques is different. The classical technique was introduced firstly by Schweppe (1970), where the non-linear estimation is solved by iterative solution of the linearized weighted least square problem (WLS). The second technique (Bonanomi,1982; Saker,1983; Zaborsky, 1980) is based essentially on non-linear transformation which lead to a linear WLS estimation. In Saker (1983), new formulation of the problem was proposed to overcome many disadvantages of the second technique. As this transformation verifies the stationary condition equations of the network through a suitable nonlinear transformation, it introduces a new approach for bad data detection, identification and correction before the estimation process. Also it introduces a powerfull tool for identifying the Topological Errors. Thus, it introduces a solution of the still persisting problem of handling topological errors in PSSSE. The main advantages of this technique are : no computation for a Jacobian matrix, complete bad-data and topological error analysis before the estimation process, and the observability problem does not represent an obstacle. These bad-data analysis is based on the fact that the bad-data indices are indicators of the matching between the measurements and the structure of the network, and the values of these indices will be affected by any topological change. This fact leads to a a quick decision about any topological change. In this paper, the authors demonstrate the cabability of the transformed measurements technique in PSSSE to handle the topological errors beside the bad-data analysis. The error analysis for the bad-data and the topological changes will be performed before the estimation process, which leads to consistant data base for PSSSE. This detection of anomalities (bad-data and topological errors) is very robust and it does not need to use any least-square estimation. Through the classical estimation, detection of bad data and topological errors is done after estimating the state vector of the system (Dopazo, 1973; Handschin, 1975; Kohlas, 1972; Merill, 1971; Mili, 1985). The residuals, the performance index and other methods, as classified in (Nour Eldin, 1985; Nour Eldin, 1986), are used afterwards to detect bad-data and topological errors. In the classical bad data analysis, the model is assumed to be free from any topological error. Also, in the classical topological error detection tests (Lutgu, 1980; Koglin, 1986), the measurements are assumed to be free from any bad data. Furthermore, the bad data analysis and TEA in these methods are treated after the estimation process. This classical treatment may lead to some convergence problems, due to the bad input information (measurements and structure) supplied to PSSSE. On the other hand, the bad data and topological error analysis based on the new technique does not require the above assumption. Fig. 1 shows schematically the different ways of the two techniques in PSSSE.

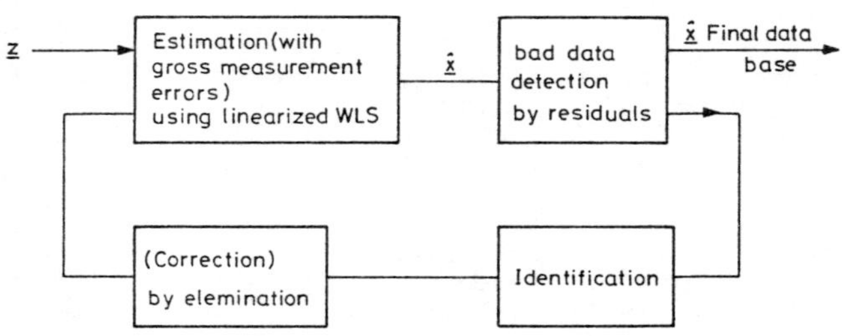

Fig. 1.a PSSSE using classical method.

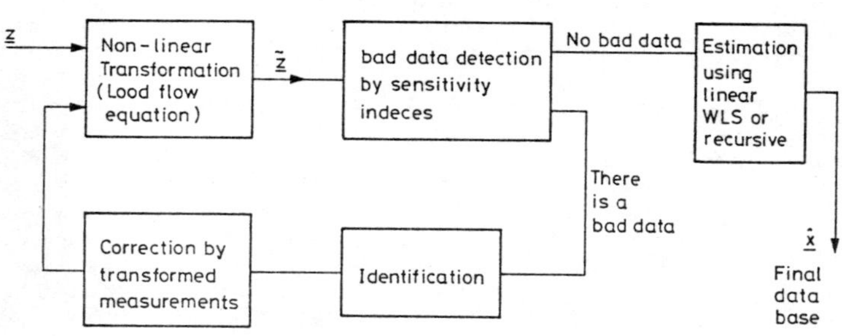

Fig. 1.b PSSSE using transformed measurements.

THE NON-LINEAR TRANSFORMATION EQUATIONS

The new formulation of the second technique (Saker, 1983; Saker, 1986) is based on a new node oriented load-flow (Nour Eldin, 1978; Nour Eldin, 1981) which is used as non-linear transformation. These transformation are performed at each node in three steps using the real time measurements of system variables available at the node j and the neighbour nodes in the following form :

P_j, Q_j : Active and reactive power injection at bus j.

V_k : Voltage magnitude of all the neighbour of bus j.

V_j : Voltage magnitude of bus j .

P_{jk}, Q_{jk} : Active and reactive power flow in the line connecting bus j and bus k.

The equations of the suitable nonlinear transformation used in (Saker, 1983; Saker, 1986) are derived with details in (Nour Eldin, 1978; Nour Eldin, 1981; Saker, 1983) and could be briefly written as :

$$V_j^{`} = E_j \sqrt{\frac{1+2P_{jj}}{2}\left(1 \pm \sqrt{1-4\,\Delta_{jj}}\right)} \qquad (1)$$

$$E_j\, e^{J\,\Psi_j^{`}} = \sum_{\substack{k=1\\k\neq j}}^{n} (Y_{jk}/Y_{jj}) \cdot V_k\, e^{J(\Theta_k-\Theta_j)} \qquad (2)$$

$$\Theta_{kj} = \tan^{-1} \frac{IM(\,V_j - \underline{E}_{jk}\,)}{RE(\,V_j - \underline{E}_{jk}\,)} \quad \text{for all } k > j \qquad (3)$$

$$\begin{bmatrix} P_{jj} \\ Q_{jj} \end{bmatrix} = \frac{1}{Y_{jj}E_j^2} \begin{bmatrix} \cos\Phi_{jj} & -\sin\Phi_{jj} \\ \sin\Phi_{jj} & \cos\Phi_{jj} \end{bmatrix} \begin{bmatrix} P_j \\ Q_j \end{bmatrix} \qquad (4)$$

$$\Psi_j = \arcsin\,(\,Q_{jj}(E_j\,/V_j\,) \qquad (5)$$

$$\Delta_{jj} = (P_{jj}^2 + Q_{jj}^2)/(\,1 + 2\,P_{jj}\,)^2 > 0 \qquad (6)$$

where

$V_j^{`}$: The transformed voltage as given by equation (1)

Θ_j : The angle at bus j.

Θ_{kj} : The angle difference on the line k-j.

E_j : Adjacent voltage of bus j as defined in (Nour Eldin, 1978).

Ψ_j, $\Psi_j^{`}$: Adjacent angles of bus j as calculated in (Saker, 1983).

$\underline{E}_{jk}$: The complex voltage drop on the line j-k, which is calculated from the power flow measurements.

Y_{jj}, Φ_{jj} : Amplitude and angle of the sum of all admittances connected to node j.

Y_{jk}, Φ_{jk} : Amplitude and angle of the admittance between bus j and bus k.

P_{jj}, Q_{jj}: The node oriented power, which are defined in (Nour Eldin, 1978; Nour Eldin 1981) as equation (4).

Δ_{jj} : The nodal load factor is defined in (Nour Eldin, 1981) as equation (6).

It was shown in (Saker, 1983) under stationary condition that :

$$\Psi_j^{`} = \Psi_j \quad \text{and} \quad V_j^{`} = V_j \qquad (7)$$

These transformed measurements have the advantage to be linearly related to bus voltages and angles. It is possible to evaluate measurements of a bus j using the informations of its neighbours only. The new technique, as shwon above, handels each node together with its neighbours and investigats whether the measurements taken from this group of neighbours match together and with the structure of the system. This matching is judged by the sensitive local bad data indices at each node R_Θ and R_V of angle and voltage respectively. Bad data identification and corrections are also possible as shown in (Saker, 1983; Saker, 1986). The sensitive local indices are defined in (Saker, 1983) as follows :

$$R_{\Theta j} = \frac{/\Psi'_j - \Psi_j/}{\sigma_\Theta} \quad \text{and} \quad R_{Vj} = \frac{/V'_j - V_j/}{\sigma_V} \qquad (8)$$

where σ_Θ and σ_V are the maximum accepted differences for Gaussian measurement errors and $R_{Vj} = 0$, $R_{\Theta j} = 0$ for the normal stationary operating point. Under the presence of additive random noise to the measured variables, the values of V'_j, V_j, Ψ'_j and Ψ_j will change in a random way. There will exist a random differences between this variables. The maximum accepted value of this differences under the assumption of Gaussian random noise is calculated by off-line simulation of statistically representative sets of measurements and the determination of the standard deviations σ_{Vj} and $\sigma_{\Theta j}$. In the presence of bad data in the measurement, the indices are greater than one, otherwise they must be less than one. The relations between the indices $R_{\Theta j}$; R_{Vj} and the system measurements are derived in (Saker, 1983). The general relation between these bad data indices and the different measurements are linear and could be written for power flow or injection measurements as :

$$RT = C\,Z + d \qquad (9)$$

where

RT : bad data index,
Z : measurement,
C : constant and d = conatant

The constants C and d were proved to be a function of the network structure (Saker, 1983). Therefore, any difference in the model from the actuale structure will affect the values of the indices (Saker, 1983; Saker, 1985) specially at the point of zero measurements, where RT=d at Z = 0 . Using this fact, one can give a quick indication about any topological error, by comparing an off-line calculated d-values with that calculated on-line. For random measurements and correct matching, the d-value will vary in a certain range as in Fig. 2. If the d-value is outside this range, a topological error is detected. The next step is to identify the different types of the topological errors.

TYPES OF TOPOLOGICAL ERRORS AND THEIR IDENTIFICATION

This section clearifies the types of topological errors and thier effect on the constant d in the index-measurement relation. The probable types of topological errors are (Lutgu, 1980) :

- A line is represented in the PSSSE model but it is actually out of service in the system.
- A line is not represented in the PSSSE model but it is actually in service in the system.
- A line is open (not open) ended in the model but it is not in the actual system.

The value of the constant d will be affected by the above three cases. Therefore, the range of the value d (for each line at both ends) is calculated off-line statistically. During the calculation of bad data indices, after the non-linear transformation, the d-values for candidates lines are compared with the computed off-line values using the next relative axis formula :

$$RTP = \frac{RT - (RTMX + RTMN)/2}{(RTMX - RTMN)/2} \qquad (10)$$

where

RTMX : The expected maximum of the d-value (calculated off-line)
RTMN : The expected minimum of the d-value (calculated off-line)
RT : R_{Θ} calculated on-line after choosing the candidates
RTP : The proposed index for Topological Error Analysis (TEA)

A topological error (a line is configured different from its model) detection will be achieved if /RTP/ 1 at both ends of the candidate line. The candidates lines for the first type of topological error are chosen such that the measured power flow is less than certain limit (this limit is equal to three times of the standard deviation of these measurements). The candidates lines for the second type can not be obtained as specified above, because the T. line is not represented in the model (but it exists in the actual network). Consequently, the corresponding flow measurement does not exist in the set of measurements. Therefore, the candidates lines for the second type are determined after handling of bad data detection. The two buses with heighest RT are chosen as two ends of a candidate line, and thier power injection are set to zero. Then a new transformation followed by a topological error detection test will be applied. Fig. 3

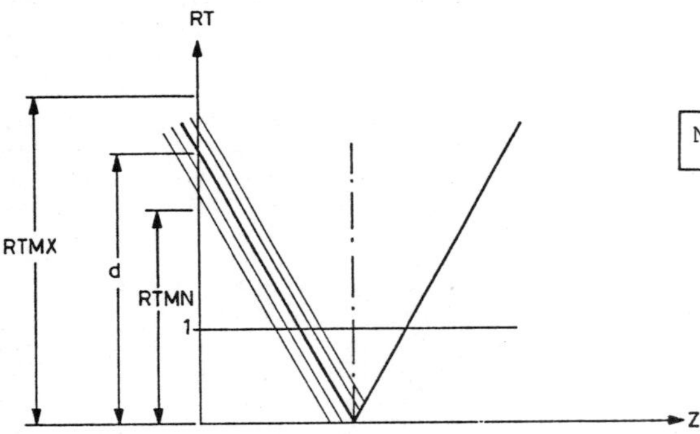

Fig. 2 The relation between the bad data indices and the power measurements.

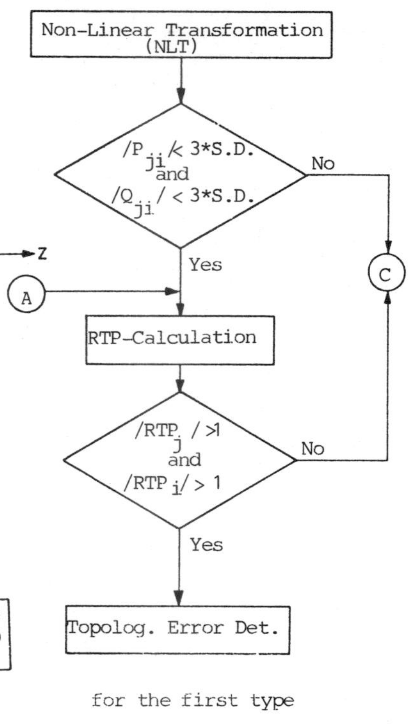
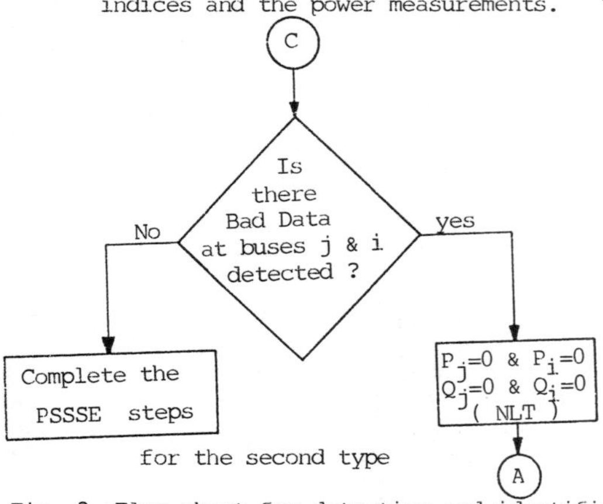

Fig. 3. Flow chart for detection and identification of topological errors

showes the flow chart of the different detection steps. The last type of the topological error is not considerd in the simulation. It could be considerd as subset from the first type, because the flow measurements will be also near zero, and the matching will not be correct at both ends of the T. line.

TEST RESULTS

The proposed algorithm is tested using the IEEE-14 bus test system as given in (Bonanomi, 1982). The topological errors are simulated in the load flow program to get the actual measurements after adding the random errors. Only sample results on the first two types are presented. Different cases for bad data in the presence of the topological error are simulated.

Table 1 showes the RTP-values for some deleted lines in the actual network (but existing in the model). The bad data indices at both ends of the line, RT1 and RT2, are compared with the corresponding limits RTMN1, RTMX1, RTMN2, and RTMX2 using the equation (10), to get RTP1 at the first end of the T.line and RTP2 at the second end. It is clear that /RTP1/ and /RTP2/° 1, which means a detection of topological error at the corresponding T. line in hand. If this detection takes place for a candidat of the first type, the topological error is identified as the T.line is out of service (i.e. the T.line is open from both ends). As an example, T. line No. 3 (first row in Table 1) is deleted from the actual network (but exists in the model). The flow measurements in this line was found to be near zero. This T. line is chosen as a candidat of the first type topological error. Therefore, the topological error test must be done. The RT-values at both ends of T. line 3 are calculated (RT1= 0.362 , RT2= 0.138). These values are compared with the off-line calculated range (RTMN1= 21.454 , RTMX1= 24.092 , RTMN2= 15.941 , and RTMX2= 17.761). This comparison is judged using equation (10) through RTP-calculation. RTP1= -16.992 and RTP2= -18.364 (/RTP1/ 1 and /RTP2/ 1). This means a detection of topological error at T. line 3, which is identified as that the line is out of service. In Table 1, similar results are shown for the lines 6, 7, 8, 12, 15, and 18 showing the robustness of detection and identification of topological errors of the first type.

TABLE 1 RTP-Values for a Topological Error from the First Type

T.L.	K1	K2	RT1	RTMN1	RTMX1	/RTP1/	RT2	RTMN2	RTMX2	/RTP2/
3	2	3	0.362	21.454	24.092	16.992	0.138	15.941	17.761	18.364
6	3	4	0.067	4.788	6.380	6.932	0.271	8.794	10.577	10.560
7	4	5	0.630	11.502	13.549	11.622	0.324	13.932	16.614	11.144
8	4	7	0.190	6.802	8.664	8.103	0.088	12.942	14.629	16.238
10	5	6	1.987	5.630	8.430	3.602	3.957	21.852	24.33	15.393
12	6	12	0.556	2.808	6.005	2.409	2.067	3.013	4.606	2.188
15	7	9	0.079	13.060	14.542	18.521	5.859	7.976	10.354	2.780
18	10	11	1.313	3.572	5.504	3.339	0.833	2.428	21.654	1.166

K1, K2 : The buses at both ends of a candidate line (CL).
RT1, RT2 : R_o at both ends of a candidate line (CL).
RTMN1,RTMN2 : The expected min. for the d_value at both ends of CL.
RTMX1,RTMX2 : The expected max. for the d_value at both ends of CL.
RTP1 ,RTP2 : The values of the proposed index for TEA at both ends of CL.

Table 2 showes RTP-values for the same type for line 3 as well as line 12 in presence of different types of bad data (injection at both ends and voltage at both ends) with different amount of bad data. It is clear from the obtained results, that the dececion is not affected by the presence of bad data, as /RTP/ is still more than one, which means a robust detection of topological error. The robustness of detection and identification of topological errors of the first type is not only insensitive to the location of the topological errors within the network but also insensitive to the presence of bad date in the set of measurements. This makes the proposed method superior to known methods as (Koglin, 1986), where bad data free measurements is essential.

Table 2 RTP-Values for a (TE) from the First Type in the Presence of Bad Data

T.L.	K1	K2	RT1	RTMN1	RTMX1	/RTP1/	RT2	RTMN2	RTMX2	/RTP2/
3	2	3	4.430	21.454	24.092	13.907	53.832	15.941	17.761	40.636
3	2	3	4.430	21.454	24.092	13.907	22.361	15.941	17.761	6.055
3	2	3	1.479	21.454	24.092	16.144	53.832	15.941	17.761	40.636
3	2	3	1.479	21.454	24.092	16.144	12.020	15.941	17.761	5.308
3	2	3	1.479	21.454	24.092	16.144	22.361	15.941	17.761	6.055
3	2	3	1.271	21.454	24.092	16.302	14.494	15.941	17.761	2.589
3	2	3	1.271	21.454	24.092	16.302	22.361	15.941	17.761	6.055
3	2	3	7.591	21.454	24.092	11.511	13.759	15.941	17.761	3.397
3	2	3	8.571	21.454	24.092	10.767	5.915	15.941	17.761	12.017
3	2	3	3.200	21.454	24.092	14.840	11.639	15.941	17.761	5.726
3	2	3	6.041	21.454	24.092	12.685	68.499	15.941	17.761	56.753
3	2	3	4.140	21.454	24.092	14.127	4.962	15.941	17.761	13.063
12	6	12	0.909	2.808	6.005	2.188	8.341	3.013	4.606	5.690
12	6	12	2.540	2.808	6.005	1.168	0.957	3.013	4.606	3.582
12	6	12	2.017	2.808	6.005	1.495	6.905	3.013	4.606	3.886
12	6	12	2.017	2.808	6.005	1.495	0.532	3.013	4.606	4.115
12	6	12	0.630	2.808	6.005	2.363	0.532	3.013	4.606	4.115
12	6	12	1.150	2.808	6.005	2.037	1.207	3.013	4.606	3.268
12	6	12	6.251	2.808	6.005	1.154	6.486	3.013	4.606	3.360

Table 3 showes RTP-values for simulated topological errors of the second type (The line exists in the actual network and is not represnted in the model). After the calculation of bad data indices, the candidate line is chosen such that it connects between two buses having the heigest RT-value. Note that this T. line may or may not be exist in the model. The power injection at the above defined buses are set to zero. The non-linear transformation is performed again, with calculation for RTP1 and RTP2 using the equation (10), based on RTMN1,RTMX1, RTMN2, and RTMX2. These values are calculated off-line with zero power injection. It is clear from the obtained results that /RTP1/ and /RTP2/ are more than one, which means a detection of topological error from the second type. If a T. line exists between the above defined buses, this means that the simulated admittance of the T.L. (in the model) is not correct.

TABLE 3 RTP-Values for a Topological Error of the Second Type

K1	K2	RT1	RTMN1	RTMX1	/RTP1/	RT2	RTMN2	RTMX2	/RTP2/
2	1	10.207	0.007	2.063	8.922	9.154	14.388	16.209	6.747
5	1	7.051	0.034	2.393	4.949	11.519	14.388	16.209	4.150
3	2	10.160	20.687	22.541	12.357	3.447	0.007	2.063	2.346
4	2	1.597	8.720	10.600	8.578	2.792	0.007	2.063	1.709
3	4	18.104	20.687	22.541	3.787	12.874	8.720	10.600	3.420
4	5	2.964	8.720	10.600	7.123	7.220	0.034	2.393	5.092
7	4	8.374	0.002	0.860	18.516	13.249	8.720	10.600	3.818
9	4	2.796	4.219	6.755	2.122	12.147	8.720	10.600	2.647
6	5	2.214	2.654	5.812	1.279	5.690	0.034	2.393	3.795
7	9	7.561	0.002	0.860	16.623	2.579	4.219	6.755	2.294
14	13	18.387	18.560	27.529	1.038	5.308	3.347	5.006	1.364

CONCLUSION

The transfrmed measurements technique in PSSSE introduces a robust and powerfull tool for topological error detection and identification without any restriction on the presence of bad data. This detection, and identification of topological errors without assumping of bad data free measurements, is achieved through the matching to the network equation at stationary condition of the network. It investigates the network structure before the estimation process is applied, and the topological error detection could be achieved even in the presence of bad data. Therefore, the new topological error analysis has essential advantages over the classical approaches.

REFERENCES

Bonanomi, P. (1982). Data Validation by Network Search Technique for Power System, Monitoring and Control. Dissertation ETH-Zürich No. 7116.

Dopazo, J.F., O. A. Klitin, and A. M. Sanon (1973). State estimation for power systems, detection and identification of gross measurement errors. Proc. 8th PICA Conference, Minneapolis. 313-318.

Handschin, E., F.C. Schweppe, J. Kohlas, and A. Fiechter (1975). Bad data analyses for power system state estimation. IEEE Trans. Power Appara. & Syst., PAS-94, 329-337.

Koglin, H. J., D. Oeding, and K.D. Schmitt (1986). Identification of topological errors in state estimation. IEE "PSMC 86" Conference, University of Durham, UK, Conference Publication No. 266, 245-251.

Kohlas, J. (1972). On bad data suppression in estimation. IEEE Trans. autom. Control, AC-17, 827-828.

Lutgu, R. L., D. F. Hackett, K. C. Liu, and D. D. Might (1980). Power system state estimation : Detection of topological errors. IEEE Trans. Power Appar. & Syst., PAS-99, 2406-2412.

Mansour, A. E. (1987). Fast and robust bad data detection, identification and correction with subsequent direct recursive state estimation. Accepted for 9th PSCC Conference in Lisbon Portugal.

Merill, H. M., and F. C. Schweppe (1971). Bad data suppression in power system static state estimation. IEEE Trans. Power Appar. & Syst., PAS-90, 2718-2725.

Mili, L., T. Van Cutsem, and M. Ribbens-Pavella (1985). Bad data identification mehtods in power system state estimation - A Comperative study. IEEE Trans. Power Appar. & Syst., PAS-104, 3037-3049.

Nour Eldin, H. A. (1978). Regions of stationary operating points in electrical power distribution network. 6th PSCC Conference, Darmstadt, West Germany.

Nour Eldin, H. A., and E. Lerch (1981). Numeerical determination of the stationary operating points in power networks. 7th PSCC Conference, Lausanne.

Nour Eldin, H. A., and M. A. M. Hassan (1985). Literature survey on bad data problem. Internal institute report 85/19, University of Wuppertal, West Germany.

Nour Eldin, H. A., and M.A.M Hassan (1986). Measurements placement for electric power system static state estimation - Part I : Basic concepts. Internal institute report 86/4, University of Wuppertal, West Germany.

Sakr, M. M. F. (1983). Static state estimation in electric power system. Dissertation University of Cairo, Egypt.

Sakr, M. M. F., A. Bahgat, and A. R. El-Shafei (1985) Dynamic state estimation in power systems with abnormalities detection, identification and correction. IEE Control 85 Conference, Cambridge, Conference Publication No.252, 1,245-251.

Sakr, M. M. F., E. Lerch, and H. A. Nour Eldin (1986). Pre-estimation and bad data detection in power systems. IEE "PSMC 86" Conference, University of Durham, UK, Conference Publication No. 266, 135-139.

Schweppe, F. C., J. Wildes, and D. P. Rom (1970). Power system static state estimation, Part I, II, and III. IEEE Trans. Power Appar. & Syst., PAS-89, 120-135.

Zaborsky, J., K. W. Whang, and K. V. Prasad (1980). Ultra fast state estimation for large electrical power systems. IEEE Trans. autom. Control, AC-25, 839-841.

A New Approach for Transient Security Assessment and Enhancement by Pattern Recognition

J.A. Peças Lopes*, F.P. Maciel Barbosa*, J.P. Marques de Sá*
J.M.G. Sá da Costa**

**Departamento de Engenharia Electrotécnica, Faculdade de Engenharia da Universidade Porto, Porto, Portugal*
***C.A.U.T.L., Instituto Superior Técnico, Lisboa, Portugal*

ABSTRACT

This paper presents a new approach for transient stability assessment and enhancement in electric power system, based on the application of pattern recognition techniques. Results obtained with the CIGRE test system are presented.

KEYWORDS

Security assessment and enhancement; transient stability; dynamic state variables; security functions; preventive control measures;

INTRODUCTION

The operation of modern power systems still a challenge to system engineers and cientists mainly due to its size and complexity. In these systems a high degree of reliability and security is expected in order to provide the customer with a continuous service of high quality and low cost. However, due to economic reasons, most power systems do not have the degree of reliability and security that permits a dispatch engineer to rely on the security of the system as a whole.
Most of the time power system operates in a normal state, in which the quality of the power delivered and the safety of the system are reached. However, power systems are continously faced with a wide variety of disturbances which may lead the system to move from a normal state to an emergency state. When the system is in an emergency state if no corrective action are taken in due time a general or partial shut down may be expected. As looking for corrective actions asks some delay, it is necessary to assess the effect of a disturbance on the system before it occurs and to take apropriate preventive actions.
Figure 1 shows an overall scheme of the major steps needed to assess system security and enhancement. The operating state of the power system is estimated by the so-called state estimator based on system measurements. Considering the estimated state and taking into account the load prediction t seconds ahead the state where the system will be in a close future. Each disturbance of a set is applied to the future state and its stability assessed. The system is said to be insecure for a given disturbance if this leads to a loss of synchrounism. Otherwise, it will be secure. If an insecure state is assessed preventive control actions are designed and presented to the system engineer.

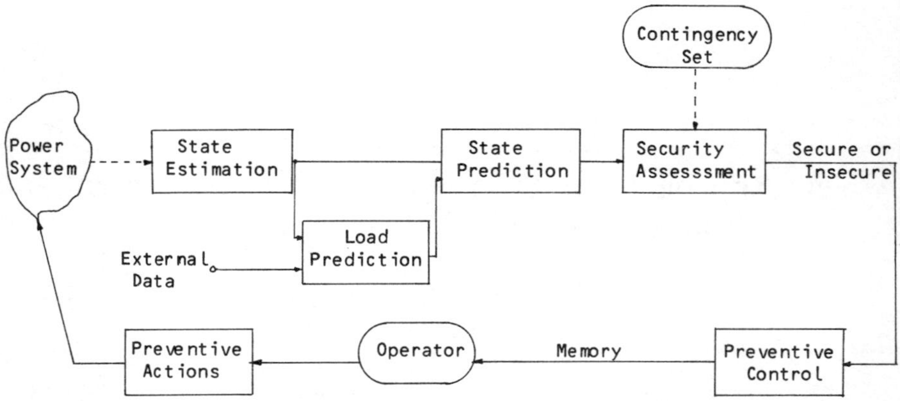

Fig. 1. Practicall scheme for transient security assessment and enhancement

Here we present a new and efficient methodology based on pattern recognition techniques for online transient stability assessment and enhancement. The system assessment is performed by evaluating a linear security function which separates the projected secure and insecure regions along the Fisher direction, being the security function of the form

$$S(\underline{X}) \begin{cases} \geq 0 & \underline{X} \text{ secure} \\ < 0 & \underline{X} \text{ insecure} \end{cases}$$

where $\underline{X}$ is the system state in a close future.
The system enhancement is achieved by changing the output powers of the generation security function moving the projected point well inside into the secure region.

POWER SYSTEM MODEL

The transient process in power systems means interaction of a large number of generating units interconnected by a transmission and distribution network.
It is widely accepted in transient stability studies of multimachine power systems the use of a simplified dynamic description of the system, which leads to a model given by a set of second order differential equations:

$$M_i \frac{d^2\delta_i}{d\,t^2} = Pmec_i - Pe_i \qquad (1)$$

$$i = 1,\ldots,m$$

where M_i is the inertia constant, δ_i the absolute rotor angle, $Pmec_i$ the mechanical input power and Pe_i the electrical output power, of the ith generator, being m the number of equivalent machines in the system.
In this model the mechanical power will be assumed constant during the transient period and the damping effect neglected.
Electrically the synchronous machine is represented by a constant voltage source E_i, in series with a transient reactance X_i. Loads are represented by constant admittances.
When using the center of inertia or center of angle (COA) as reference frame, the new speeds and angles are defined as $\tilde{\omega}_i = \omega_i - \omega_o$ and $\theta_i = \delta_i - \delta_o$, with the center of inertia δ_o given by :

$$\delta_o = \frac{1}{MT} \sum_{i=1}^{m} M_i \delta_i$$

and $\omega_o = d\delta_o/dt$ with $MT = \sum_{i=1}^{m} M_i$.

Therefore the swing equations in the new coordinate system are

$$\dot{\theta}_i = \tilde{\omega}_i \tag{2}$$

$$\dot{\tilde{\omega}}_i = \frac{Pmec_i - Pe_i}{M_i} - \frac{1}{MT} Pcoa \tag{3}$$

where

$$Pcoa = \sum_{i=1}^{m} Pmec_i - Pe_i$$

TRANSIENT STABILITY ASSESSMENT BY PATTERN RECOGNITION

The design of pattern recognition systems involves several stages. It begins with class definition and the formation of a set of labelled samples (training set) of those classes represented in a suitable measurement space. The design is finished when a classifier is derived which assigns a unique lable to new samples. The designed classifier is function of a small set of variables usually called features, resulting from a feature selection stage.

In our case, we are faced with a two-class pattern recognition problem and the aim is to design a classifier which represents the boundary between the stable and unstable regions.

Primary Variables Choice

As any other physical system, a power system can be fully described by a set of primary variables which are usually called measurements. The number of measurements depends on the size of the power system and on the detail with which one wishes to reproduce the system behaviour.

The sucess of the application of pattern recognition techniques depends significantly on the primary variables choice. It is an intuitive idea that the measurements to be chosen should be directly related to the physical causes behind the transient stability phenomena. In this way it is expected to have initial high discriminating capabilities between the two classes. As this work was developed in order to assess the transient stability and to enable the definition of systematic procedures for the transient stability enhancement, the primary variables choice was guided by the following criteria :

1 - good information about the system dynamics;
2 - to be directly or indirectly related with parameters that enable to develop preventive controling measures;
3 - to be low cost computational variables, obtained from the pre-fault load flow conditions or calculated at t_o^+, in order to minimize the classification time.

Having in mind the above considerations and the good hints given in the works of Gupta (1975), Pavella (1982) and Sá da Costa (1982), the following variables were chosen :

- the accelerating powers, Pac_i, calculated just after the fault occurence (t_o^+), given by :

$$\frac{Pmec_i - Pe_i(t_o^+)}{M_i} \qquad i = 1,\dots,m \tag{4}$$

- their derivatives, $\dot{P}ac_i$, relative to the rotor angle θ_i, given by

$$- \frac{1}{M_i} \frac{\partial\, Pe_i(t_o^+)}{\partial\, \theta_i} \qquad (5)$$

$$i=1,\ldots,m \quad i \neq k$$

where k is the nearest machine of the default;

- the center of angle power calculated at t_o^+ , $Pcoa(t_o^+)$;
- the total inertia constant, MT;
- the ratio $Pcoa(t_o^+)/MT$;
- the total load level consumption of the system, PL_i, just before the contingency occurence.

Training Set Generation

The design of a pattern recognition system requires a representative set of samples, usually termed as training set. The samples in this set are generated through off-line simulations and for each contingency to be considered its stability analysed and labelled as stable or unstable.
In order to perform the training set generation an algorithm which takes into account the forecast load configurations, the most probable unit comittment alternatives and economical dispatch was developed. The true stability classification for each contingency was performed using the transient energy method as described in Athay and co-workers (1980), being the stability region evaluation computed through the potential energy boundary surface (PEBS) method.

Feature Selection

In order to choose from a set of measurements the ones that have better discriminatory power a feature selection stage must be performed. This stage reduces the dimensionality of the measurement space by selecting the most significant primary variables, removing any redundant information and at the same time rearranging them in decreasing order of their discriminant power.
In our case, the feature selection was performed by means of a single ranking procedure using a separability F measure, as used in Sá da Costa (1982).
The final choice of the variables to be used in the classifier as features was performed through a forward search procedure, having in mind the minimization of the classification error when the classifier is applied to the training set.

SECURITY FUNCTION DESIGN

Having in mind the practicability of the transient stability enhancement project a linear classifier was chosen to be the best structure for the security function. Several approaches are available and have already been used to derive linear classifiers when dealing with this type of problem, such as : the perceptron, least square methods, linear programing methods, Bayes linear decision rules.
Considering the deterministic procedure used in the training set generation it would be dificult to estimate suitable multidimensional distributions in order to use statistical methods in the security function design. Therefore, a combination of a deterministic and a statistical procedures were used to obtain the linear classifier that assures the best separation between the two (stable and unstable) classes and consequently the minimization of the expected probability of error. This methodology consists in finding an optimum linear transformation that converts the n sample points, represented by an $\underline{X}$ vector in the feature space, into n points orthogonally projected over a single direction, through a mathematical relation of the type :

$$Y = \underline{A}^T \underline{X} \qquad (6)$$

where Y is the new sample point and $\underline{A}$ is the transformation vector.
The determination of this optimum space direction is obtained through the calculation of the $\underline{A}$ vector that maximizes a criterion function $J(\underline{A})$ given by:

$$J(\underline{A}) = \frac{\underline{A}^T \underline{S}_B \underline{A}}{\underline{A}^T \underline{S}_W \underline{A}}$$

This criterion corresponds to the maximization of the ratio of the between-class scatter matrix $\underline{S}_B$ to the sum of the within-class scatter matrices $\underline{S}_W$.
Accordingly with Duda (1973) the maximization of $J(\underline{A})$ leads to:

$$\underline{A} = \underline{S}_W^{-1}(\underline{m}_s - \underline{m}_u) \tag{7}$$

where $\underline{m}_i$ and n_i are respectivelly the mean vectors and the number of sample points of each class, being i=s,u (stable, unstable).
In order to have the security function written in the canonical form presented before an independent term a_o must be added to the relation (6). As all sample points are projected in a single direction it is then possible to estimate undimensional distributions for the projected points in each class. Assuming normal distributions the independent term a_o is calculated from the following equation:

$$\frac{1}{2}(a_o - m_s)^2 \frac{1}{\sigma_s^2} - \frac{1}{2}(a_o - m_u)\frac{1}{\sigma_u^2} + \frac{1}{2} \ln(\frac{\sigma_s^2}{\sigma_u^2}) = \ln(\frac{P(\omega_s)}{P(\omega_u)}) \tag{8}$$

which uses the well-known formula of the quadratic Bayesian classifier with estimated "a priori" probabilities $P(\omega_s)$ and $P(\omega_u)$ (Fukunaga, 1972). (m_s, σ_s^2) and (m_u, σ_u^2) are the estimated parameters of each class distribution.

TRANSIENT STABILITY ENHANCEMENT

Transient stability of power systems can be basically enhanced by three preventive kind of measures: by changing the output powers of the generators, by load shedding or by changing the transmission network structure. The present methodology is based on the first method to perform the preventive control. Here, the preventive controling measures to be taken are based on the idea that in order to make stable an unstable configuration it is only necessary to move the projected operating point along the Fisher direction in order to attain a prespecified target point well inside the stable region. This target point is defined by imposing the value ε that the security function $S(\underline{X})$ will have at the new operating state as result of the application of the preventive control measures.
However, ε must have a relative small value in order to avoid to make large changes in the economical output powers of each machine.
The required shifting of the security function $S(\underline{X})$ can be calculated by:

$$\Delta S(\underline{X}) = |S(\underline{X})| + \varepsilon \tag{9}$$

In terms of the security function $\Delta S(\underline{X})$ can be expressed as:

$$\Delta S(\underline{X}) = \sum_{i=1}^{n_1} \frac{\partial S(\underline{X})}{\partial V_i} \Delta V_i \tag{10}$$

where n_1 is the number of features used by the classifier, V_i and ΔV_i are respectivelly the ith feature and its change.
As the features can be expressed in terms of the pre-fault values of the mechanical input powers $Pmec_i$ and internal rotor angles θ_i, if we assume that the other

system variables will remain constant, equation (10) will be replaced by:

$$\Delta S(\underline{X}) = \sum_{i=1}^{m} \frac{\partial\ S(\underline{X})}{\partial\ Pmec_i} \Delta Pmec_i + \sum_{i=1}^{m} \frac{\partial\ S(\underline{X})}{\partial\ \theta_i} \Delta\theta_i \tag{11}$$

Practically speaking the required $\Delta S(\underline{X})$ can be obtained by decreasing and increasing respectively the output powers of the most unstable generator (i) and another machine (k), while keeping power balance. The most unstable generator is identified by the accelerating power method at t_o. The choice of the machine (k) that will increase its output power is made taking into account the maximum generating capabilities, the accelerating power at t_o^+ and the additional generating economic costs in choosing that generator.
The mechanical power to be exchanged between generators j and k, in order to assure the power system transient stability for a specified contingency is obtained from equation (12), where the changes induced in the rotor angles by the changes of the output powers of generators i and k are expressed by the well-known decoupled load-flow equation (Stott, 1974):

$$\underline{\Delta P} = \underline{H}\ \ \underline{\Delta\theta}$$

leading to a more suitable form of equation (11), i.e.:

$$\Delta S(\underline{X}) = \left(\frac{\partial\ S(\underline{X})}{\partial\ Pmec_k} - \frac{\partial\ S(\underline{X})}{\partial\ Pmec_i} \right) \Delta Pmec +$$

$$+ \sum_{j=1}^{m} \frac{\partial\ S(\underline{X})}{\partial\ \theta_j} (Z_{jk} - Z_{ji}) \Delta Pmec \tag{12}$$

where Z_{jk} are the elements of matrix $\underline{Z}$ obtained from the inversion of $\underline{H}$ and $\Delta Pmec = |\Delta Pmec_i| = \Delta Pmec_k$. The elements of the $\underline{H}$ matrix were calculated using the reduced pre-fault admittance matrix and the pre-fault rotor angle values.

NUMERICAL RESULTS

Security Functions Performances

The described Pattern Recognition approach to the transient stability assessment is illustrated here for the CIGRE test system, with four three-phase fault type contingencies. Due to space limitations its data is not presented here, but can be obtained from request.
For each one of the contingencies a training set was generated with the algorithm refered before, considering 38 different load levels with the total active load varying from 845 MW to 1850 MW. The contingencies data as well as the number of secure and insecure operating points of each contingency training set are given in table 1.

Table 1 Contingencies Data

Contingency nº	Bus nearby default	Line switched	Elimination time (s)	Nº of stable op. points	Nº of unstable op. points
1	1	1-3	.32	183	138
2	3	3-9	.33	236	83
3	6	4-6	.41	174	144
4	3	3-4	.36	203	128

The variables selected to be used by each security function are presented in decreasing order of their F values at table 2.
The performance of each security function was quantified through the training and test set errors evaluation. In order to theoretically validate the experimental results obtained, the upper (ε_u) and lower (ε_l) Chernoff bounds on the error probability were calculated. Table 3 presents the global results concerning the training and test set errors for each contingency, as well as the Chernoff bounds.

Table 2 Features to be Used by each Security Function

Contingency nº	Selected features
1	Pac_1, PL, $\dot{P}ac_4$, Pac_2, Pac_4, Pac_6, Pcoa, $\dot{P}ac_2$, $\dot{P}ac_5$, $\dot{P}ac_3$
2	Pac_4, Pac_3, PL, $\dot{P}ac_4$
3	Pac_6, Pac_4, Pac_7, Pcoa, Pcoa/MT, $\dot{P}ac_4$, $\dot{P}ac_2$, Pac_3
4	Pac_3, Pac_7, Pcoa/MT, Pac_4, Pac_5, $\dot{P}ac_1$, Pac_2, $\dot{P}ac_2$, $\dot{P}ac_4$

Table 3 Training and Test Errors and Chernoff Bounds on the Probability of Error

Contingency nº	Training error %	Test error %	ε_u %	ε_l %
1	3.12	4.36	11.88	1.41
2	4.44	5.08	16.27	2.65
3	3.77	4.08	9.05	0.82
4	3.93	4.89	10.58	1.12

The training set errors presented are optimum ones, in the sense that the separation threshold were calculated through Bayesian decision rules based on the minimization of the probability of error. However, in some cases it may be desirable to increasing these errors, introducing a bias on the classifier in order to reduce the false dismissals (insecure pattern classified as secure) errors.

Implementation of Preventive Controling Measures

This section presents the numerical results obtained with the preventive control methodology, for a given operating state characterised by the output powers presented in table 4 and assuming that contingency nº 3 is the most probable to occur, due to some external adverse conditions.

Table 4 Generator Output Powers Before and After Preventive Control

Generator	Output power (MW) Before		After
1	253.81		253.81
2	201.90		201.90
3	291.11		291.11
4	355.00		355.00
5	193.11		193.11
6	208.20	------→	195.15
7	219.32	------→	232.37

If contingency nº 3 takes place the corresponding security function gives $S_3(\underline{X})=-1.493$. The presented negative value alerts the operator to the potential unstability of this operation state. The dispatch engineer must then demand to the preventive security control system the actions to be taken in order to get out of the alert state.

In this case the preventive control strategy suggests that the equivalent machine nº 6 should exchange with machine nº 7 13.05 MW in order that the security function value becomes equal to 0.4. Table 4 summarizes the changes in the generators output powers before and after the implementation of the preventive control strategy. Machine nº 6 was the one to be chosen among other candidates, because of the lowest increase in the generating economic costs of this solution versus the other possible ones. The 0.4 target security function value was chosen in order to assure that the new system operating point was going to be defenitly stable.

Given the pre-fault condition and for the used test system less than one second was needed to classify an operating state and define the preventive control measures to implement in case of an alert situation. This time was evaluated in a CDC CYBER 720/170 computer.

The proposed methodology has already been extended to several simultaneous contingencies, using an iterative procedure, in order to establishe a total preventive scheme.

CONCLUSIONS

A new approach to transient security assesssment and enhancement of power systems based on pattern recognition techniques has been developed. The security function adopted is based on a set of primary variables which are directly or indirectly related with the parameters that enable to develop preventive control measures. This function separates the projected secure and insecure regions along the Fisher direction and gives the power system degree of security.

Preventive control measures are defined based on a prespecified value of the security function, which gives directly the changing to be made in the output powers of the generators.

Numerical results have shown the efficiency of the method.

The proposed methodology constitues part of an intelligent security system being developed to solve the transient security assessment and enhancement problem of large power systems, which takes into account heuristic knowledge given by the design engineers and operators through pattern recognition, problem solving and knowledge representation techniques.

REFERENCES

Athay, T., Sherkat, V. R., Podmore, R., Virmani, S., and Puech, C. (1980). Transient energy stability analysis, Conference on System Engineering for Power: Emergency Operating State Control - Section IV, Davos, Switzerland, 1-19.

Duda, R. O., and Hart, P. E. (1973). Non parametric techniques. Pattern Classification and Scene Analysis, John Wiley. pp. 114-118.

Gupta, C. L., and El-Abiad, A. H. (1975). Transient stability analysis of power systems by pattern recognition. Proceedings Midwest Power Symposium, Akron.

Fukunaga, K. (1972). Linear classifiers. Introduction to Statistical Pattern Recognition, Academic Press. pp. 90-100.

Ribbens-Pavella, M., Murthy, P. G., and Horward, J. L. (1982). Transient stability index for on-line stability assessment and contingency evaluation. Electrical Power and Energy Systems, 4, nº 2, 91-99.

Sá da Costa, J. (1982). Application of Pattern Recognition to Transient Security Assessment in Power Systems. Ph. D. thesis, UMIST, Manchester. pp. 20-26.

Stott, B. (1974). Review of load-flow calculation methods. Proceedings IEEE, 62, nº7.

Model-Based Failure Diagnosis in Power Plants

Pentti Lautala, Tuula Ruokonen, Martti Välisuo, Juhani Hyvärinen

Helsinki University of Technology,
Control Engineering Laboratory

ABSTRACT

This paper describes methods, applications and experiences dealt with during the research project Failure Diagnosis and Condition Monitoring in Power Plants. The diagnosis algorithms developed are model-based, i.e. they are based on monitoring the process using process measurements and models describing the quantitative operation of the process. A model is here understood as being any description of the operation of the process and the relations between the measurements. The stucture of the model is chosen to suit the process monitored and its failure mechanisms in the best possible way: a dynamic or static mathematical model, a characteristic curve or a presentative value or function. The methods are also designed to be as fault selective as possible in order to make the location more effective.

KEYWORDS

Failure diagnosis, fault detection, power plant, dynamic model, static model, characteristic curve.

INTRODUCTION

The nature of the work of the operator in industrial processes has changed with the introduction of the modern digital process control systems and video control rooms. The control system measures and processes a lot of data and the operator supervises the plant by scanning the displays. In a normal situation the system handles most of the necessary functions automatically, but when there is a disturbance the system does not give the operator sufficient support: often we get a lot of alarms, most of which are irrelevant. The solution is an automatic failure diagnosis system that is part of or parallel to the process control system.

The main task of a real-time failure diagnosis system is to monitor the plant constantly and to detect, locate and even predict disturbances and failures. There are numerous diagnosis methods with specific applications. One traditional method is the vibrational analysis of rotating machines. A mathematical model of a process may reveal new diagnosis methods.

In the research project we have studied application of model-based methods to two power plants owned by the finnish power company Imatran Voima: a coal power plant in Naantali and a peat power plant in Jyväskylä.

MODEL-BASED REASONING

The idea of model-based failure detection is to check whether the measured variables are consistent according to the model. It is a good idea to divide the process into small subprocesses and to describe them with submodels. Submodels are more attractive than full models, because they are easier to construct and to update them. The submodels also make it easy to locate a fault: if the inconsistence of some measured variables is detected by a certain submodel, there is a fault in the respective subprocess. Some model types and diagnosis principles are listed below.

Dynamic model:
- a matched filter is constructed and the difference between the process output and the filter output is observed
- a state estimator is constructed and the estimated state is observed
- a state estimator is constructed and statistical properties of the innovation signal are observed. (Willsky, 1976)

Static model:
- a presentative value, e.g. efficiency, is calculated and observed
- the model is presented as an equation or as a characteristic curve and the difference between the process output and the model output is observed
- the model is presented as a characteristic curve and the difference between the reference curve and a recently estimated curve is observed.

It is essential to know the most common failures and to understand the failure mechanisms. It is therefore necessary to consult with the operators and other experts at the very beginning of a diagnosis system design project.

STRUCTURE OF THE SYSTEM

The purpose of a real-time failure diagnosis system is to help the operator in a failure or some other unusual situation and possibly to take action in certain urgent situations. The system contains data about the process state at that moment and about the process in general, and it provides the operator with analysed information so that he can make decisions and take action. The operator must be able to decide what kind of help he requires from the system.

Since the present-day process control systems do not seem to be suitable for diagnostic purposes, a separate failure dignosis and condition monitoring system is necessary. A hierarchical structure, as in Fig. 1, has many theoretical and practical advantages. The

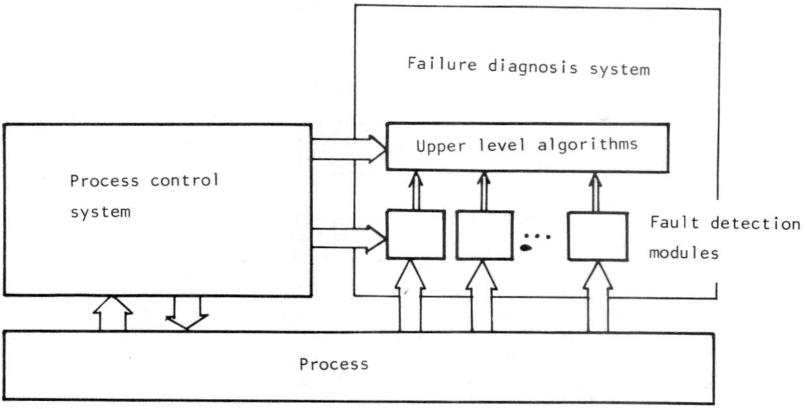

Fig. 1. Data processing in a failure diagnosis and condition monitoring system.

lower level consists of mutually independent modules, each of which monitors one device or function. The information given by the modules is combined and interpreted on the upper level. This kind of structure makes it easy to construct and expand the system. The independence of the modules makes it possible to use quite different methods in the same system. (Ruokonen, Lautala, 1986)

On the upper level the faults are identified and located by the means of inference rules. For this purpose an expert system has useful features: most expert systems have an easy-to-use user interface, they are capable of explaining how they get their results and their knowledge base can be expanded gradually. On the other hand, it may be difficult to make an expert system run in real time. However, the real time requirement is less critical on the upper level of the diagnosis and the inference task is reduced because the lower level modules pre-process the data.

THE SIMULATORS

A simulator is an efficient tool for studying a system in which the operator is a significant decision maker. In most cases the diagnosis system gives the operator advice and alarms, and he then makes the final decisions. Of course, in clear and sudden situations automatic actions are also possible. This means that the interface between the algorithms and the operator is such a significant part of the diagnosis that the merits of a diagnostic algorithm cannot reasonably be tested without it.

Testing the diagnosis system with a real system or with a simulator is easier than testing a process control system because it can be done in an open loop. With a real process the difficulty lies in fault generation, which can be done very easily on a simulator.

A few simulators have been constructed in the research project for different subprocesses in power plants, viz. feedwater valves, coal

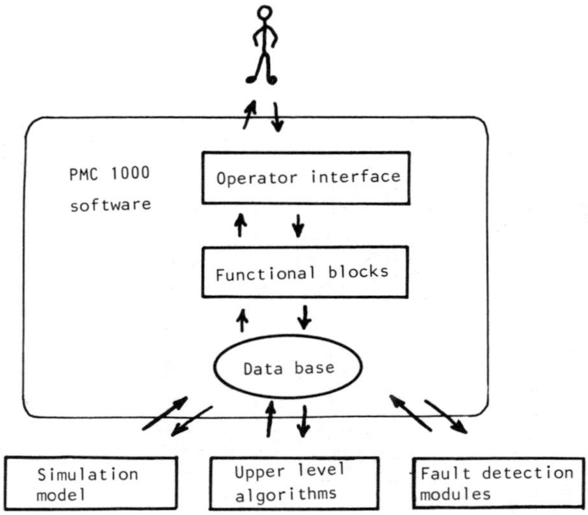

Fig. 2. Block diagram of the simulator.

burning and peat burning. The simulators have the structure presented in Fig. 2. PMC 1000 process monitoring and control software by HP was used. The simulators include a process model or they utilize recorded measurements. (Lautala, Välisuo, 1986)

DYNAMIC MODELS IN FAULT DETECTION

General

When model-based methods are applied to a slow process, the model must include the dynamics of the process. Fig. 3 shows a block diagram of a fault detection algorithm based on a dynamic model. It includes a matched filter that gives an estimate of the process output. The signal that indicates the fault, i.e. the test signal, is the low-pass filtered difference between the output and the estimate. Another possible test signal is the integral of this difference. It indicates the slow disturbances. (Iserman, 1984), (Lautala, Välisuo, 1986).

Application to Coal Milling and Burning

A prototype fault diagnosis system was constructed for a counter pressure coal power plant with 86 MW electric and 175 MW heat power in Naantali, Finland. The hardware consists of a MicroVAX I microcomputer with a signal processor, fast AD converters and a graphic colour terminal. The diagnosis algorithms include vibration analysis, life time calculation, effectivity calculation and model-based diagnosis algorithms. The system is now int the process of being installed, so no results can be presented yet. Some of the model-based algorithms have been tested in a simulator and the simulation results are presented here.

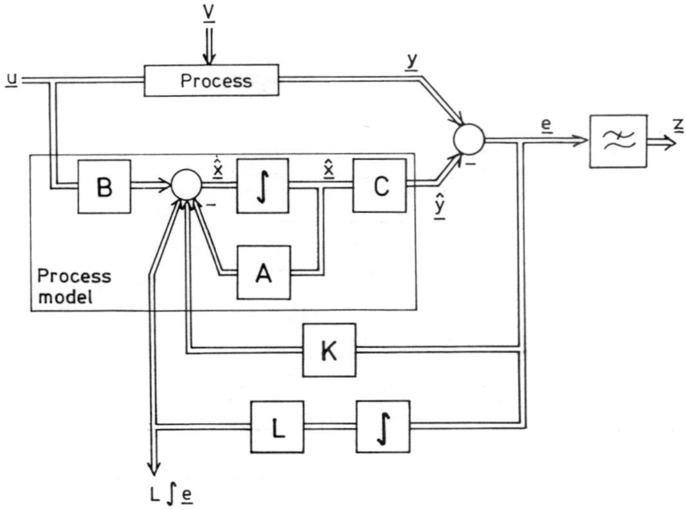

Fig. 3. Fault detection algorithm.

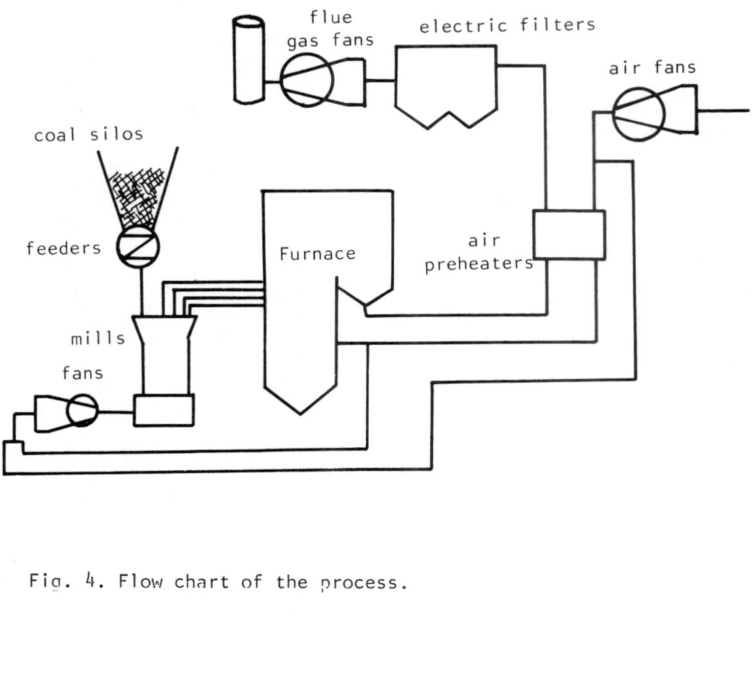

Fig. 4. Flow chart of the process.

A simplified process flow chart is shown in Fig. 4. The plant has three parallel fuel supply lines and mills. Each mill feeds four pipelines transferring the pulverized coal to the furnace. The continuously adjustable feeders supply coal from the silos to the mill. The coal flow is not measured. The coal flow from the mill to the furnace is controlled by the mill air. The plant has two air lines with two air fans and two preheaters. About 20 % of the air is used as mill air and the rest is used as secondary burning air. In the furnace the pulverized coal burns and the heat released is transferred mainly to the boiler and the superheaters. Two flue gas lines exist in the system. After the superheaters the flue gas goes through the preheaters and the electric filters to the stack. The model used here for the flue gas lines does not include the preheaters or the electric filters.

The measurements available for failure diagnosis in the fuel supply and the furnace are
- speeds of the feeders
- flows of the mill air
- air flow to the furnace
- pressure differences over the mills
- temperatures in the mills
- pressure in the furnace
- oxygen content in the flue gas

The following simulated faults are possible in the simulator. The feeder does not feed the right amount of coal to the mill. The milling is disturbed e.g. because of wet coal and some of the burners do not get enough coal. Some of the burners stop.

These faults can be detected and located if the coal amounts in the mills can be measured or estimated and if the fuel power released in the furnace is known. These variables are not directly measurable and matched filters are constructed to track the flow resistances of the mills, which are in proportion to the coal stored in the mills, and to track the oxygen consumption. It has been shown that the oxygen consumption is a good measure of the fuel power (Kortela, Lautala, 1981). The normal alarms of the separate signals can also be used in the diagnosis.

Next we shall look at the operation of the simulator in a fault situation. Fig. 5 shows some trends and Table 1 the output of the alarm printer. The dynamics of the simulation model is scaled so that the model is twice as fast as the real process.

To begin with the process is in a steady state. The coal feed rates of both mills are 60% and the oxygen content of the flue gas is 4%. Then the operator increases the coal feed rate of mill 1 to 70%. This action also changes the the transport air flow and the total air flow because of the feedforward loops. The figure shows the normal responses of the pressure difference and the fuel power.

At time 19:52:00 an abrupt disturbance takes place in mill 1: the coal flow into the mill decreases though the speed of the feeder remains constant. Several variables change and at 19:55:50 the fault diagnosis system locates a fault in mill 1. If the operator wants to know more details, he gets the following information: The test variables of mill 1 and the oxygen consumption have exceeded the lower limits and the oxygen content has exceeded the upper limit, which is 6.5%. From the trends he can see that the situation is getting worse.

By the time the process reaches a steady state, the fuel power has decreased 12 MW and the test variable of mill 1 indicates that the disturbance equals a decrease of 17 percentage points in the feed rate. The operator decides to adjust the secondary air flow to cut down the high oxygen content. The oxygen content alarm ceases, but the test variables remain unchanged, because the process is still inconsistent with the model used in the fault detection, due to the disturbance.

Table 1. Alarms during the simulation:		
Variable limit alarms:		
h:min:s		
19:54:10	Mill 1 test variable	Low advisory
19:54:20	Fuel power test variable	Low advisory
19:55:50	Mill 1 test variable	Low critical
19:56:00	Flue gas oxygen	High advisory
19:56:50	Fuel power test variable	Low critical
20:00:00	Flue gas oxygen	OK
Identified faults:		
19:55:50	Too little coal in mill 1.	

STATIC MODELS IN FAULT DETECTION

The process can be described by a static model if it is fast or if we just want to examine the steady state. A static model is usually sufficient for monitoring w.g. valves, pumps and fans.

A static model describes the relation between two or more simultaneously measured variables. The sampling interval is not critical. The statistical properties of measurement noise change when nonlinear calculations are made, so attention should be paid to the filtering.

A static model can be constructed on the grounds of physicallaws. The quantitative relations of the input and output variables are calculated from the physical dimensions of the process. Another approach is to use the physical laws in order to solve the form of the model and to estimate the parameters from some sufficiently rich measurement data series.

A characteristic curve is one kind of static model. It is a variable that is measured or calculated from the measurements and that is presented as a function of the operating point. The characteristics of valves and pumps are familiar examples. A characteristic curve can be stored as an array in the computer memory.

Checking the difference between the model output and the process output is one way of monitoring the process. Another way is to estimate continuously the present characteristic curve and to compare it with the reference curve.

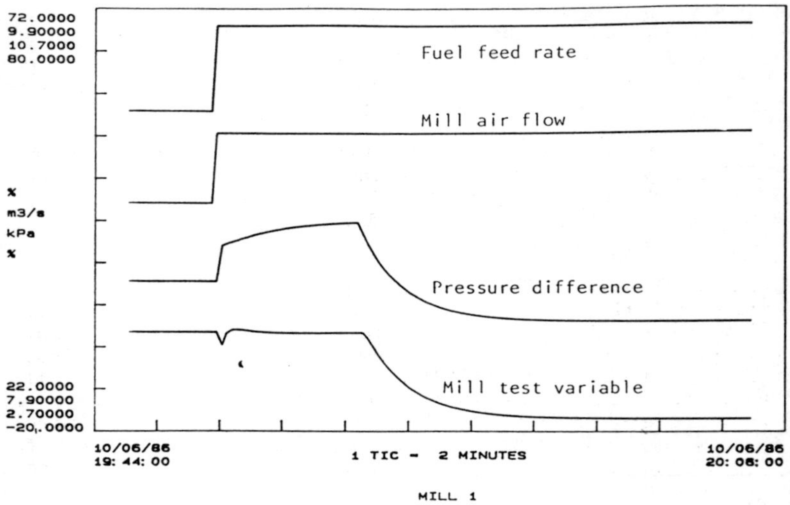

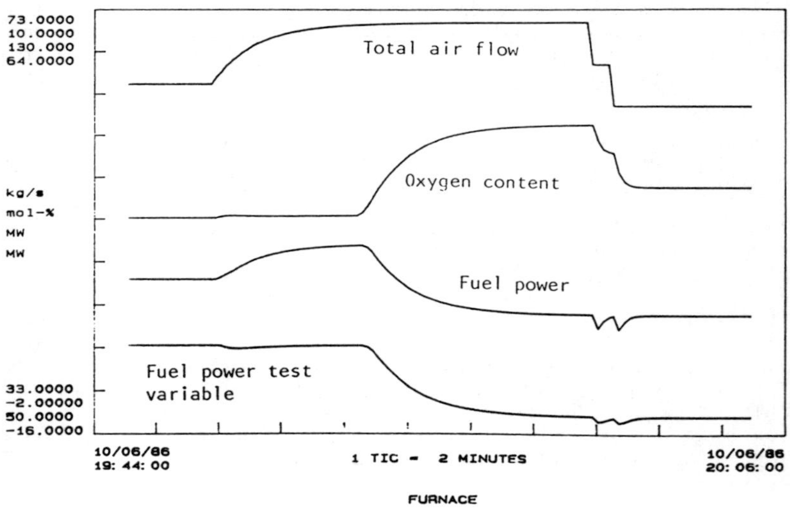

Fig. 5. The trends of some important variables during the simulation described in the text.

Application to burning air fans

Fans are common components in power plants and are often located at critical points in the process. They are also remarkable power consumers. It is thus profitable for economic and other reasons to monitor their condition. Methods applying characteristic curves have appeared to be effective for this purpose.

At the Rauhalahti peat power plant two forced draft fans were examined. They are of axial type and the air flow rate is controlled by changing the blade angle. The fans are placed parallel in the air line.The gas flows, pressure differences over the fans, power consumptions and the blade angles are measured.

The characteristic curves issued by the fan manufacturer plot the dependence of the pressure difference and the power consumption on the flow rate and the blade angle. These curves were approximated with suitable polynomials. The flow rate was determined from both of these polynomials and could thus be calculated from the other measurements. (Fig. 6). The flow rates calculated both from the pressure difference and from the power consumption are approximately equal but they both differ from the measured value by a factor of about two. The reason for this is that the blade angle measurement was inaccurate.

In fault diagnosis the error between the calculated and measured flow rates can be used as a measure of the fan's condition. To some extent it also seems possible to use the fan and its characteristic curve to measure the flow rate through the fan. It might be particulary useful to measure the flue gas flow rate in this way.

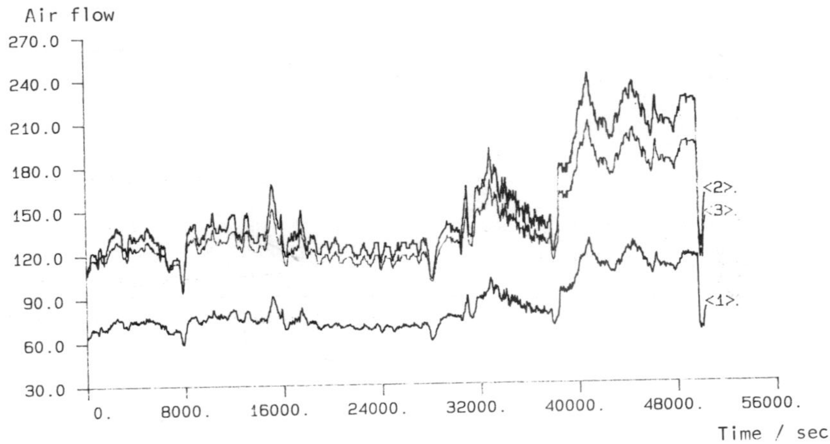

Fig. 6. Comparision of the measured and calculated values of the air flow.

1. Measured air flow.
2. Air flow calculated from the pressure difference and the blade angle.
3. Air flow calculated from the fan power consumption and the blade angle.

CONCLUSION

Fault diagnosis is becoming more and more important as an aid to the process operator. It gives him information about the process state,advice and instructions and takes some urgent actions automatically.

Data processing and memory capacities are nowadays so cheap that it is profitable to apply failure diagnosis methods even to small plants. Microcomputer based systems in particular have made this possible. Failure diagnosis will in the future be integrated into the process control system.

REFERENCES

Iserman, R. (1984). Fehlerdiagnose mit Prozessmodellen. Technisches Messen 51. Jahrgang, Heft 10, 345-355.

Kortela, U., Lautala, P. (1981). A New Control Concept for a Coal Power Plant, 8th IFAC World Congress, Kyoto, Japan.

Lautala P., Välisuo, M. (1986) A Simulator for a Fault Diagnosis System. Report 69, Helsinki University of Technology, Control Engineering Laboratory.

Ruokonen, T., Lautala, P. (1986) Model-based Reasoning in the Failure Diagnosis of Industrial Processes. Report 68, Helsinki University of Technology, Control Engineering Laboratory.

Willsky, A., S. (1976) A Survey of Design Methods for Failure Detection in Dynamic Systems. Automatica 12.

Reactor Noise Analysis Applying Multivariable AR & ARMA Methods in NPP PAKS

J. Bokor*, S. Veress** and I. Nagy***

Computer and Automation Institute, Hungarian Academy of Sciences, 1111 Budapest, Kende u.13-17, Hungary
***Institute for Computing Technique, 1751, Budapest, P.O. Box 65, Hungary*
****Nuclear Power Company, Paks, P.O. Box 71, Hungary*

ABSTRACT

This paper presents the application of multivariable AR and ARMA methods in reactor noise analysis and surveillance systems. Using the identified models of noise signals, the detection of changes in the operation of the nuclear power plant or in its components can be performed either in the model space or computing the spectral characteristics, in the frequency domain. An example for detecting abnormal vibration of a control rod is investigated.

KEYWORDS

ARMA models, signal processing, identification, spectral analysis.

INTRODUCTION

Due to the increasing demands on reliability and safety of technical plants and their components, the development of monitoring and surveillance systems are receiving an increasing interest. This holds as well for advanced processes with highest demands on reliability and safety, like aeronautics and nuclear power plants (NPP). One of the important development in surveillance and diagnostics in NPP-s is based on the reactor noise analysis.

In the reactor noise analysis informations are obtained from fluctuating part (as opposed to stady state values) of the plant signals, in order to

- monitor the reactor components by verifying conformity with expected status, detection and diagnosis of anomalies,
- determine physical parameters and/or general dynamic characteristics of the reactor,
- provide informations about normal or abnormal operation of the plant or in its components.

The overall task of noise monitoring and surveillance system can usually be devided into three main areas.

Signal processing

Reactor noise signals are generally modelled by stochastic processes or by stochastic, fields when the spatial distribution is to take into considerations. In this modelling approach the noise can be characterized either in the time or in the frequency domain. The purpose is to convert the noise signals to a useable form (parametric models, spectral densities, e.t.c) in which the relevant signal properties can be distinguished.

Pattern recognition

The purpose is to detect statistically significant changes in noise signal characteristics.

There are several approaches to the detection of changes based on modelling of noise signals.

In case of known model structures and parameters, the detection of abrupt changes can be performed by applying fault sensitive filters, Jones (1973), whiteness and chi-squared tests of the residuals of the appropriately designed Kalman-filter, Mehra and Peschon (1971), applying a bank of Kalman-filter with multiple hypothesis testing (the multiple model approach), see e.g. Willsky (1986), or the generalized likelihood ratio test, Willsky and Jones (1974).

An other approach is based on the detection of changes in the space of identified model parameters, see Isermann (1984). The advantage of this method appears in the case, when the model structure of the plant generating the signals is known and there exists a bijective map between the (unmeasurable) physical parameters and the estimated model parameters.

The third approach considers the situation where neither the model structures nor the relationship to specific physical parameters are known (this is very often the case in reactor noise analysis). The detection of changes is based on detection of significant changes in the identified model structures and/or parameters.

Using the above methods, the change detection can be performed under the given conditions either in time-domain, in parameter space or in the model-space.

In reactor noise analysis, however, the most widely used methods are based on change detection in the frequency domain, using the spectral characteristics of the signals. The estimation of spectral characteristics are based on nonparametric, like Fast Fourier Transform (FFT) methods and parametric, like Maximum Entropy Method (MEM), AutoRegressive (AR) and AutoRegressive Moving Average (ARMA) spectral estimation methods, see e.g. Haykin (1979). Efficient procedures for extracting relevants features of spectral functions and for performing discriminant analysis based on these features are now available.

Diagnostics

The informations provided by noise analysis and pattern recognition are used to diagnose the causes of detected changes and to inform operating personnel about suspected anomaly. Recently, artificial intelligence tools and the emerging technologies of expert systems are also utilized in the realization of the noise diagnostic systems.

This paper discusses the signal processing methodology applied in the noise analysis and surveillance system of NPP paks. In the processing of noise signals, the black-box situation will be considered, and the methods for identification of parametric models will be reviewed in the next paragraph. In accordance with this approach, the detection of (not abrupt) changes can be performed in the model/parameter space, and also in the frequency domain using the spectral functions computed from the identified models.

The last paragraph investigates an application example, where the abnormal vibration of a control rod is detected using multivariable AR models of in-core neutron noises.

SIGNAL PROCESSING METHODOLOGY APPLIED IN REACTOR NOISE ANALYSIS

The methods applied in reactor noise analysis will be rewieved. We focus our attention to the application of parametric signal processing and spectral analysis methods, thus classical nonparametric, like correlation and FFT-based spectral analysis methods will not be considered here.

Reactor noises can be regarded as multivariable stochastic processes modelled as the output of a linear filter driven by vector valued white noise sequence. Assume, that y_t is a p-dimensional discrete-time Gaussian process having rational Hermitian spectral density $\phi(z)$, $z = \exp(i\omega)$. Assume that $\phi(z)$ is positive definite for all $\omega \in [-\pi,\pi]$, then applying spectral factorization, one can obtaine a transfer matrix representation of y_t:

$$y_t = T(z)e_t, \quad e_t \sim N(0,R), \tag{1}$$

where e_t is Gaussian white noise sequence and $T(z)$ is a rational transfer matrix. Applying the results of stochastic realization theory, $T(z)$ can be realized in state-space, matrix fraction or ARMA forms. Since the use of transfer matrix and ARMA models are superior in the spectral analysis context, we consider these parametrizations only. ARMA models can be written in the form

$$A(D)y_t = C(D)e_t \tag{2}$$

where $A(D)$, $C(D)$ are coprime nonsingular polynomial matrices with $A(o) = C(o) = I_p$, where I_p is a p×p dimensional unit matrix and D is the delay operator.

The polynomial matrices A,C of the above monic ARMA models can be derived from the Hankel-matrices associated with the transfer matrix $T(z)$, see Gevers (1986) or from the constructibility matrix associated with $T(z)$, see Bokor and Keviczky (1987). The (McMillan) degree of an ARMA model is defined by $n = \deg\det A(D)$.

The irreducible ARMA models in Eq.2 are generally not uniquelly identifiable except they are in canonic or pseudocanonic form, see Gevers and Wertz (1984) for this important issue. The structure of canonic and pseudocanonic ARMA forms are uniquelly defined by the socalled structural indices, n_i, $i = 1,\dots,p$, $\Sigma_i^p n_i = n$.

AR models are obtained from Eq. 2 if specifically $C(D) \equiv I_p$. The discussion of the structure and the computation of McMillan degree of AR models can also be found in Gevers (1986).

It should be noted, that in signal processing applications the use of fully parametrized AR models ($n_1 = n_2 = \dots = n_p$) is the most common one.

In the remaining of this paragraph we shortly review the structure and parameter estimation methods associated with the above model classes available in the "method-base" of the noise analysis system. The statistical tests that can be used to detect changes in the models will also be revealed.
Since the computation of spectral functions from the identified models is fairly straightforward, these will be omitted here.

Transfer matrix models

The parameter estimation can be solved by the application of Maximum Likelihood (ML) method. It is also possible to apply various statistical tests to the residuals to validate the model.

ARMA models

In this noise analysis system the ML and Extended Least Squares (ELS) methods are applied for parameter estimation of ARMA models.

Using ML method, the structure estimation (determination of the structural indices from the observations) can be solved by applying the information criteria BIC or Ø, see e.g. Hannan and Kavalieris (1984).

Using ELS method, the procedure suggested by Bokor and Keviczky (1982) can be applied. This procedure fully exploits the structural properties of monic canonical or pseudocanonical ARMA forms and applies an F-test to the residuals.

AR models

A variety of Least Squares (LS)-type method has been elaborated for parameter estimation of AR models, or for estimation of AR parameters in ARMA models. Using fully parametrized AR models the parameter estimation can be solved recursively for model structures with increasing degree. This can be solved by applying fast lattice filters, see e.g. Whittle (1963), Haykin (1979), Friedlander (1982).

Since the lattice-filters are recursive in structure, it is easy to apply criteria like Final Prediction Error (FPE) or AIC for structure estimation. The methods available in the noise analysis system are based on the Whittle--algorithm, on the lattice algorithm for MEM and a recently developed lattice algorithm for ARMA parameter estimation.

Statistical test applied to detect changes in the model structures and parameters

Having a bank of reference models representing "normal" situations, the models identified in the current situations are compared to detect changes in the structure and/or in the parameters. Since the computation of complete structure re-estimation is very time consuming, usually the parameters are only reestimated in the previously identified model structure. For this reason it is useful to apply statistical hypothesis testing to obtain information about changes in the structure. Associated with the ML method, the authors elaborated the Lagrangian Multiplyer (LM) and Likelihood Ratio (LR) tests for multivariable models. Related references are Postkitt and Tremayne (1981), Veres (1985,1986).

Utilizing the asimptotic properties of ML estimators, the estimated parameters can be compared to the particular reference model parameters to detect changes in the parameter space. The estimate of covariance matrix of the parameters is computed in the parameter estimation procedure, and can be used to define distance measures to perform discriminant analysis.

In the common practice the identified models are used to compute the spectral characteristics (auto- and cros,power spectral densities /APSD, CPSD/, phase, coherence, noise power contribution functions, e.t.c.) of the noise signals. These are further used for feature extraction and pattern recognition in the frequency domain.

AN APPLICATION OF AR AND ARMA METHODS IN THE NOISE ANALYSIS AND SURVEILLANCE SYSTEM OF NPP PAKS

The application of the methods described in the previous paragraph will be illustrated by an example of detecting abnormal vibration of a control rod of a nuclear reactor unit operating in NPP Paks.
The reactor units are VVER-440 (PWR) type reactors producing 1760 MW thermal and approximately 440 MW electric power. The six loop primary coolant circuits, the arrangement of reactor pressure vessel, main coolant pumps, main valves, steam generators, the pressurizer and the pipings are shown on Fig. 1.
The reactors are equipped as standard instrumentation with ionisation chambers allocated in the biological shild around the reactor units and with assemblies of in-core Self Power Neutron Detectors (SPNDs). The SPNDs are located in tubes at various radial directions, see Fig. 2. Each tube involves 7 detectors in different vertical positions. Some of the ionizations chambers are connected to noise preamplifyers, others are connected to a current-to-frequency converters. The Noise Analysis and Surveillance (NAS) system uses these signals after frequency--to-voltage converters, (IK, JK signals), leaving the original control function of these chambers unchanged.
The reactor inlet and outlet pressure (PB, PK signals) and the reactor core temperature fluctuations (TH signals), are also measured applying piezo-electric pressure transducers and standard thermocouples respectively. Accelerometers are mounted for vibration monitoring and analysis in the primary system (R and S signals). The arrangement of the above detectors can be seen on Fig. 1 and Fig. 2. The signals of selected detectors are processed by the NAS system after appropriate amplifying, filtering and A/D conversions.

The purpose of the following stochastic modelling and subsequent frequency domain analysis was to find appropriate fault signatures usable to detect abnormal vibrations of control rods. In this experimental situation it was known in advance that one of the control rod was damaged. This situation was also studied by Glöckler at al. (1986) using FFT-based spectral analysis.
For the identification of AR or ARMA models the output variables were selected first. The signal of three SPND detectors allocated at the same vertical but different radial positions and the signal of one accelerometer mounted at the control rood driving mechanism (CRDM), denoted by SO, were chosen as output variables. The signals of SPNDs are denoted by SP4, where the number indicates their vertical position, and their radial positions is defined by two coordinates in accordance with Fig. 2.

The measurements were performed under various reactor conditions and vertical control rod positions as shown in Table 1. The position of the damaged control rod is represented by the midle rod position. The structure and parameter estimation was performed by the procedure described in Bokor and Keviczky (1982), using 1024 samples obtained by 64/s sampling frequency. Perhaps surprisingly, the AR model class with structural indices $n_1 = n_2 = n_3 = 4$, $n_4 = 6$ proved to be the best one for all experimental situation. The ordering of the output variables can be seen on Fig. 4. The parameter estimates changed significantly in each case.

The frequency functions APSD, CPSD, phase, coherence and noise power contribution (NPC) were computed from the identified models.
The APSDs representing the "normal" situation is shown on Fig. 3. Analysing the APSDs of ex-core signals on Fig. 3, one can observe the resonance peaks in the frequency interval 0.4-0.55/s not existing in the normal case. Performing also Signal Transmission Path (STP) analysis based on the partial coherence and NPC functions on Fig. 4, see Oguma and Turkcan (1985), one can recognize the effect of source noise of the in-core neutron flux fluctuations to the acceleration of vibration of the CRDM. The source noise of in-core neutron flux fluctuation is the abnormal vibration of control rod in the reactor core. Consequently the CRDM vibration and neutron flux fluctuation have a common source noise due to the

the abnormal vibration of control rod in the above frequency range. This shows that anomalies in the control rod vibrations can be detected by analysing in-core neutronflux and/or control rod driving mechanism vibration signals.
Evaluating the values of in-core neutronflux APSDs, the location of the damaged control rod can also be guessed.

CONCLUSION

The paper considered the application of AR and ARMA modelling of reactor noises in the NAS system of NPP Paks. The advantage of this parametric approach in processing the noise signals can be utilized in obtaining informations about normal or abnormal operation of the plant, or in fault detection and diagnostics.

REFERENCES

Basseville, M. and A. Benveniste (1986). Detection of Abrupt Changes in Signals and Dynamical Systems. Springer Verlag, Berlin Heidelberg, New York, Tokyo.

Bokor J. and L. Keviczky (1982). Int. J. Control, Vol. 36, No. 3, 461.

Bokor J. and L. Keviczky (1987). Int. J. Control, Vol. 45, No. 3, 861.

Friedlander, B. (1982). IEEE Trans. on Inform. Theory, IT-28, 639.

Gevers, M.R. and V. Wertz (1984). Automatica, Vol. 20, No. 3., 333.

Gevers, M.R. (1986). Int. J. Control, Vol. 43, No. 6, 1745.

Glöckler, O., G. Por and J. Valko (1986), Proc. 6th Power Plant Dynamics, Control & Testing Symposium, Knoxwille.

Hannan, E.J. and L. Kavalieris (1984). Adv. Appl. Prob. 16, 492.

Haykin, S. (1979). Nonlinear Methods in Spectral Analysis, Springer Verlag Berlin,Heidelberg.New York, Tokyo.

Iserman, R. (1984). Automatica, Vol. 20, Ns. 4, 387.

Jones, H.L. (1973). PhD Thesis, Dept. of Aeronautics, M.I.T. Cambridge, Mass.

Mehra, R.K. and J. Peshon (1971). Automatica, Vol. 7, 637.

Oguma, R. and E. Türkcan (1985). Progress in Nuclear Energy, 15, 863.

Postkitt, D.S. and Tremayne (1981). Ann. Statistics, 9.

Veres, S. (1985). Proc. of the 7th. IFAC Symp. on Ident. and System Parameter estimation, York, England.

Veres, S. (1986). Asymptotic distribution of likelihood ratios for overparametrized ARMA processes to appear in Journal of Time Series Analysis.

Whittle, P. (1963), Biometrika, 50.

Willsky, A.S. (1976). Automatica, Vol. 12, 601.

Willsky, A.S. and H.L. Jones (1976). IEEE Trans. Autom. Control, AC-21, 108.

Number of Experiment	1.	2.	3.	4.
Reactor power [%]	63	60	58	23
Positions of control rods [cm]	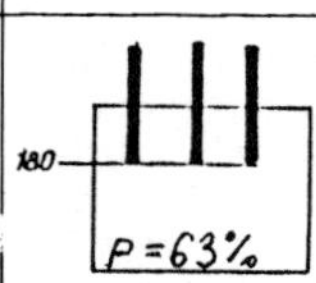	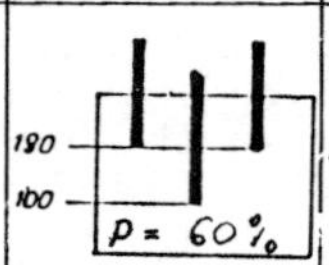	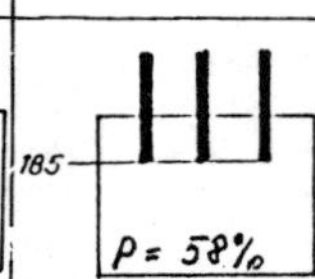	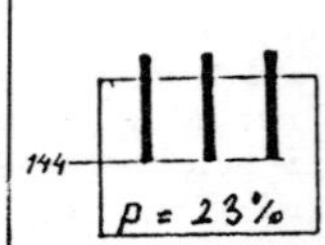

TABLE 1. Experimental conditions

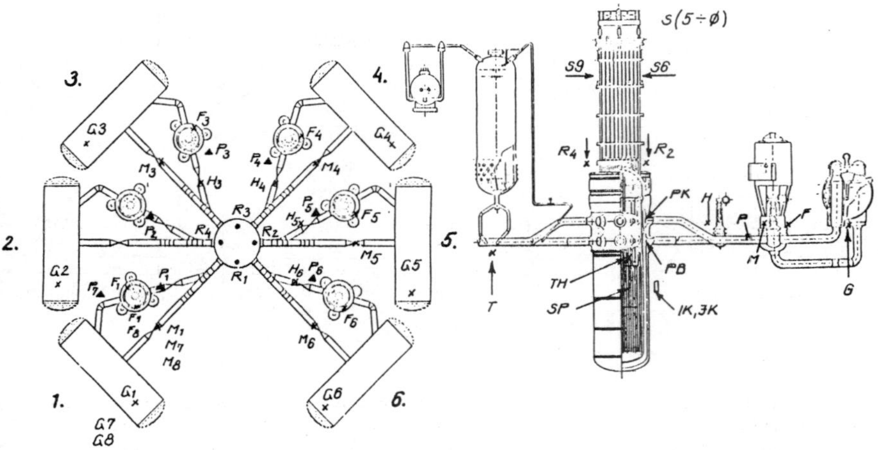

Fig. 1 Arrangement of accelerometers, pressure transducers, SPNDs, ionisation chambers and thermocouples used by noise analysis and surveillance system, NPP "PAKS"

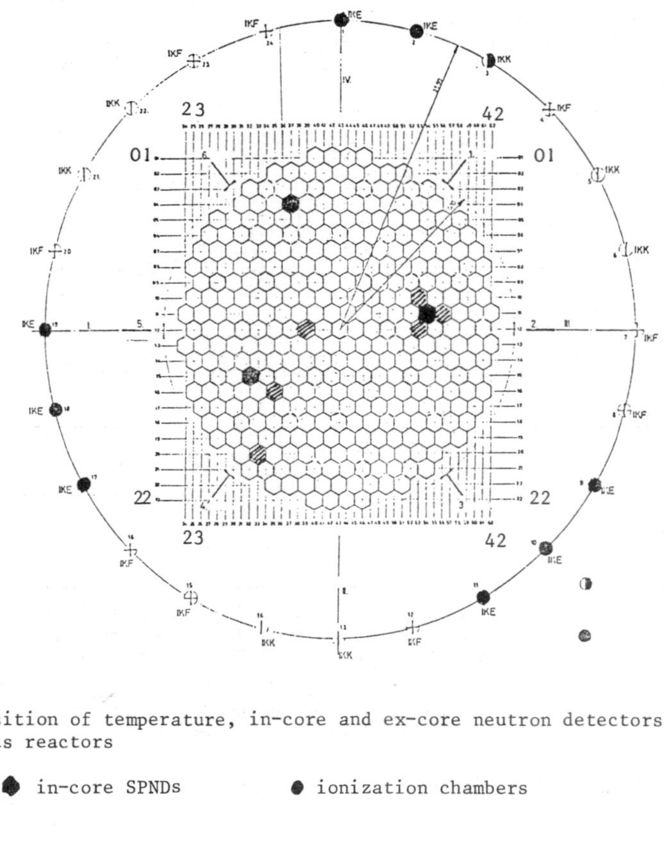

Fig. 2 Position of temperature, in-core and ex-core neutron detectors in NPP Paks reactors

● in-core SPNDs ● ionization chambers

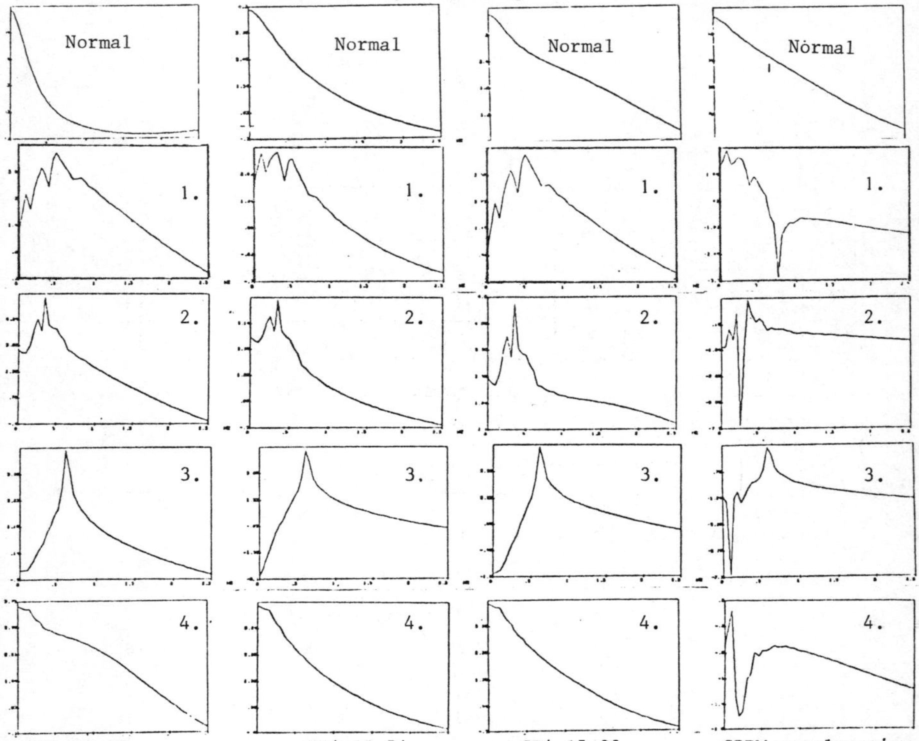

SP4 04-37 SP4 11-54 SP4 15-32 CRDM acceleration

Fig. 3. APSDs of in-core neutron flux and CROM acceleration signals.

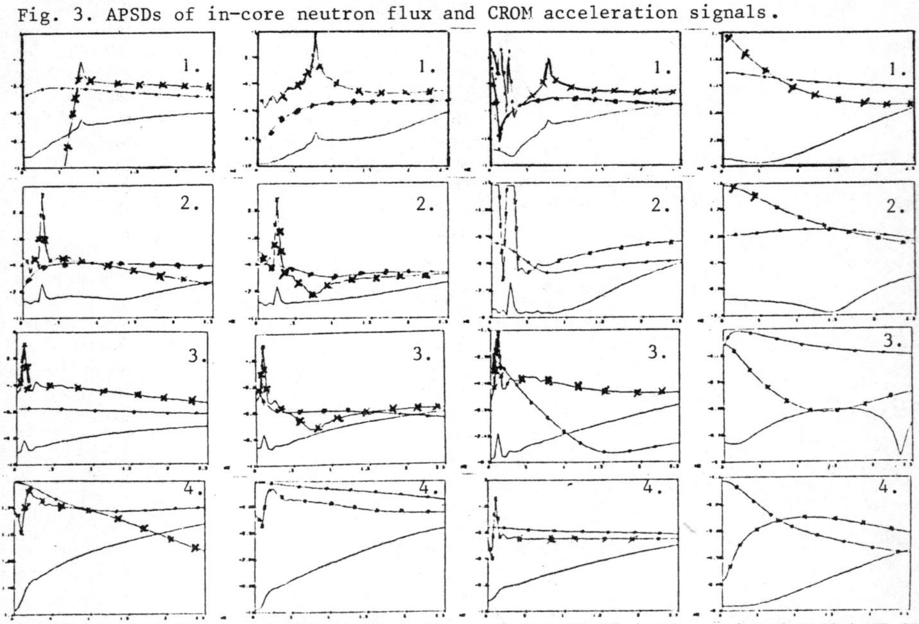

i:SP4 04-37 j:SO i:SP4 11-54 j:SO i:SP4 15-32 j:SO i:SP4 11-54 j: 15-32

Fig. 4. Partial coherence —, NPC(i,j) ‑•‑•‑, NPC(j,i) ‑x‑x‑ functions.

Robust Observer of a Synchronous Generator

W. Mielczarski, P. Woźniak

Institute of Electrical Power Engn.,
Lódź Technical University, Lódź, Poland

ABSTRACT

The problem of the non-linear synchronous generator state observation is presented. After a non-linear transformation of the state we can obtain a model of the system in the observer canonical form which allows the design of observer with parameters dependent on outputs of the system. This approach improves some properties of the observers, however does not assure their robustness. A concept of proportional observers with an additional integrating component (called hereafter a proportional-integrating observer or shortly PI-observer) is presented and their robustness is discussed. Typical transient response patterns obtained by means of simulation are presented using the PI-observer and comparisions are made with similar results obtained by proportional observers .

KEYWORDS

Robust observers; non-linear observers; power system; synchronous generator modeling .

INTRODUCTION

Many control applications require that the state of a dynamic system is available at all instants of time. The state observation problem concentrates on design of an auxilliary dynamic system known as observer. The observer theory for linear systems has been developed since 1963 in many different directions reviewed by O'Reilly (1983). The problem of a non-linear observer was considered by Roitenberg (1970), Thou (1973), Kou and co-workers (1975), Kochne (1982), Bestle and Zeitz (1982), Li and Tao (1986). Mielczarski (1987) also proposed the novel method of non-linear observer design which proves to be the most general one .

In this work the following non-linear system is considered

$$\begin{aligned} \dot{x} &= f(x,u) \\ y &= h(x,u) \quad , \end{aligned} \qquad (1)$$

where $x \in R^n$, $y \in R^m$, and relevant non-linear observer

$$\begin{aligned} \dot{z} &= f(z,u) + K\,(h(x,u)-h(z,u)) \\ v &= h(z,u)\ , \end{aligned} \qquad (2)$$

where $z \in R^n$, $v \in R^m$, K is (m x n) gain matrix. The error equation for the observer can be written in the form

$$\dot{e} = f(z,u) - f(x,u) + K\,(h(x,u)-h(z,u)) \qquad (3)$$

Before we formulate the theorem proved by Mielczarski (1987) we must define for the observer (2) the matrix R_s

$$R_s = A - K\,C \qquad (4)$$

where A and C are jacobian matrices for f and h respectively .

Theorem (Mielczarski 1986,1987)

For the non-linear system (1) the relevant non-linear observer defined by equation (2) is stable, i.e. the solutions of the error equation (3) are asymptotically stable in the Lyapunov sense, if in a neighbourhood G of the origin the following conditions hold :

1. $f_i(x,u)$ and $h_i(x,u) \in C^k$ for (t_0, ∞) in G ,
2. the non-linear system (1) is locally observable ,
3. the eigenvalues of R_s are chosen such that
   $Re(\lambda_i(R_s))$, i = 1,2,...,n ,
4. $|\Delta x_i| < \eta_x$ and $|\Delta u_i| < \eta_u$ in G .

A number of non-linear observers for a synchronous generator and their properties were presented and discussed by Mielczarski (1987, part II). The matrix K of these observers is assumed to be stationary and does not depend on the output of the system (1). It is obvious that non-linear observer in which the elements of the gain matrix R_s are functions of tha output should have better dynamic properties. Such an observer may be obtained by non-linear transformation of the state proposed by Bestle and Zeitz (1983). For a better understanding, we review briefly their results .
Bestle and Zeitz (1983) considered the non-linear system similiar to (1)

$$\begin{aligned} \dot{x} &= f(x,t) \\ y &= h(x,t)\ , \end{aligned} \qquad (5)$$

which is assumed to be observable and transformable to the observer canonical form

$$\dot{x}^* = E_n\, x^* - a^*(x^*_n)\ , \qquad (6)$$

where the matrix

$$E_n = \begin{vmatrix} 0 & . & . & 0 \\ 1 & . & & 0 \\ . & . & & . \\ 0 & . & .1 & 0 \end{vmatrix} \qquad (7)$$

The non-linear and time-variable transformation (and its inverse) of the system (1) into the form (6) is denoted by

$$x^* = W^{-1}(x\ ,t) \qquad (8a)$$

$$x = W\ (x^*,t) \qquad (8b)$$

Because the non-linear terms a* in the canonical form (6) was assumed to depend on the output only, the x* can be reproduced exactly by the observer

$$\dot{z}^* = E_n z^* - a^*(y,t) + K (y-z_n) \quad (9)$$

with the gain vector

$$K^* = (p_o I_n + p_1 M + \ldots + p_{n-1} M^{n-1} + M^n)\, q(z,t) \quad (10)$$

where : I_n denotes identity matrix $(n \times n)$,

$$N(\partial h/\partial x)^T = \partial/\partial t\, (\partial h/\partial x)^T + f^T(\partial/\partial x(\partial h/\partial x))^T + (\partial h/\partial x)^T \partial f/\partial x \quad (11)$$

is the differential operator, and observability matrix has the form

$$Q(x,t) = \begin{vmatrix} (\partial h/\partial x)^T \\ N(\partial h/\partial x)^T \\ \cdot \\ N^{n-1}(\partial h/\partial x)^T \end{vmatrix} , \quad (12)$$

the vector $$q(x,t) = Q^{-1} |0\ 0\ \ldots\ 1|^T = Q^{-1} C^{*T} \quad (13)$$

and the differential operator M

$$M\, q(x,t) = -\partial/\partial t\, q(x,t) - (\partial/\partial x\, q(x,t)) f + \partial f/\partial x\, q(x,t) . \quad (14)$$

The computation of the non-linear transformation $W(x^*,t)$ requires the integration of a system of non-linear partial differential equations

$$\partial W/\partial x(x,t) = |\, q(x,t)\ Nq(x,t)\ \ldots\ N^{n-1}q(x,t)\, | \quad (15)$$

$$\partial a^*/\partial x^*_n(x,t) = -\, |\partial W/\partial x^*|^{-1} N^n q(x,t) \quad (16)$$

The error of the observer (7) possesses a linear time-invariable dynamic matrix $E_n - K C^*$ and it can be designed by the eigenvalue assignment method (i.e. by computation of $p_o, p_1, \ldots$ coefficients values for given eigenvalues $\lambda_1, \lambda_2, \ldots$) .

NON-LINEAR OBSERVERS FOR A SYNCHRONOUS GENERATOR

The following model of a synchronous generator is considered

$$\dot{x} = f(x) + B u \qquad (17)$$
$$y = x_1 ,$$

where $x^T = |\, x_1\ x_2\, | = |\delta\ \omega|$,

$$u = |\omega_b\ P_m| ,$$

$$B = \begin{vmatrix} \omega_b & 0 \\ 0 & \gamma T_m \end{vmatrix} ,$$

$$f(x) = |\omega_b x_2\ \ P_m - a_1 \sin(x_1) - d\, x_2\, |^T ,$$

δ - the load angle |rad|, ω - the angular velocity, P_m - control variable proportional to mechanical power on the generator shaft, a_1 - constant proportional to the maximum value of electrical power output, d - damping coefficient. All variables, except δ , are in |p.u.|. Values of all parameters for the study system are given in the Appendix .

The non-linear observer based on Mielczarski´s theorem for the model (15) is in the form (named hereafter PN1)

$$\dot{z} = f(z) + Bu + K\,(y-z_1) \quad , \tag{18}$$

where $K = |\,k_1\ k_2\,|^T$ is constant gain vector for given eigenvalues λ_1, λ_2. Using the method proposed by Bestle and Zeitz (1983) and developed by Li and Tao (1986) we constructed another observer (called hereafter PN2) in the form

$$\dot{z} = f(z) + Bu + K(z_1)\,(y-z_1) \quad , \tag{19}$$

where the gain vector elements are functions of z_1 given by :

$$\begin{aligned} k_1 &= p_1 + d \\ k_2 &= (p_o - p_1 d + d^2) - a_1\cos(z_{1o}) \quad , \end{aligned} \tag{20}$$

where p_o and p_1 are such that eigenvalues of (19) equal λ_1 , λ_2 .

The next step of our work was to establish the PI-observer. The proportional-integrating observer for linear sytems was proposed by Wojciechowski (1978) and developed for non-linear systems by Mielczarski (1986) .
The PI-observer for non-linear system (1) has the form

$$\dot{z} = f(z,u) + Bu + K\,(h(x,u)-h(z,u)) - gw \tag{21a}$$
$$\dot{w} = h(x,u) - h(z,u) \tag{21b}$$

The equation for error (defined by (3)) dynamics can be written as follows

$$\dot{e} = f(z,u) - f(x,u) + K\,(h(x,u)-h(z,u)) - gw \tag{22}$$

It can be shown (Mielczarski, 1987) that the error equation (21b) satisfies the assumption of the Theorem if the eigenvalues of matrix $R_s{}^*$ are chosen to satisfy the inequality

$$\mathrm{Re}(\lambda_i(R_s{}^*)) = \lambda < 0 \quad , \quad i = 1, 2, \ldots, n \tag{23}$$

where

$$R_s{}^* = \begin{vmatrix} A - K\,C & g \\ -C & 0 \end{vmatrix}^{(n+1)\times(n+1)} , \tag{24}$$

A and C are jacobian matrices for f and h respectively , K and g are vectors of coefficients . The system (21)-(22) describing the PI-observer has 2n unassigned elements , n of the vector K and n of the vector g. We have used n+1 elements to assign eigenvalues of $R_s{}^*$ and the other n-1 to improve robustness of the PI-observer .

SIMULATION

In order to confirm the usefulness of the presented observers' design methods, a few typical transient response patterns have been simulated for the PN1 and PN2 observers and they are presented for comparative purpose. All transients are ploted on the phase plane. The stable point $\delta = 0.89\ |\mathrm{rad}|$ and $\omega = 1$ is chosen and it points the origin of the coordinate system.
The trajectories of the generator and PN1 and PN2 observers are ploted on Fig.1 to show the influence of vector K dependency upon the generator output. The necessary relations were obtained using equations (10)-(14). It is apparent that trajectory of the PN1 observer keeps closer to the trajectory of the generator.
This improvement does not suffice to keep both trajectories close

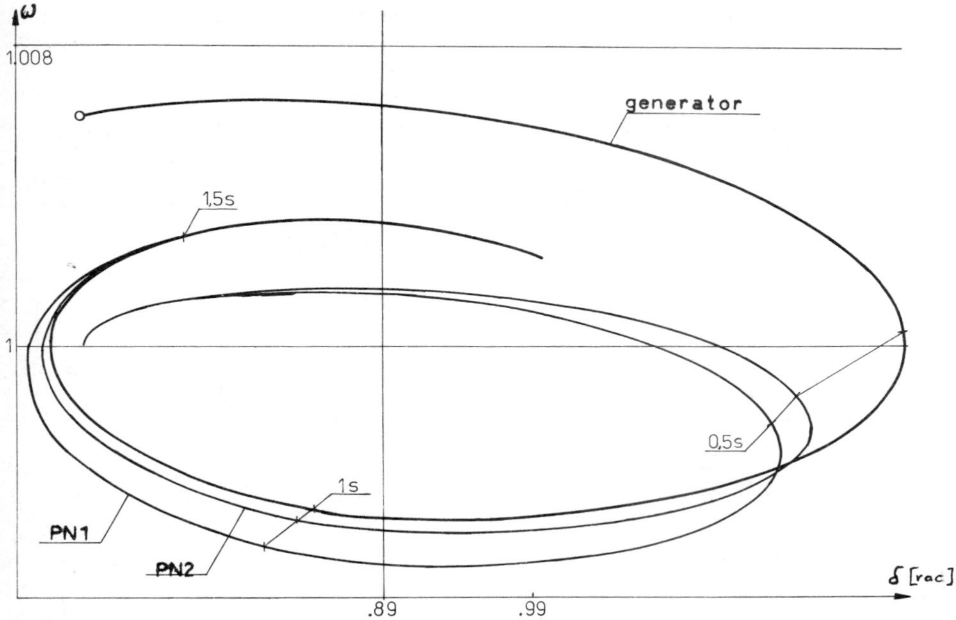

Fig. 1. Trajectories of synchronous generator and PN1, PN2 observers; $\lambda_1 = \lambda_2 = -5$, $x_1(t_o) = z_1(t_o) = z_1^*(t_o) = 0.69$ $x_2(t_o) = 1.005$, $z_2(t_o) = z_2^*(t_o) = 1.000$

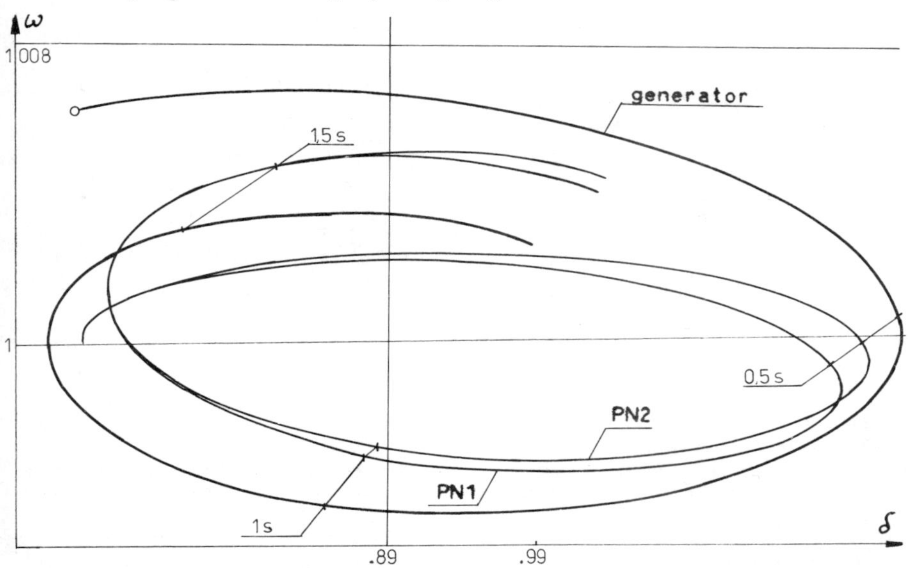

Fig. 2. Trajectories of synchronous generator and PN1, PN2 observers; $\lambda_1 = \lambda_2 = -5$, $a_1^* = 0.95\ a_1$, $x_1(t_o) = z_1(t_o) = z_1^*(t_o) = 0.69$, $x_2(t_o) = 1.005$, $z_2(t_o) = z_2^*(t_o) = 1.000$

when the difference between coefficients in the generator model and in the observer occurs. Fig. 2 shows trajectories of observers PN1 and PN2 even for small change of a_1 constant, denoted as a_1^* for both types of observers, do not track the synchronous generator trajectory.
In such case the only solution is application of the PI-observer . It allows to obtain estimation of the inaccessible state variable (in the sample system it is $x_2 \triangleq \omega$) withot an error. The PI-observer defined by equations (21)-(22) shows a robustness for a wide range of a model constants variation. On the fig. 3 an example transient response pattern is presented for $a_1^* = 0.75\ a_1$.

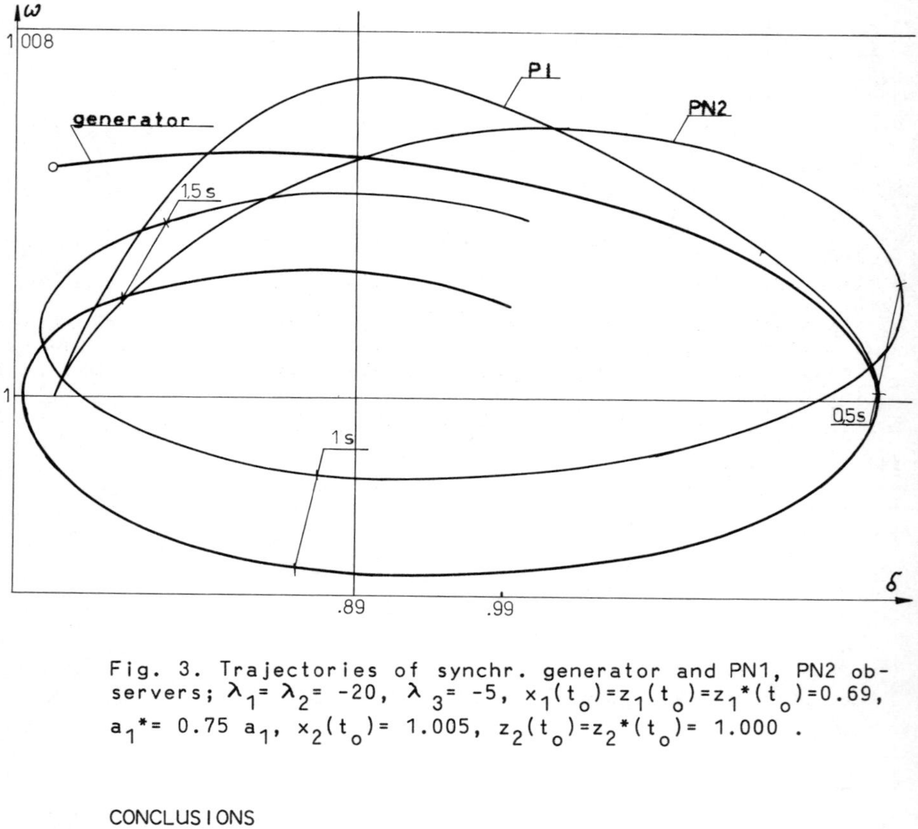

Fig. 3. Trajectories of synchr. generator and PN1, PN2 observers; $\lambda_1 = \lambda_2 = -20$, $\lambda_3 = -5$, $x_1(t_o) = z_1(t_o) = z_1^*(t_o) = 0.69$, $a_1^* = 0.75\ a_1$, $x_2(t_o) = 1.005$, $z_2(t_o) = z_2^*(t_o) = 1.000$.

CONCLUSIONS

Presented simulation results confirm that both types of non-linear proportional observers (i.e. PN1 and PN2) are stable in a vast region and for the wide range of eigenvalues. The PN2 observer has better properties because of smaller overregulations during transients (fig.1).
Any deviation of the observers' parameters values from the real ones in the synchronous generator causes a steady-state error of the inaccesible state vactor variable estimation for both PN1 and PN2 observers (fig.2).
The PI-observers proposed by authors assure good estimation even for big deviation of the a_1 constant in the sample system (fig.3) .

APPENDIX

The sample system data for turbogenerator (S_n = 432 MVA) (Mielczarski ,1986)

$\omega_b = 314.159$, $a_1 = 0.094$, $d = 0.877$, $P_m = 0.950$.

REFERENCES

Bestle, D., Zeitz, M. (1983). Int. J. Control, 38, 419-431 .
Koehne, M. (1982). Modelling and simulation of nonlinear and adaptive observers. Proceedings 10th World Congress, Montreal, Canada, August 8-13, 1982, 1-8 .
Kou, R., Elliott, S. and Tarn, J. (1975) . Inf. Control, 29, 204-216.
Li, C.W., and Tao, L.W. (1986) . Int. J. Control, 44, 1703-1713 .
Mielczarski, W. (1986) . The state observers of a synchronous generator. monography Technical Univ. of Łódź Press, vol. 84, (in Polish) , (1987) . Observing the state of a synchronous generator. Part I. Theory. Part II. Applications . Int. J. Control, Accepted for publication in 1987 .
O'Reilly, J. (1983) . Observers for Linear Systems, Academic Press, London .
Roitenberg, Y.Y. (1970) . SIAM J. Control, 8, 338-345 .
Thau, E. (1973) . Int. J. Control, 17, 471-479 .
Wojciechowski, B. (1978) . The analysis and synthesis of proportional-integrating observers of scalar linear continuous, stationary, linear systems. Ph. D. dissertation, Silesian Technical Univ. , (in Polish) .
Zeitz, M. (1979) . Nichtlinear Beobachter, Regelungstechnik, heft 8, 241-272 .

III. Robust, Fault-Tolerant and Intelligent Controllers

Towards a Global Approach for the Design of Reliable Control Systems

M. Mariton, P. Bertrand

Laboratoire des Signaux et Systèmes CNRS - ESE
Plateau du Moulton, 91190 Gif-sur-Yvette, France

0. Introduction

To achieve high performance requirements, modern spacecraft and aircraft require sophisticated control functions performed by digital fly-by-wire systems. A typical example is provided by the X-29 or Rafale aircraft which were made open-loop unstable to ensure higher manoeuvering capacities and which rely on feedback control to guarantee stability.

It is then easy to understand that such aircraft become vulnerable to incidents like actuator or sensor failure and that **fault tolerance** becomes a key Flight Control System property.

To obtain this property, the occurrence of failures must be included in the control design model, and the purpose of this paper is to introduce a suitable mathematical framework that reflects both the performance and reliability points of view. Using this framework, some preliminary results are reported towards a global approach which includes in the same analysis the three basic steps of a design, namely components selection, placement, and control law synthesis.

The rest of the paper is organized as follows. Starting from a simple duplex example, the proposed model is first introduced. It is then shown how the optimal control laws designed for this model can achieve both the reliability and performance objectives. Next, the selection and placement of failure-prone actuators is considered: a direct measure of actuator quality is defined as the attainable extremal value of an averaged cost function and an indirect measure, called the degree of controllability, is introduced through the notion of stochastic ε-controllability (it would also have been possible to consider sensors with measures like a degree of observability or the error variance of the optimal Kalman filter). Finally, an example illustrates the proposed ideas and some perspectives are outlined.

1. A model for the design of Reliable Control Systems

A basic ingredient of reliability is redundancy as analyzed by (Von Neumann 1956): to ensure a satisfactory performance level after the occurrence of some failures, it is necessary to have components organized in some redundancy levels. Consider the example of a duplex system

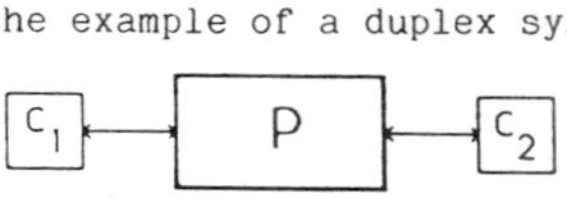

Figure 1 - Duplex Control System

Denoting by λ the failure rate and by μ the maintenance rate of C_1 and C_2, one can associate with Figure 1 four regimes, or modes,

$$\begin{array}{llllll} r=1 & C_1 & C_2 & r=3 & C_1 & \mathcal{C}_2 \\ r=2 & \mathcal{C}_1 & C_2 & r=4 & \mathcal{C}_1 & \mathcal{C}_2 \end{array}$$

and a matrix Π of transition rates to describe the random jumps of the regime $r(t)$ in $S=\{1,2,3,4\}$

$$\Pi=\begin{bmatrix} -2\lambda & \lambda & \lambda & 0 \\ \mu & -(\lambda+\mu) & 0 & \lambda \\ \mu & 0 & -(\lambda+\mu) & \lambda \\ 0 & \mu & \mu & -2\mu \end{bmatrix}$$

Mixing the dynamics of the controlled plant and this description of failures as a jump process, hybrid Jump Linear models are obtained: to the usual **plant state,** $x(t) \in \mathbf{R}^n$, one appends a discrete regime variable, called the **plant mode,** $r(t) \in S = \{1,2,\dots,N\}$ (N=4 for the above example).

The simplest model for the evolution of the plant mode and the plant state consists of a Markov chain on S and a linear ordinary differential equation:

$$\mathrm{Prob}\{r(t+\Delta) = j \,|\, r(t)=i\} = \begin{cases} \pi_{ij}\Delta + o(\Delta) & i \neq j \\ 1+\pi_{ii}\Delta + o(\Delta) & i=j \end{cases} \tag{1}$$

with $\Pi=(\pi_{ij})_{i,j=1,N}$ the matrix of transition rates,

$$\begin{cases} \dot{x}(t) = A(r(t))x(t) + B(r(t))u(t) \\ y(t) = C(r(t))y(t) \end{cases} \tag{2}$$

with $u \in \mathbf{R}^m$ and $y \in \mathbf{R}^r$ the control and observation vectors respectively. Matrices A, B and C depend on the mode $r(t)$ to describe the way failures affect the plant state dynamics.

Similar models were introduced in discrete-time by (Birdwell 1978, Chizeck 1982) and it is believed that (1), (2) provides a suitable mathematical framework for the analysis of reliability issues in control system design. This claim is illustrated in the following sections.

2. Dynamic reliability and stochastic optimality

The model (1), (2) incorporates a description of failures, and therefore optimal control laws based on this model should provide fault tolerance together with performance.

Consider for example an averaged quadratic cost function

$$J = \frac{1}{2} E\{ \int_{t_o}^{t_f} (x'Qx + u'Ru)dt \,|\, t_o, x(t_o), r(t_o)\} \tag{3}$$

where E denotes mathematical expectation and $Q(\geq 0)$ and $R(>0)$ are mode dependent weighting matrices $(Q=Q(r(t)),\ R=R(r(t))$ since the designer may affect different weightings to the control authority and output excursion depending on the failure regime.

It was shown by (Sworder 1969) and (Wonham 1971) that the control law that minimizes (3) when the plant state and mode are both measured is

$$u^*(t) = -R_i^{-1} B_i' \Lambda_i(t)\, x_i(t) \qquad \text{when } r(t)=i \tag{4}$$

for the Λ_i's, i=1 to N, solutions of coupled Riccati equations

$$i=1,N \begin{cases} -\dot{\Lambda}_i = A_i'\Lambda_i + \Lambda_i A_i - \Lambda_i B_i R_i^{-1} B_i' \Lambda_i + \sum_{j=1}^{N} \pi_{ij}\Lambda_j + Q_i \\ \Lambda_i(t_f) = 0 \end{cases} \tag{5}$$

From (4) and (5), it is clear that this optimal feedback has two desirable properties from a reliability point of view: **Reconfiguration** and **Anticipation.** Once the mode jumps from i to j, say following an actuator failure, the gain is switched from $-R_i^{-1}B_i'\Lambda_i$ to $-R_j^{-1}B_j'\Lambda_j$; this is the Reconfiguration property. Anticipation over the occurrence of failures is also obtained since the coupling term $(\sum\pi_{ij}\Lambda_j)$ in (5) accounts for eventual transitions to degraded regimes, the more likely a failure in mode i (π_{ij} large), the more it affects the current Riccati matrix Λ_i ($\pi_{ij}\Lambda_j$ large).

It was shown in (Mariton 1986a) how to accommodate structural constraints while preserving these qualitative properties: optimization was performed when either the state or the mode is not measured ($u^* = U(y,r)$ or $u^* = U(x)$) and a significant improvement of performance was obtained with respect to the stabilizing synthesis of (Siljak 1980).

3. Components selection and placement

Two approaches are proposed for the selection and placement of failure-prone actuators, based on direct and indirect measures. Sensors could also have been considered in a dual manner.

a. A direct approach

The minimal value of the cost (3) in a given class of admissible control can be interpreted as a direct measure of the quality of an actuator configuration (= selection + placement).

Fortunately, the choice of a configuration is very often restricted by some design constraints or commonsense a priori limitations, so that the problem can be restricted to a few concurrent possibilities.

Consider for example the case where one has to decide between two choices, say A and B. Two optimal syntheses are then performed

$$J^*_{A \text{ or } B} = \text{Min } (3) \tag{6}$$

s.t. (1),(2) and actuator configuration A or B; and the decision is based on the attained values J^*_A and J^*_B: one will prefer configuration A to configuration B when $J^*_A < J^*_B$.

A related approach is (Montgomery 1983): the performance is analyzed a posteriori using a Jump Linear (JL) model like (1), (2) but the control synthesis is a classical Linear Quadratic (LQ) one. Thus the control law obtained does not include a description of failures and there appears a twist between the design phase (deterministic LQ models) and the analysis phase (stochastic JL model). The approach proposed above seems more consistent since it uses the JLQ model throughout; moreover, the example presented in section c below demonstrates that the best configuration choice significantly depends on the control law used, thus making the twist in Montgomery's proposition only more sensitive.

A forthcoming paper (Mariton 1987) will be entirely devoted to this direct approach, and we now examine a complementary indirect approach in more detail.

b. An indirect approach

The notion of stochastic ε-controllability introduced in (Mariton 1986b) can be used as an indirect measure of the quality of an actuator configuration. The definition is as follows:

"An initial state x_o is said to be stochastically ε-controllable with probability ρ over the time interval $[t_o,t_f]$ if there exists a control law $u(t,x)$ such that

$$\text{Prob}\{ \| x(t_f) \|^2 \geq \varepsilon \mid x(t_o) = x_o\} \leq 1-\rho \tag{7}$$

where $x(t_f)$ indicates the value at $t=t_f$ of the trajectory starting at x_o for $t=t_o$ and under the control u".

It was possible to obtain sufficient conditions for controllability of Jump Linear models:
"An initial state x_o of the Jump Linear system (1),(2) is stochastically ε-controllable with probability ρ over $[t_o,t_f]$ if the following conditions are satisfied

(i) There exist N matrices $K_i(t)$, defined over $[t_o,t_f]$, bounded, symmetric and positive definite, solutions of the set of Bernoulli equations

$$i=1,N \quad \dot{K}_i+A_i'K_i+K_iA_i-K_iB_iB_i'K_i + \sum_{j=1}^{N} \pi_{ij}K_j=0, \quad K_i(t_f)=In/\alpha_i \tag{8}$$

for $\alpha_i>0$, i=1 to N.

(ii) x_o satisfies $\quad x_o' K_i(t_o)x_o \leq (1-\rho)\ \varepsilon/\alpha_i$ (9)

when $r(t_o)=i$".

The proof can be found in (Mariton 1986b), and consists of explicitly exhibiting the control law required by the definition, using the martingale inequality to obtain the bound (7).

Here attention is focused on condition (ii). For fixed ε, ρ and α_i, i=1 to N, and for the K_i's solutions of (8), (9) defines an ellipsoid in $\mathbf{R}^n$: all the x_o which are inside the ellipsoid characterized by

$$x_o' K_i(t_o) = (1-\rho)\varepsilon/\alpha_i \tag{10}$$

can be driven in an ε-neighbourhood of the origin, with probability ρ, over $[t_o,t_f]$. The size of this ellipsoid thus measures some "relative controllability" of the system: the larger the ellipsoid the more controllable the system.

A related approach is (Van der Velde 1982) where a deterministic minimum energy problem was used to obtain a Degree of Controllability from the size of the ellipsoid of states that can be driven to zero with a given energy in a fixed time. The originality of the proposed approach is that it includes a description of failures: through the K_i, i=1 to N, solutions of (8), the Jump Linear model (1), (2) influences the size of the ellipsoid (10). However, to go from (10) to a more aggregated measure of controllability, one can follow a path similar to (Van der Velde 1982).

Denoting by V_E the volume of the ellipsoid and V_S the volume of the largest space contained in the ellipsoid, the Degree of Controllability DC is defined as

$$DC = (V_S + (V_S/V_E)(V_E-V_S))^{1/n} \tag{11}$$

The V_S term is predominant; it translates the smallest distance from which x_o can be driven into an ε-ball around the origin with probability ρ in a fixed time t_f. When there is a balance among the actuators ($V_E=V_S$), V_S completely characterizes DC. However, a second term is added to increase DC when there exist some directions in $\mathbf{R}^n$ which are more controllable than others. The volume between the sphere and the ellipsoid V_E-V_S is scaled by V_S/V_E so that the relative measure given by DC is limited to between 0% and 200%. The n-th root finally ensures that DC is proportional to the linear dimensions of the state space.

The above definition for DC is, of course, somewhat arbitrary, and other expressions could have been chosen. The difficulty here is to aggregate the information contained in the n-th dimensional ellipsoid (n can be quite large) into a scalar measure. Some indications are inevitably lost in such a process and, together with DC, other measures should be deduced from (10). For the example studied below (§c) the least controllable direction was computed to give another valuable indication apart from DC.

The volumes V_E and V_S are given in terms of the eigenvalues of $K_i(t_o)$ when $\pi(t_o)=i$: if $\lambda_1,\lambda_2,\ldots,\lambda_n$ are the eigenvalues of $K_i(t_o)$ and

$\lambda_{min}=\text{Min}\{\lambda_1,\lambda_2,\ldots,\lambda_m\}$, one gets $V_E = a\sqrt{\prod_i \lambda_i}$ and $V_S = a\sqrt{\lambda^n_{min}}$ (12)

with $d=(1-\rho)(\varepsilon/\alpha_1)$, the scale factor a is $a=\pi d$ for $n=2$, $a=\frac{4\pi}{3}d^{3/2}$ for $n=3$ ...

c. Example

Approaches a and b will be illustrated for the flexible system of Figure 2 whose force actuators (f_1 and f_2) can be mounted on any of the three masses.

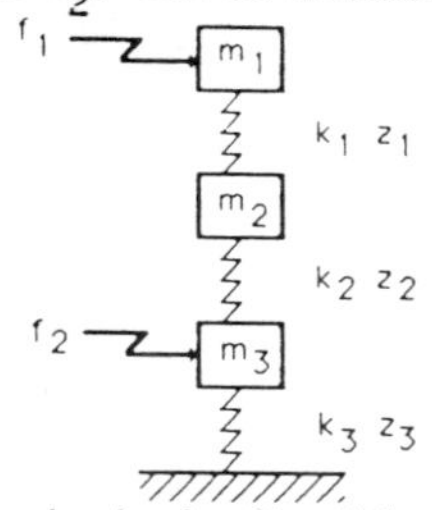

Figure 2 - A simple flexible structure

Attaching numerical values to the model parameters m_i, k_i, z_i, i = 1 to 3, one obtains a state-space model as

$$\dot{x} = \begin{vmatrix} 0. & 1. & 0. & 0. & 0. & 0. \\ -20. & -0.5 & 20. & 0. & 0. & 0. \\ 0. & 0. & 0. & 1. & 0. & 0. \\ 10. & 0. & -20. & -0.5 & 10. & 0. \\ 0. & 0. & 0. & 0. & 0. & 1. \\ 0. & 0. & 10. & 0. & -20. & -0.5 \end{vmatrix} x + \begin{vmatrix} 0. & 0. \\ 1. & 0. \\ 0. & 0. \\ 0. & 0. \\ 0. & 0. \\ 0. & 1. \end{vmatrix} u \quad (13)$$

with $x'=[x_1,\dot{x}_1,x_2,\dot{x}_2,x_3,\dot{x}_3]$ and $u'=[f_1,f_2]$ where x_i is the position of the ith mass. For a different configuration of the actuators the control matrix in (13) is modified by moving ones and zeros, and an actuator failure produces a zero at the corresponding entry.

The average life-time of the ith actuator is $1/\pi_{ii}$ and no maintenance is allowed. Excluding the catastrophic failure case $(B(r=4)=0)$, the matrix of transition rates for the three remaining models is

$$\Pi = \begin{vmatrix} -(\pi_1+\pi_2) & \pi_1 & \pi_2 \\ 0. & 0. & 0. \\ 0. & 0. & 0. \end{vmatrix} \quad (14)$$

so that the initial mode $r(t_o)=1$ is transient with two failed absorbing modes $r=2$ and $r=3$.

To illustrate approach a, consider the two configurations in Figure 3

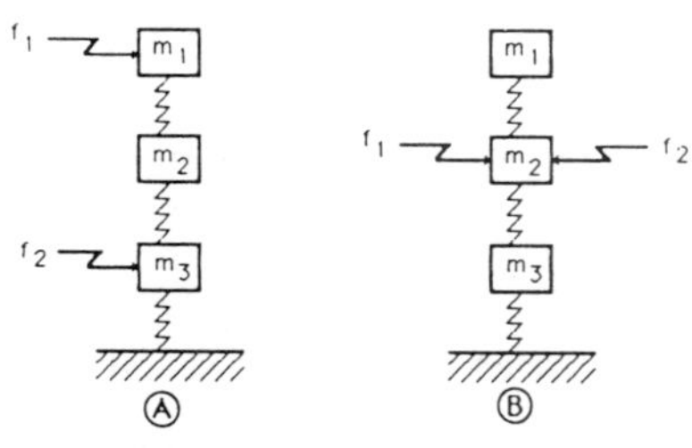

Figure 3 - Two possible actuator configurations

For a state feedback with switching gains, the optimal solution (4) leads to a minimal cost index expressed by

$$J^* = \frac{1}{2} x'_o \Lambda_1(t_o) x_o \qquad (15)$$

where the Λ_i's, i=1 to 3, are solutions of (5). Through the JL model (1),(2), and the couplings in (5), this value of J^* contains the influence of failures and of the actuator configuration.

For the two choices of Figure 3, the cost (15) is plotted for different values of π_1 and π_2 as a function of t_f on Figure 4 to 6 when x_o is distributed over a sphere of radius $\sqrt{2}$ (J^*=trace Λ_1).

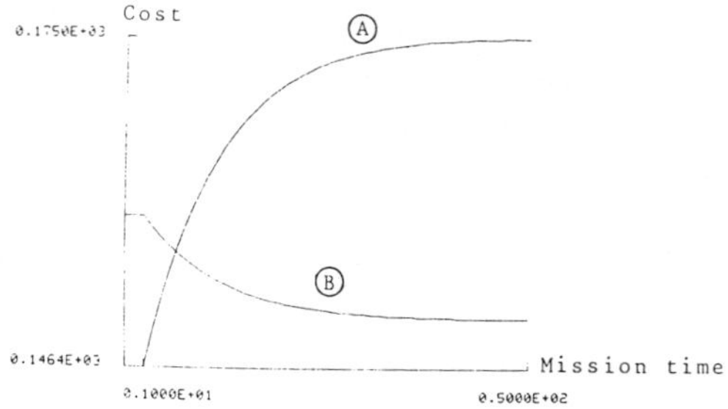

Figure 4 - Configuration A / B , π_1=1. and π_2=0.

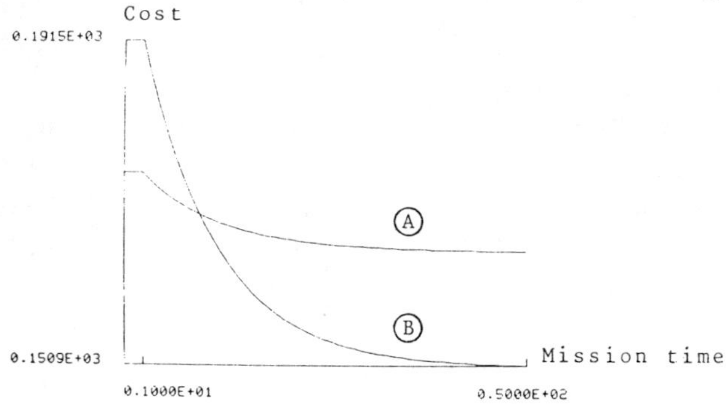

Figure 5 - Configuration A / B , π_1=2. and π_2=0.5

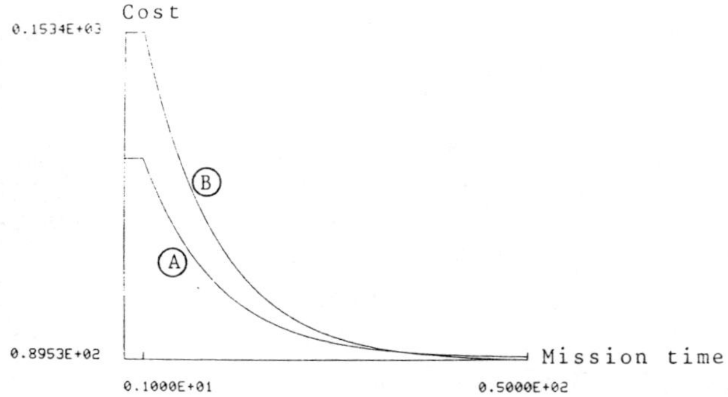

Figure 6 - Configuration A / B , π_1=1. and π_2=1.

The striking fact is that the best choice depends on the mission duration t_f: for large t_f $J^*_B < J^*_A$ and for small t_f $J^*_A < J^*_B$. The interpretation is that for short missions failures are unlikely, and one should prefer to spread the actuators over the flexible structure to improve performance (choose A), while for long missions failures will eventually occur, so that actuators clustered in the middle of the structure are preferable to improve redundancy, hence reliability (choose B).

The power of the proposed modelization is thus demonstrated: thanks to the model (1),(2) the cost function (3) reflects both the Performance and Reliability requirements and gives a quantitative support for the trade-off between these two aspects of the control system design.

The influence of the control law actually used on the choice of a configuration was overlooked in (Montgomery 1983). Its importance is shown in Figure 7, where the overcost $J^*_B - J^*_A$ is plotted as a function of t_f for the proposed JLQ regulator and the LQ regulator of Montgomery.

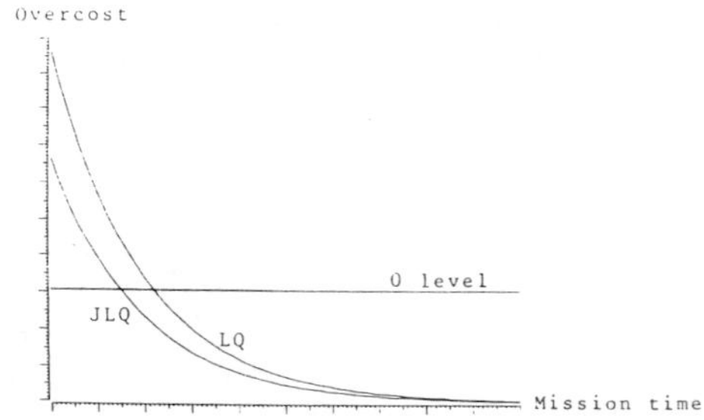

Figure 7 - Influence of the JLQ/LQ synthesis

The point to stress now is that the mission time after which the best choice becomes B is shorter with the JLQ regulator. Again this influence reflects the modelization of failures in the JL setting: since the JLQ regulator "knows", through the model (1),(2), that failures will occur, it favours the configuration where actuators are most redundant for shorter missions than the LQ regulator which is "blind" to failures.

For the illustration of approach b, consider three concurrent configurations as in Figure 8.

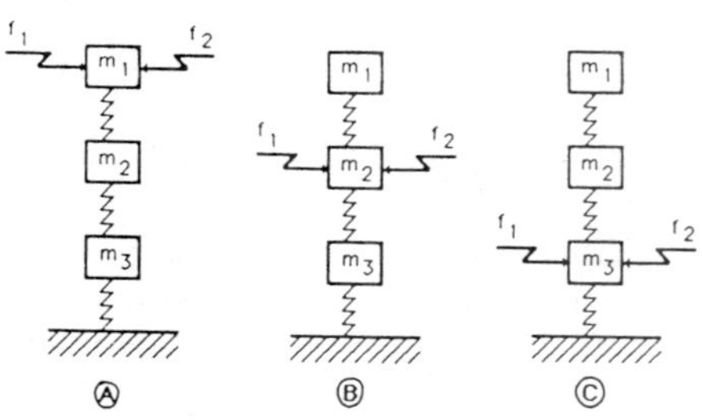

Figure 8 - Three double-actuator configurations

Matrices K_1, K_2 and K_3 were computed for configurations A , B and C with $\rho=0.9$, $\varepsilon=0.1$, $\alpha_1=\alpha_2=\alpha_3=0.1$ and $t_f=1$, and from the value of $K_1(t_o)$ the Degree of Controllability DC given by (11) is displayed in the following table, together with the least controllable direction.

Configuration A	VE = 1383.2
Least controllable Direction (0.02, 0.02, 0.06, 0.06, 0.77, 0.63)'	VS = 114.8 DC = 2.46

Configuration B	VE = 1384.3
Least controllable Direction (0.62, 0.59, 0.01, 0.10, 0.61, 0.60)'	VS = 153.5 DC = 2.60

$\pi_1 = \pi_2 = 1.$

Configuration C	VE = 1379.5
Least controllable Direction (0.75, 0.61, 0.19, 0.13, 0.08, 0.03)'	VS = 114.6 DC = 2.45

Results are qualitatively similar to the results obtained with the direct approach. A very valuable piece of information also proposed is the least controllable direction. In some applications, physical constraints will impose the initial state to belong to a subspace of $\mathbf{R}^n$, and it will be important to ensure that the chosen configuration keep the least controllable direction as orthogonal as possible to this subspace.

4. Conclusion and perspectives

The reported results illustrate the efficiency of the proposed model (1),(2), and it is believed that they constitute the initial steps towards a global approach in the design of reliable control systems. The three basic parts of the design, components selection, components placement, and synthesis of the control law itself, are thus integrated to reflect simultaneously the Performance and Reliability points of view.

Future research will consider refinement of this formulation. For example, continuous degradations of components will be described by a Markov chain on a continuum, and performance will be evaluated beyond the mean value of the cost function to obtain the distribution of the realized cost.

References

BIRDWELL, J.D., 1978, "On reliable control system designs", Ph.D. Dissertation, Mass. Inst. Technology, report n° ESL-TH-821.

CHIZECK, H.J., 1982, "Fault-tolerant optimal control", Ph.D. Dissertation, Mass. Inst. Technology, report n° 903-23077.

MARITON, M. and P. BERTRAND, 1986a, "Improved multiplex control systems: dynamic reliability and stochastic optimality", Int. J. Control, vol.44, n°1, pp.219-234.

MARITON, M., 1986b, "On controllability of linear systems with stochastic jump parameters", IEEE Trans. Automatic Control, vol.AC-31, n°7, pp.680-683.

MARITON, M. and P. BERTRAND, 1986c, "Reliable flight control systems: components placement and feedback synthesis", submitted for publication.

MONTGOMERY, R.C., 1983, "Reliability considerations in the placement of control systems components", Proc. AIAA Guidance and Control Conf., Gatlinburg.

SILJAK, D.D., 1980, "Reliable control using multiple control systems", Int. J. Control, vol.31, n°2, pp.303-329.

SWORDER, D.D., 1969, "Feedback control for a class of linear systems with jump parameters", IEEE Trans. Automatic Control, vol.AC-14, n°1, pp.9-14.

VANDERVELDE, W.E. and C.R. CARIGNAN, 1982, "A dynamic measure of controllability and observability for the placement of actuators and sensors on large space structures", Mass. Inst. Technology, report n° SSL-2-82.

VON NEUMANN, J., 1956, "Automata studies", C.E. Shannon and E.F. Moore, eds., Annals of Mathematics, n°34, Princeton University Press, Princeton.

WONHAM, W.M., 1971, "Random differential equations in control theory", in Probabilistic Methods in Applied Mathematics, A.T. Bharucha-Reid, ed., vol.2, Academic Press, New York.

Reliable Controllers for Serially Interconnected Systems

L. Bakule

Institute of Information Theory and Automation, Czechoslovak Academy of Sciences, 182 08 Prague 8, Czechoslovakia

ABSTRACT

A reliable stabilization technique using a multi-controller configuration with passive redundancy is derived, where the serial interconnection structure of subsystems is employed to decompose the system into the individual subsystems. The problem of designing of a pair of decentralized controllers is considered in such a way that a controller acting alone or their sum stabilizes the plant. Supposing strongly stabilizable plant it is derived that using decomposition the decentralized controllers can be designed both for fully decomposed and for nondecomposable dynamical requirements. One of two controllers can be designed arbitrarily. The stabilization technique is applied to reliable robust tracking for serially interconnected systems.

KEYWORDS

Linear systems; reliable stabilization; serially interconnected systems; multi-controller configuration; robustness; decomposition; decentralized controllers.

INTRODUCTION

Feedback control problems for interconnected plants must often be solved in a decentralized way, i.e. the controller consists of several control stations which operate independently on the corresponding subsystems of the plant. The design of such controllers represents a problem because the solution must satisfy the struc-

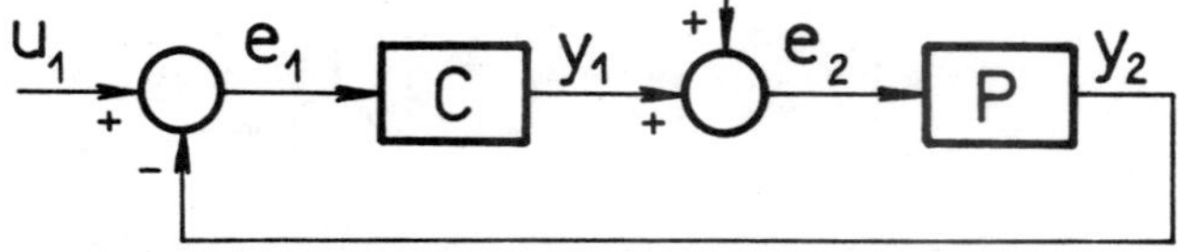

Fig. 1. Unstable plant P stabilized with controller C.

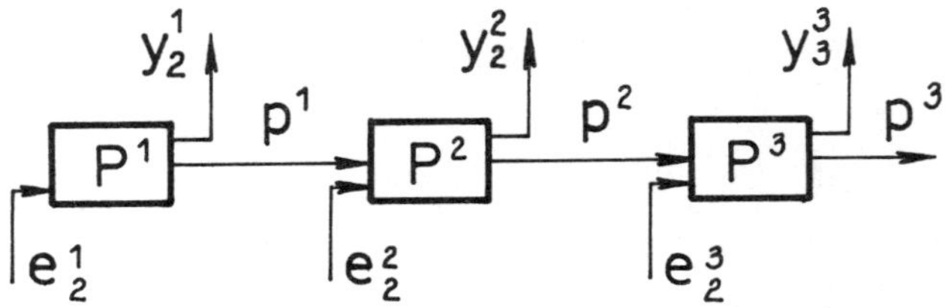

Fig. 2. Serially interconnected subsystems.

tural constraint on the controller structure. This is true even if the plant is completely known. The reduction of complexity and dimensionality is one of the main problems in the decentralized control design. If the plant has subsystems-interconnection structure, a complete design problem can be decomposed into subproblems, so that the controllers can be designed nearly independently by means of subsystems only. Such a decomposition has been described for instance in Bakule and Lunze (1985, 1986).

The problem of determining a feedback controller C for unstable plant P to stabilize it has been studied for many years, see Fig.1. This paper deals with the dual of the simultaneous stabilization, i.e. reliable stabilization which is considered for serially connected subsystems, see Fig. 2. A multi-controller configuration with passive redundancy is used for each subsystem P^i, where a plant has the form $P=P^1...P^N$, see Fig. 3. Considering a plant P parallel structure of sontrollers C_1, C_2 such that the closed loop system is stable, as well as either one of controllers is equal to zero. The motivation for this problem statement is as follows: The problem was formulated by Siljak (1978, 1980) and used for reliable stabilization by Vidyasagar and Viswanadham (1985). In the normal mode, both controllers are operational and the system is stable. Should either controller fail the system is still stable, even though other properties mifht be affected adversly. Such scheme is reliable against a single controller failure. The failed controller is replaced by the zero controller, i.e. the zero transfer matrix plus a constant (step) disturbance at the output. The concept of a multi-controller configuration with passive redundancy differs from the concept of a multi-controller configuration with active redundancy, see e.g. Vidyasagar and Viswanadham (1985).

The objective of this paper is to present conditions on P employing the serial connection structure that ensure the existence of decentralized controllers C_1^i, C_2^i for all i. The results state that suit-

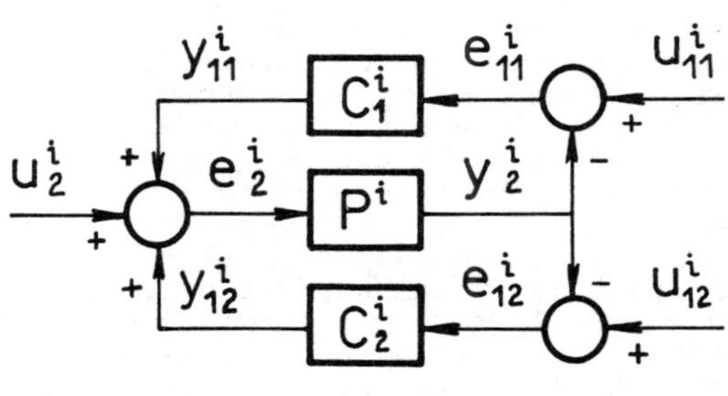

Fig. 3. A multi-controller configuration with passive redundancy for P^i.

able controllers C_1^i, C_2^i exist if P^i are strongly stabilizable, i.e. P^i can be stabilized using a stable controller.

PROBLEM FORMULATION

Suppose a given unstable plant in the form $P=P^1 \ldots P^N$, where P^i are strongly stabilizable, i.e. they can be stabilized by a stable controller. Denote R(s) the field of rational functions with real coefficients, S denote the subset of R(s) consisting of proper stable rational functions. M(R(s)) denote the set of matrices whose elements belong to R(s), and M(S) introduce analogously. Suppose compatible dimansions P, $C \in M(R(s))$ and denote $e=(e_1, e_2)^T$, $u=(u_1, u_2)^T$ in Fig. 1

$$e = H(P, C)u ,$$

where

$$H(P, C) = \begin{bmatrix} (I + PC)^{-1} & -P(I + CP)^{-1} \\ C(I + PC)^{-1} & (I + CP)^{-1} \end{bmatrix} .$$

The closed loop system is stable and C stabilizes P if $H(P,C) \in M(S)$ P is strongly stabilizable if the intersection $S(P) \cap M(S)$ is nonempty.

The Decentralized reliable stabilization problem (DRSP) can be formulated. Suppose $P \in M(R(s))$ for $P=P^1 \ldots P^N$, find if possible the decentralized controllers C_1^i, C_2^i such that

(i) $C_1^i \in S(P^i)$,

(ii) $C_2^i \in S(P^i)$,

(iii) $C_1^i + C_2^i \in S(P^i)$, and

(iv) supposing C_1^i, C_2^i as $C_j^i = (\tilde{D}_{cj}^i)^{-1} \tilde{N}_{cj}^i$, where $\tilde{D}_{cj}^i$, $\tilde{N}_{cj}^i$ are for each i left-coprime for j=1, 2, then D_{cj}^i are right-coprime.

The conditions (i)-(iii) mean that each of controllers and their parallel combination stabilize P^i. The condition (iv) means that the controllers have no internal duplication of poles in the closed right half-plane, i.e. this condition says that the parallel combination is stabilizable and detectable.

Further, extend the reliable stabilization technique to reliable regulation for serially interconnected subsystems.

PROBLEM SOLUTION

Because of the serial connection of the subsystems, the problem of designing the complete decentralized controller is decomposed into N subproblems. This decomposition can be performed either by the decomposition of dynamical requirements on the global system into the independent subproblems (Bakule and Lunze, 1985), or the dynamical requirements cannot be decomposed into independent subproblem requirements (bakule and Lunze, 1986). Because these dynamical requirements concern the closed loop stability, the asymptotic regulation in the closed loop system and the dynamic I/O-behaviour, the problem solution is restricted on the existence conditions for reliable stabilization only. It corresponds with the decomposition of dynamical requirements on the global serially interconnected system into the dynamical requirements on the individual subsystems. Further, the case when dynamical requirements cannot be decomposed concerns mainly the I/O-behaviour or robustness. This can be considered for serially connected subsystems as a specific tracking problem with a given reference signal from the (i-1)-th subsystem when considering the i-th subsystem. Denote $P_c=P_c^1...P_c^N$ the closed loop system $P=P^1...P^N$. Concerning the requirements on the closed loop system the followimg Lemma can be used.

<u>Lemma 1.</u> The closed loop system $P_c=P_c^1...P_c^N$ satisfies the requirements: 1) stability, 2) asymptotic regulation iff all isolated closed loop subsystems P_c^i are stable.

Proof. It is given in Bakule and Lunze (1986).

Lemma 1 point out that the solution of the DRSP is transformed into the solution of the reliable stabilization problem on the subsystem level. Following the decomposition principle (Callier and Desoer, 1982) which says that if C_1^i stabilizes P^i and C_2^i stabilizes $P^i(I+C_1^iP^i)^{-1}$ implies that $C_1^i+C_2^i$ stabilizes P^i. Therefore, introduce The Decentralized modified reliable stabilization problem (DMRSP). Suppose $P\in M(R(s))$, $P=P^1...P^N$ is a serially connected plant, find if possible pairs of decentralized controllers such that

(i´) $C_1^i \in S(P^i)$

(ii´) C_2^i stabilizes both P^i and $P_1^i=P^i(I+C_1^iP^i)^{-1}$.

Following the results by Vidyasagar and Viswanadham (1985) for centralized reliable stabilization, it holds for the DMRSP the following.

<u>Theorem 1.</u> Given a unstable plant $P=P^1...P^N$ M(R(s)) with serial connection structure, the DMRSP has a solution iff all plants P^i are strongly stabilizable, Moreover, if P^i are strongly stabilizable, then for any $C_1^i \in S(P^i)$ M(S), there exists a corresponding C_2^i such that (i´) and (ii´) hold.

Proof. Because of the serial connection structure of the plant P, the assumption of strongly stabilizable P^i and the Lemma 1, the global problem is decomposed into the independent subsystems problems. The assertion for independent subsystem problems is proved in Vidyasagar and Viswanadham (1985).

Let us deal with reliable robust tracking for serially connected plants. Tracking a given reference signal means that one-directional effect occurs. Suppose that for the i-th plant P^i the feedback subsystem $P_c^{(i)}=P_c^1...P_c^{(i-1)}$ with an output $r^{(i)}$is a system its output is the reference signal into the subsystem P^i, $P_c^{(i)}=(\tilde{D}_r^i)^{-1}\tilde{N}_r^i$, where $\tilde{D}_r^i$, $\tilde{N}_r^i$ are left-coprime. To present the results for reliable tracking a decentralized multi-controller configuration for serially connected plants introduce the basic concepts.

Given a plant $P^i \in M(R(s))$, a basic neighborhood of P^i is a set $N(P^i)$ in M(R(s)) of the form

$$N(P^i) = \left\{ N_1^i(D_1^i)^{-1}: \quad \left\| \begin{bmatrix} N_1^i - N^i \\ D_1^i - D^i \end{bmatrix} \right\| < \varepsilon \right\},$$

where (N^i, D^i) is a r.c.f of P^i. A robustness against perturbations in P^i means that there is N(Pi) such that any property (stability, tracking) holds for all plants P^i in $N(P^i)$. The controller C^i solves the robust tracking problem for P^i if

(i) $C^i \in S(P^i)$,

(ii) $(I+P^iC^i)^{-1}(\tilde{D}^i_r)^{-1}\tilde{N}^i_r \in M(S)$,

(iii) the properties (i), (ii) are robust against perturbations in P^i.

The following result is used for reliable robust tracking. It is proved in Vidyasagar and Viswanadham (1985).

Lemma 2. Let $(\tilde{D}^i,\tilde{N}^i)$ ba any l.c.f. of P^i and let α^i_r denote the largest invariant factor of $\tilde{D}^i_r$. Suppose $\alpha^i_r I$, $\tilde{N}^i$ are right-coprime. Then C^i solves the robust tracking problem iff $\alpha^i_r C^i$ stabilizes P^i/α^i_r. All solutions are given by $(\alpha^i_r)^{-1}S(P^i/\alpha^i_r)$.

The robust tracking can be refomulated as a stabilization of a modified plant. It means that the reference input $(\tilde{D}^i_r)^{-1}\tilde{N}^i_r$ is applied to both controllers C^i_1, C^i_2 with modified plant P^i.

Theorem 2. Suppose a plant $P=P^1...P^N$ and for P^i the reference input generator $(\tilde{D}^i_r)^{-1}\tilde{N}^i_r$ and strongly stabilizable plant P^i/α^i_r, where α^i_r is the largest invariant factor of $\tilde{D}^i_r$. Suppose further that C^i_1 solves the robust tracking problem for P^i. Then there exist a C^i_2 such that C^i_1, C^i_2 together solve the reliable robust tracking problem for P^i. If this problem is solved for all i it is solved for the plant P.

Proof. It follows from the decomposition of the robust tracking problem for the plant P into the robust tracking problems for P^i, from the Theorem 1 and Lemma 2 immediately. For serially connected subsystems into the plant P, the result is obtained by the sequential application of the results for the plant P^i from the i-th to the (i+1)-th subsystem for all i.

CONCLUSION

The problem of designing a pair of decentralized controllers for a given serially connected subsystems acting either alone or together and stabilizing the stongly stabilizable plant is studied. The decomposition of the reliable stabilization and robust tracking problems into the independent subproblems enables to present the procedures to solve these problems at the subsystem level only.

REFERENCES

Bakule, L., and J. Lunze (1985). Completely decentralized design of decentralized controllers for serially interconnected systems. 30th Internat. Wissenschaft. Kolloquium ., 1 , Ilmenau (GDR), 15-18.

Bakule, L., and J. Lunze (1986). Sequential design of robust decentralized controllers for serially interconnected systems. Problems Contr. Inform. Theory ,Vol. 15(2), 167-177.

Callier, F. M., and C.A.Desoer (1982). Multivariable Feedback Systems. Springer, Berlin.

Hodžič, M., and D.D. Šiljak (1984). An LQG design for large sparse systems. Proc. 9th World IFAC Congress, vol. 10, Budapest, 293-298.

Minto, K. D., and M. Vidyasagar (1986). A state-space approach to simultaneous stabilization. Control-Theory and Advanced Technology, Vol. 2, No.1, 39-64.

Özguner, Ü., and W. R. Perkins (1978). Optimal control of multilevel large-scale systems. Internat. J. Control, vol 28, No.6, 967-980.

Šiljak, D. D.(1978). On reliability of control. Proc. Conf.Dec. Control, San Diego, 687-694.

Šiljak, D. D.(1980). Dynamic reliability using multiple control systems. Internat.J.Control, vol.31, 303.

Vidyasagar, M., and N. Viswanadham (1982). Algebraic design techniques for reliable stabilization. IEEE Trans.Aut.Control, vol. AC-27, 1085.

Vidyasagar, M., and N. Viswanadham (1985). Reliable stabilization using a multi-controller configuration. Automatica, Vol.21, 599-602.

Multivariable Control Systems Design: A case Study of Robust Control of Nuclear Power Plants

Dj. B. Petkovski

Fac. of. Tech. Sciences, V. Vlahovića 3, Novi Sad, Yugoslavia

ABSTRACT

The computer aided robust multivariable control system design procedure is given. The procedure allows the possibility of adding extra controllers which can be used to improve the robustness characteristics of the nominal system. New robustness results in output feedback control system designs are presented which lead to reduction of conservatism in robustness tests. A mathematical model of a real physical system, a nuclear power plant, is used to demonstrate the power of the proposed methodology for stability robustness analysis and robust multivariable control systems design.

KEYWORDS

Output feedback control; time domain robustness criteria; design of robust control systems; stability robustness; nuclear power plant.

INTRODUCTION

Theoretical models used for controller design are most often simplified by neglecting nonlinearities, time-varying parameters and higher order dynamics. Another important theme in system theory is the preservation of various system-theoretic properties in the face of perturbations in the system model. In this light the problem of multivariable closed-loop system analysis and design in the presence of the process perturbations as well as modelling uncertainties, has attracted much recent attention.

One of the most basic issues that arise in this class of problems is the stability robustness of control system, i.e., its ability to maintain stability in the face of uncertainty and perturbations. This, of course, is due to the fundamental importance of the issue. A practical control system is usually unacceptable if small system perturbations and modelling errors result in system unstability. If the robustness evaluation come up with embarassingly small marginess, a design adjustment procedure to improve robustness margins would be very desirable. The first step in robust controller design for multivariable control systems is the development of nonconservative criteria for robustness analysis. With few exeptions, recently proposed methods to assess stability robustness of multivariable control systems are well suited for weakly structured uncertainties because they require only a matrix norm to bound the uncertainty. In other words, in these approaches the class of "allowable perturbations" is completely instructured except for the definition of the applicable norm. Such robustness criteria tend to be very conservative for highly structured uncertainties because they fail to exploit the addition structure.

To overcome these difficulties new approaches have been proposed in the literature (see for example Lehtomaki et al. 1981, Postlethwaite et al. 1981, Yamen et al. 1984). However, these techniques have been proposed in the frequency domain. Having in mind that most of the modern methods for multivariable control system design are based on the time domain methodology and use state space mathematical models, the time domain robustness criteria are usuallu better suited for this task. In this paper it is shown that the usefulness of the time domain robustness test for physical systems can be greatly enhanced by a judicious choice of robustness criteria combined with physical knowledge of specific modelling errors and possible perturbations, using the state-space model of the system.

Simplified mathematical models give rise to the elegant classical as well as modern controller design procedures, but in practice, the implementation of control systems is still concerned with tuning of control system and over-night redesign procedures, what could happend because of varying nature of open-loop system dynamics. In addition, as mentioned, if the robustness evaluation come up with embarassingly small margins, a design adjustment procedure to improve robustness margins would be very desirable. In this paper, a new procedure for improving the feedback properties, including robustness, of multivariable control system, is also proposed. The procedure allows the possibility of adding extra controllers which can be used to "fine-tune" the closed loop behaviour of the nominal control system.

To illustrate the concepts developed in this paper, a 15-order mathematical model of a nuclear power plant is used throughout.

PROBLEM FORMULATION

It is assumed that the process to be controlled can be adequately represented in the region of nominal operation point, by the controllable linear state model

$$\dot{x}(t) = Ax(t) + Bu(t) \tag{1}$$

where $x(t)\varepsilon R^n$ is a state vector, and $u(t)\varepsilon R^m$ is a control vector. The information available to the controller is assumed to be

$$y(t) = Cx(t) \tag{2}$$

where $y(t)\varepsilon R^r$ is an output vector.

The control u(t) is assumed to be a direct feedback from the output y(t), namely,

$$u(t) = Ey(t) \tag{3}$$

where E is a time-invariant gain matrix.

There are straightforward methods to find output feedback control law in order to achieve satisfactory performances of the nominal closed-loop system. The precise meaning of the term "satisfactory" depends upon the actual application. One rudimentary criterion is to require that the closed-loop poles be at preassigned locations in the complex plane. The pole-assignment approach has been studied by several researchers (see e.g. Fallside, 1977). In more complex situations, it is usually necessary to resort to more sophisticated design procedures based on optimal control theory. The most common optimal control criterion functions are based simply on the minimization of time integrals of quadratic forms in the state (or output) and input variables (see e.g. Kvakernaak, 1972).

We assume that the system, eqns. (1)-(3), is stabilizable and that by one of the above approaches the feedback matrix E has been selected so that the nominal closed loop system is asymptotically stable. In this case, we can define the following Lyapunov matrix equation.

$$(A+BEC)^T P+P(A+BEC) = -Q \qquad (4)$$

where Q is positive definite matrix, $Q=Q^T>0$.

As known, mathematical models are usually inexactly identified, and at best an approximation of reality. In addition, it is a common procedure in engineering practice to work with mathematical models that are simpler, but less accurate, than the best available models of a given physical process. Finally, the systems themselves are subject to unpredictable changes with time, so that their is always a discrepancy between the physical reality and the mathematical models.

The class of systems considered throughout this paper is a perturbed version at the nominal system, eqns. (1)-(3) satisfying,

$$\dot{x}(t) = Ax(t) + Bu(t) + \Delta Ax(t) + \Delta Bu(t) \qquad (5)$$

$$u(t) = ECx(t) \qquad (6)$$

where A, B, E and C are the same as in the nominal system; so that all modelling errors and parameter variations are lumped into the matrices ΔA and ΔB.

When modelling errors and large parameter variations are presented in the system dynamics, then generally nothing can be inferred about the solution of the changed situation, and it is clear that a satisfactory notion of robustness for multivariable dynamic feedback systems must be able to characterize the magnitude of the arbitrary perturbations which may be tolerated without unstability.

As in the case of observers of Kalman filters implementation as well as decentralized control systems (Petkovski 1984, 1985), for output feedback designs no guaranteed robustness properties hold. Because the robustness evaluation can come up with embarassingly small marginess, a "designer adjustment procedure" to improve robustness would be very desirable.

In what follows both isuess, robustness evaluation and design of robust output feedback control systems will be addressed.

ROBUSTNESS CHARACTERIZATION

There are two main approaches which have been applied to robustness characterization of multivariable control systems:

(i) the frequency domain approach (e.g. Lehtomaki et al, 1981, Kouvaritakis, 1985, which is based on the transfer function representation of a system, and

(ii) the time domain approach (e.g. Patel et al, 1980, Petkovski 1984, 1985, Yedavallie, 1985), which is based on a state space representation of a system.

This section will study the robust stability problem for a state space representation of a system using a time domain approach. The above problem has been extensively studied for the case of unstructured perturbations (only a bound on the norm of the perturbation matrices ΔA and ΔB, eqn. (5) are given) as well as structured perturbations (the structure of perturbations in ΔA and ΔB is specified and the bounds on such structured perturbations are given).

For example, the following recent result has been obtained for the unstructured case:

Theorem 1. (Patel et al, 1980) The perturbed system (5) and (6) is stable if

$$\bar{\sigma}\,(\Delta A + \Delta BEC) < \frac{1}{\bar{\sigma}(P)} = \mu_1 \tag{7}$$

where $\bar{\sigma}(\cdot)$ denotes largest singular value of $(\cdot)$ and P satisfies the Lyapunov equation

$$(A + BEC)^T P + P(A + BEC) + 2I = 0 \tag{8}$$

For the structured case, the following result was recently obtained:

Theorem 2. (Yedavallie, 1985) Assume that the elements of E,

$$\Delta E = \Delta A + \Delta BEC \tag{9}$$

are restricted so that

$$|\Delta E_{ij}| < \varepsilon_{ij} \tag{10}$$

and let $\varepsilon \overset{\Delta}{=} \max\limits_{i,j} \varepsilon_{ij}$. Then the perturbed system (5) and (6) is stable if

$$\varepsilon < \frac{1}{\bar{\sigma}|(|P|U)_s|} = \mu_2 \tag{11}$$

where P satisfies (8), U is a matrix with elements $u_{ij}=\varepsilon_{ij}/\varepsilon$, $|(\cdot)|$ denotes modulus matrix, i.e., matrix formed by taking moduli of elements of $(\cdot)$ and $(\cdot)_s$ denotes symmetric part of the matrix $(\cdot)$, $(\cdot)_s = \frac{1}{2}((\cdot) + (\cdot)^T)$.

To mitigate the conservatism of the existing robustness criteria we shall combine the information concerning the nature of the pertubations which can physically occur, and mathematical characterization of the perturbation matrices. Namely, in many practical situations, it may happen that there exists some a priori knowledge of the perturbations which may be encountered. In this case there is a need for a more detailed description of the "allowable perturbations", and we define the perturbations in certain directions which are more appropriate from a physical standpoint. This leads to decomposition of the perturbations into two components, one of which lies along the given direction in the space of all perturbation matrices. In particular, define the matrices ΔA and ΔB as

$$\Delta A = \eta(t)\, A^1 + \Delta A^1 \tag{12}$$

$$\Delta B = \eta(t)\, B^1 + \Delta B^1 \tag{13}$$

where (t) is a scalar function, and ΔA^1 and ΔB^1 represent the perturbations in the system dynamics, which lie out of the directions A^1 and B^1, respectively.

Now, suppose that the perturbation matrices ΔA and ΔB lie entirely along the directions A^1 and B^1, respectively, that is $\Delta A^1=0$ and $\Delta B^1=0$. The following theorem gives the sector (η_{min},η_{max}), i.e., the bounds on the scalar function $\gamma(t)$ such that the perturbed system remains stable.

Theorem 3. If the following inequalities are satisfied

$$\lambda^{-1}_{min}(\bar{W}) = \eta_{min} < \eta(t) < \eta_{max} = \lambda^{-1}_{max}(\bar{W}) \tag{14}$$

for all $t\varepsilon[0,\infty)$, where $\eta(t)$ is memoryless, time-varying nonlinearity, and

$$\bar{W} = (P(A^1 + B^1EC))_s \tag{15}$$

and where the matrix P is positive definite solution of (8), $\lambda_{min}(\cdot)$ and $\lambda_{max}(\cdot)$ denote minimum and maximum eigenvalue of the matrix $(\cdot)$, respectively, and $(\cdot)_S$ denotes symmetric part of the matrix $(\cdot)$; then the perturbed system (5) and (6) with ΔA and ΔB defined with (12) and (13) for $\Delta A^1=0$ and $\Delta B^1=0$, is asymptotically stable.

Proof: All proofs are omitted, and are available from the author.

In essence, Theorem 3 shows that if $\eta(t)\varepsilon(\eta_{min},\eta_{max})$ then the perturbes system remains asymptotically stable. However, the robustness results of Theorem 3 have been restricted to the case when the perturbations in the system dynamics lie entirely along the given directions in the space of all perturbation matrices. If the scalar function $\eta(t)$ is selected in the sector (14), than this selection uniquely determines the bounds on the matrices ΔA^1 and ΔB^1, which represent all the perturbations which lie out of the direction A^1 and B^1, respectively. The bounds on ΔA^1 and ΔB^1 can be established using any of the existing robustness criteria dealing with unstructured perturbations.

As mentioned for output feedback designs no guaranteed robustness properties hold. If the robustness evaluation come up with embarassingly small marginess, a "design adjustment procedure" to improve robustness margins would be very desirable. This problem is addressed in the next section.

FINE TUNING OUTPUT FEEDBACK CONTROL

In this section a general procedure is developed for improving the feedback properties, including robustness of multivariable, output feedback control systems. The procedure allows the possibility of adding extra controller which can be used to "fine-tune" the closed-loop behaviour of the nominal system. In this case, the system (1)-(3) is represented as

$$\dot{x}(t) = Ax(t) + B(u(t) + \bar{u}(t)) \tag{16}$$

$$y(t) = Cx(t) \tag{17}$$

where u(t) is the nominal control law defined with (3) and

$$\bar{u}(t) = Fx(t) \tag{18}$$

defines the additional control law, which can be used to "fine-tune" the closed-loop behaviour of the nominal control system.

In what follows a theorem is given which completely characterizes the class of feedback maps that do not "interfere" with the stabilizing actions of the nomiral control law u(t), eqn.(3). This leads to a "recipe" for addition controller design.

Theorem 4. If the output feedback closed-loop system

$$\dot{x}(t) = (A + BEC)x(t) \tag{19}$$

is asymptotically stable, then the closed-loop system

$$\dot{x}(t) = (A + BEC + \bar{B}EC)x(t) \tag{20}$$

is also asymptotically stable for all $\bar{B}$ defined by

$$\bar{B} = -P^{-1}C^TE^TW \tag{21}$$

where W is arbitrary symmetric, non-negative definite matrix, and P is the positive definite solution of the Lyapunov matrix equation (4).

Now the calculation of the additional control law $\bar{u}(t)$ is straightforward. The objective is to define (18) so that the system (16) and the system (20) are "near" each other. Define the trajectory error vector as

$$e(t) = x^*(t) - x(t), \quad r(0) = 0 \tag{22}$$

where the notation the star (*) is used to indicate properties associate with the system (16). Letting

$$r(t) = (BF - \bar{B}EC)x(t) \tag{23}$$

the equation

$$\dot{e}(t) = (A+(B+\bar{B})EC)e(t) + (BF-\bar{B}EC)x(t), \quad e(0) = 0 \tag{24}$$

is expressed as

$$\dot{e}(t) = (A+(B+\bar{B})EC)e(t) + r(t) \tag{25}$$

It is obvious that if $r(t)=0$, then $e(t)=0$ for all $t\varepsilon[0,\infty)$, since $e(0)=0$. However, in general case $r(t)\neq 0$ and therefore approximation must be made. There are a variety of criteria one can choose to minimize in such a case. One obvious possibility is the minimization of

$$\min_F ||(BF-\bar{B}EC)x(t)|| \tag{26}$$

where $||(\cdot)||$ denotes the ordinary euclidian quadratic norm. Remembering that

$$\min_F ||(BF-\bar{B}EC)x(t)|| < \min_F ||(BF-\bar{B}EC)|| \; ||x(t)|| \tag{27}$$

holds for all $x(t)\varepsilon R^n$, the problem reduces to determine

$$\min_F ||BF-\bar{B}EC|| \tag{28}$$

Now, the calculation of additional control law is straightforward. When B is of full column rank then F is defined by

$$F = (B^T B)^{-1} B^T \bar{B}EC \tag{29}$$

Furthermore, when B is not of full column rank, F may still be selected with

$$F = B^\dagger \bar{B}EC \tag{30}$$

where $B^\dagger$ is a Moore-Penrose inverse of B.

Based on the proposed method the computer aided design procedure can be readily introduced. The following steps could be the design guideliness:

1. Solve the output feedback control problem in order to satisfy certain design performances.
2. Compute the resulting closed loop system matrix and the stability robustness measures. If the stability robustness of the resulting output-feedback system does not satisfy the design objectives procede the fine tuning output feedback control design.
3. Determine the design freedom matrix W, eqn.(21) and compute $\bar{B}$,(21).
4. Compute the additional output feedback gain, eqn. (29).

Computationally, the proposed method is very feasible, as it implies the numerically stable procedure. The resulting control scheme is simple and practical and the controller acts directly on the available system outputs.

NUCLEAR POWER PLANT EXAMPLE

System Description

To illustrate the theoretical concepts developed in this paper a mathematical model of the nuclear power plant of Krško, Yugoslavia, is used through.

Developing a good model for a complex system, as is nuclear power plant is usually a major effort. The application of the design methods proposed in Section 2. demands a linear model of the processes.

The common way of developing linear models is to linearize a set of nonlinear equations. The full nonlinear model of the considered nuclear power plant is represented by the system of equations

$$\dot{x}(t) = f(x(t),u(t),t) \tag{31}$$

$$y(t) = g(x(t)) \tag{32}$$

This model was linearized and reduced to 15 order system, which was later used for design purpose. The model includes the main physical processes, as well as the main control system. A very simplitied diagram ot the nuclear power plant ot Krško, as well as a detail description ot the physical system considered are given in (Arsov, 1987).

Control System Performance

An output feedback regulator based on LQ methodology with C_{10x15} was used.

A set of the weighting matrices Q and R was found by trial and error, bearing in mind the physical properties of the system, until satisfactory performance was obtained for a variety of conditions of several power levels. The selection of the weighting matrices Q and R, and the study of the trade-off between the different control systems, have been studied in (Arsov, 1987).

The robustness properties of the considered output feedback control systems when large perturbations occur in the system model have been evaluated using the results of Theorem 1, for the case of unstructured perturbations, i.e. for $A^1=0$ and $B^1=0$. Theorem 1 (see eqn.(7)) gives the following bound

$$\mu_1 = 0.3077087x10^{-3}$$

The result shows that the robustness criteria in terms of matrix norms are very conservative, and for the considered nuclear power plant system the result is of limited practical use. The measure of the "size" of the perturbations used in this approach does not distinguish the "structure" of the perturbations.

To realistically evaluate the robustness of the output feedback design considered, we shall use some a priori knowledge of the modelling uncertainties about the design model, as well as perturbations in the system matrix. Certain elements in the system matrix A are considered to be fixed. Therefore, as a first approximation the perturbation direction A^1 of the system matrix A was obtained by simply setting some of the elements of the matrix A to zero. For the control matrix $B^1=0$ was chosen. From Theorem 2 it follow that the perturbed system will remain stable for

$$\mu_2 = 0.1111084x10^{-1}$$

see eqn.(11). In a similar way from Theorem 3 it follows that the perturbed system will remain stable as long as $\eta(t)\varepsilon(\eta_{min},\eta_{max})$ for all $t\varepsilon[0,\infty)$, see eqn.(14), where

$$\eta_{min} = -0.7409858x10^{-1}; \quad \eta_{max} = 0.1111084x10^{-1}$$

As can be seen Theorem 3 provides broader robustness bounds than the measures given by Theorems 1 and 2.

To improve the robustness properties of the considered control system we will employ the design procedure proposed in Section 4. In this case the design matrix was chosen as

$$W = wI_{3x3}$$

The robustness results for different values of the scalar parameter w are given in Table 1.

TABLE 1. Robustness Results for Different Values of the Design Parameter w.

wx10	$\mu_2 x10^{-1}$	$\eta_{min} x10^{-1}$	$\eta_{max} x10^{-1}$
0.0	0.1111084	-0.740958	0.1111084
0.2	0.1248644	-1.240193	0.1248644
0.4	0.1311395	-1.566816	0.1311395
0.6	0.1342954	-1.725147	0.1342954
0.8	0.1360616	-1.776713	0.1360616
1.0	0.1370938	-1.772159	0.1370938
2.0	0.1383690	-1.557309	0.1383690
4.0	0.1374662	-1.273983	0.1374662
6.0	0.1365171	-1.14779	0.1365171
8.0	0.1358391	-1.079351	0.1358391

It is shown that $w \approx 4$ attains the most robust design. Therefore, the proposed computer aided design procedure can be used to select an appropriate value of the design parameter w to attain a robust design.

ACKNOWLEGEMENT

This work was supported in part by the U.S. - Yugoslav Joint Fund for Scientific and Technological Cooperation in cooperation with DOE under Grant PP-727.

REFERENCES

Arsov, Lj. (1987). Ph.T. University of Novi Sad, Yugoslavia (in SerboCroation)

Fallside, F. (1977). Control System Design by Pole-Zero Assignment. Academic Press, New York.

Kouvaritakis, B., and H.Latchman (1985), Int. J. Countrol, 41, 1381-1412.

Kvakernaak, H., and R.Sivan (1972). Linear Optimal Control Systems. Wiley-Interscience, New York.

Lehtomaki, N., N.Sandell, and M.Athans. (1981), Proceedings 20th I.E.E.E. Conf. Decision and Control, San Diego, CA.

Patel, R.F., and M.Toda.(1980), Proceedings Joint Automatic Control Conference, paper TD8-A.

Petkovski,Dj. (1984), R.A.I.R.O. Automatique/Sustems Analysis and Control, 18, 315-327

Petkovski, Dj., and M.Athans. (1985).Elect.Power Sys. Research, 9,253-262

Petkovski, Dj. (1985) IEE Proc., 132, Pt.D., 53-60.

Postlethwaite, I., J.M.Edmunds, and A.G.MacFarlane.(1981). I.E.E.E. Trans. Autom. Control, 26,32.

Yaman, A., B.Manousiouthakis, and P.Putz. (1984). Int. J.Control, 40,603.

Yedavallie, R.K.(1985).Int.J.Control, 42, 1507-1517.

Experimental Design of Reliable Controllers for Industrial Robots

M. Armada* & C.M. Vargas-Gomez*

**Instituto de Automatica Industrial (C.S.I.C.)*
C.N.III Km. 22,800 Arganda del Rey, Madrid, Spain

ABSTRACT

An experimental set-up for designning reliable controllers for industrial robots whose joints are driven by DC motors is presented. This experimental testbed is composed of the following elements : DC motor with its attached analogic speed regulator, which is connected to a microprocessor based boards where the digital controller is implemented; and by the other side there is a dynamometer brake with its associated electronics that permits to emulate under real time conditions the load as it is to be seen by the DC drive when it is used to actuate a given joint of a robot arm. When this load conditions are furnished to the drive then the digital control algorithms can be tested and it is possible to improve its performance by means of parameter tunning and/or by trying other control strategies.

KEYWORDS

Automatic control; electronic instrumentation; DC motors; robotics; reliability.

INTRODUCTION

It is widely recognised the increasing use of DC servomotors as prime movers of industrial and/or teleoperated robots. When DC motors are employed as joint actuators the control engineer must to abord the selection problem of the different elements that constitutes the servomechanism : motor, speed/torque reduction, velocity and position sensors, amplifiers, and the computer architecture for the digital part of the controller, based on the available information about the load, work cicle, temperature, friction,...

Actually the common practice when designing the position servomechanism for the robot arm joints consists in taken only into account the extreme conditions of the working cicle, and then to adjust the servoamplifier and the parameters of the digital controller in order to fulfil them. However this way of doing yields to a far from optimal performance when the manipulator is working out of such a limiting values because the system is overdimensioned. For this reason, in the author's opinion, a testbed for analysing the performance of the whole set : motor + load + amplifier + digital controller, should constitutes an invaluable tool so that we can understand and improve the design of high reliability, very accurate robot joint controllers, reaching as a consequence to the optimization of its time response.

The aim of this work is to present an experimental set up for designing reliable controllers for robots whose joints are driven by DC motors. This experiemntal arrangement provides the designer with high flexibility and accesibility to the most relevant variables and parameters that influence the dynamic behaviour of the robot axes.

The experimental DC motor set-up was originated by the need to design and adjust the drivers of our actually evolving robotic system intended for performing inspection and maintenance operations in the hostile environment of nuclear power plants. Such a robot system responds to the concept of telerobot, in the sense that it must perform operations both in robotic and teleoperated mode. The robotic system is composed by a high complexity mechanical manipulator, a local controller structure based on a multiprocesor system (16 bits) and a remote teleoperation station build up around a minicomputer(32 bits) with real time operating system, where off-line programming at the time of on-line programming capablities are provided.

The major component of the experimental set-up is a dynamometer brake with its own, high performance, associated electronics that permits to control and/or monitoring speed, torque and power.

Our basic idea in this experimental arrangement consists in employing the magnetic powder dynamometer brake to simulate under real time conditions the load as it is to be seen by the DC drive when it is used to actuate a given joint of the robot arm. When this load conditions are furnished to the drive, we can test the digital controllers that are programmed in the microcomputer with the aid of a microcomputer developement system and then we are in conditions of improving its performance by means of parameter tunning and/or by trying other control strategies.

THE ROBOT JOINT LEVEL CONTROL PROBLEM

A robot manipulator is composed by a series of links connected together by joints (revolute or prismatic) which permits relative displacements between adjacent links and consequently leads to a movement of the manipulator's end effector in the space volume known as robot workspace. There are two main facts that explains the widespread use of robots in industry. The mechanical flexibility just cited is the first one. The second one is based on the programming flexibility provided by the control computer.

In order to completely exploit the flexibility of the robotic devices it is very important that the two sides perform accordingly well. In other case it maybe for instance that an excellent software for programming cannot be enough for good performance if the mechanical part of the robot (the manipulator) is not properly controlled.

Our main interest in this work is to investigate how to improve the manipulator dynamic behaviour, or what is the same, the robot joint level control problem. Our research is directed not only to improve the reliability of a single joint controller, but to multiaxis controllers that take into account the mutual influences (couplings) of the different links of the robot when moving in the workspace (Armada, 1986a).

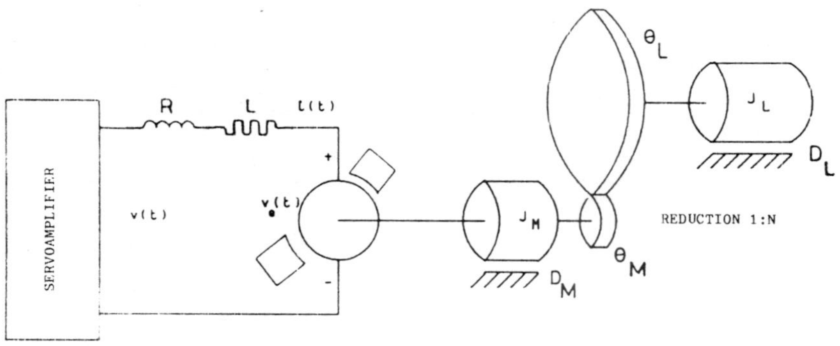

Fig. 1.- DC motor with coupled load.

Figure 1 shows a typical disposition of a permanent magnet DC motor with coupled load (the robot link) through a reduction mechanism. The electrical equation whose solution give us the time evolution (transient response) of the current delivered by the transistor chopper as a function of the input voltage (velocity set point), is as follows,

$$v(t) = R*i(t) + L*di(t)/dt + K_E*w_M(t) \qquad (1)$$

where : i(t) instantaneous current (A); R (Ohms); L (H); K_E

voltage constant (volt/rad/seg); w_M speed (rad/seg).

D'alambert principle permits the obtention of the mechanical equation,

$$K_T*i(t) = (J_M + J_L/N^2)dw(t)/dt + (D_M + D_L/N^2)w(t) + T_F + T_L/N^2 \quad (2)$$

where D_M, D_L are in oz-in-seg/rad; T_L, T_F are in oz-in; J_M, J_L are in oz-in-seg^2/rad; K_T, torque constant (oz-in/A).

This equations are the starting point to undertand the influence of different parameters on the joint drive performance.

EXPERIMENTAL SET-UP FOR DESIGNNING RELIABLE CONTROLLERS FOR ROBOTS

The developement, production, maintenance and use of DC motors as primary actuators or as movement transmision elements requires the exact knowledge of its electromechanical characteristics and of its working conditions.

When DC motors are used as prime movers it is neccesary to know his power, torque and velocity. Sometimes, depending on the specific problems posed by a particular application it is enough to measure and test some of this magnitudes in any (little) points over the characteristic curves. However when we are dealing with more complex problems like those of robots, it may be neccesary to draw out a good number or a complete family of curves.

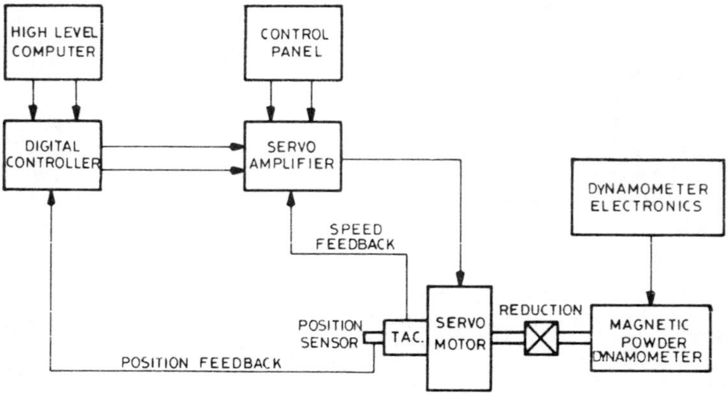

Fig. 2.- Block diagram of the experimental testbed.

To obtain the required information the DC motor drive under research can be connected to a dynamometer (eddy current or magnetic powder) brake and then it can be subject to different workcicles. Because of the eddy current dynamometer is not very adequate at low speeds we have chosen for our testbed a magnetic powder one that can furnish the maximum braking torque from 0 rpm up to 10000 rpm, what means in practice the possibility of measuring the stall torque and a sufficient speed range.

The dynamometer brake used in our testbed is controlled by a special purpose high performance electronics that facilitates the monitoring of speed, power and torque and also permits to programming workciles both in velocity and torque modes under PID control, with a wide range of tunning capabilities. Also it is possible to obtain graphical information (speed/torque/current curves) by means of a plotter connection.

The experimental set-up for designing reliable controllers for robots is composed by the following elements (see Fig. 2 and 3) : 1) Magnetic powder dynamometer and related electronics; 2)DC motors associated with servoamplifiers and sensors; 3) Digital processor. The DC motors and the dynamometer are connected by a flexible coupling and due to the universal structure of the set-up it is straighforward to change from one motor to another.

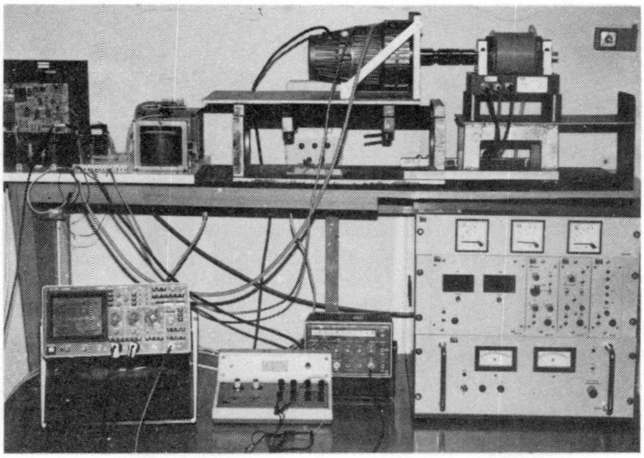

Fig. 3.- Experimental set-up for designing controllers for robots.

Operating Principle

The basic operating principle of the experimental testbed is as follows. The servoamplifier which supplies to the DC motor the driving voltage is fed by the output of a D/A converter what constitutes the speed set point. At the same time the spped of the motor is feedback to the servoamplifier in order to stabilize the servosystem. The position is read by an appropriate device (for instance an incremental optical encoder) and feedback to the digital controller. Also the loads solicited torque is measured in two ways : train gage sensors located at the dynamometer and voltage drop across a shunt resistance in the servoamplifier.

The solicited torque can be feedback to improve the dynamic response of the servosystem and/or to change the control mode from speed to torque. The analogic signal of the torque is amplified and then acquired by the digital processor throughout an A/D converter.

The position of the DC drive is controlled by means of an algorithm stored in the control computer. We use 16-bit microprocessors with mathematical co-processors (floating point accelerators).

After a careful analysis of the dynamic of our robot we obtain by means of simulation the workcicle that the individual joints of the robot must be able to cope under different load conditions. Then this information is used to program the dynamometer so that we can "emulate" in real time conditions in our testbed the individuals joint workcicle. When doing so we can monitor the behaviour of the interest variables and we are in conditions of analyse, evaluate and/or change the parameters of the digital controller in order to improve its performance and consequently to obtain more reliable controllers.

PERFORMANCE ANALYSIS AND EVALUATION OF ROBOT JOINT DRIVES

The experimental set-up can be employed to measure a number of interest parameters that greately influence the time response of the servosystem under design. Magnitudes like the electrical (T_E) and mechanical (T_m) time constants, torque and velocity contants (K_T, K_E), inertia (J), Coulomb friction (D), load and friction torque (T_L, T_F), are easily tested by means of adequate experiments.

Additionally it is easy to see the behaviour of the DC drive variables like speed or current when a load is applied to the motor shaft. One of the most interesting features provided by the servoamplifier is that the speed of the motor is maintained at the desired set point independently of the load torque perturbations at the expense of providing enough current to overcome the extra solicited torque. However a carefull analysis shows that this is an ideal situation and that an important transient (up to 50%) exists over a certain (not neccesarily short) time interval. Fig. 4 shows the influence of a load torque step (T_L = 0.84 Nm.) at regulating speed of 400 rpm. The velocity peak is about 300 rpm during 100 ms .Fig. 5 shows the time response of the servosystem with and without load, where it can be noticed that the load torque filters out the overshot, but increases the rise time.

When we close the position loop it maybe that this speed perturbations can originate position errors, mainly at high speeds. In this case adequate digital compensation must to be envisaged, implemented and evaluated with the aid of the experimental testbed.

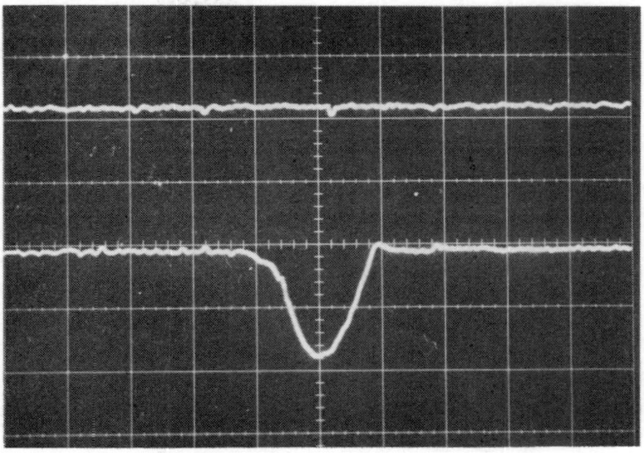

Fig. 4.- Influence of a load torque step on the speed regulation.

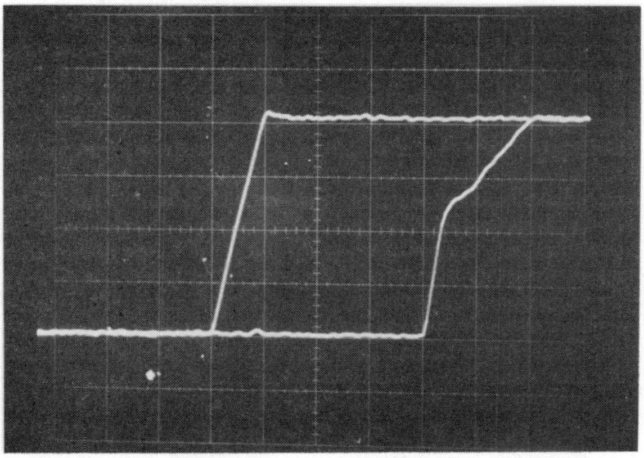

Fig. 5.- Comparison of servosystem time response.

CONCLUSION

An experimental set-up designing reliable controllers for robots whose joints are driven by DC motors has been presented. The main element of this testbed is a magnetic powder dynamometer brake that can be regulated to emulate in real time conditions the variable load that must be driven by the DC motor based servosystem. This experiemtal set-up is employed actually to design and test the digital control strategies neccesary to obtain a high reliability controller for the robotic system intended for work in the hostile environment of nuclear power plants actually under developement in our laboratory.

REFERENCES

Armada, M. (1986a).Multivariable frequency domain methods for industrial robot controllers analysis and design. *Proc. 16th ISIR, Brussels*.

Armada, M. (1986b). Control de posicion de motoores de corriente continua. *Automatica e Instrumentacion*.

Kuo, B,C. & Tal, J. (1978). *DC motors and control systems*. SRL P.C., Champaing, Illinois.

Sen, P.C. (1981). *Thyristor DC drives*. John Wiley & Sons, New York.

Optimization Control and Reliability of Steel Grinding Process

R. Pasquino

Faculty of Eng., University of Salerno, 84100, Italy

Abstract

This paper deals with a detailed analysis of optimization and reliability control of traditional steel grinding processes. The analysis is realized, on the basis of stochastic calculations, after a detailed insight into the production system.

1. Introduction

One of the relevant problems in the analysis of production systems consists in determining the gap between previsions and production related to the actual optimization of the system itself in terms of minimization of the power spent in connection with the maximization of the product and minimization both of the power spent and of the uncertainty of the production process.

In general an accurate analysis should be able to take into account in details the random behaviour of the production system or, in other words, the fluctuations of the production around some mean value.

I consider in this research work the above outlined problem related to the analysis of traditional steel grinding systems. I recall that a stochastic model for the description of the performances of steel grinding machining was proposed in a recent paper [1] as the natural generalization to the stochastic case of the well known model by Mayne and Malkin [2-4]. Further studies have been proposed in paper [5]. Such a model consists of an analytic expression which relates the metal removal rate and the power spent for such a process to the relevant technological parameters such as the diameter and depth of the grinding wheel, velocities and wheel surface and of the workpiece and depth of cutting. In addition, the model considers the random effects due to the natural fluctuations induced to the model by real working conditions. On the other hand the man-machine interaction is not directly considered.

I consider, in this paper, the following problems:

- Defining a suitable objective functional defined by the need of maximizing the removal rate, minimizing the power spent and optimizing the reliability of the system.

- Defining the reliability of the system in terms of damage probability due to burning.
- Analysing the influence of the technological parameters upon the aforementioned functional.

The next section provides a brief description of the model; the third section deals with the analysis of the optimization and reliability problem, and the fourth section with the actual application to a real production system.

2. On the Stochastic Models of the Grinding Machining Process

I refer to the model proposed in reference [1], which provides the metal removal rate, the power spent for the removal and the limit burning power, respectively Y, W and Wb, by the following equations:

$$Y = (a_1 + \varepsilon_1\alpha_1(\omega))hv \tag{1}$$

$$W = (a_2+\varepsilon_2\alpha_2(\omega))Y + (a_3+\varepsilon_3\alpha_3(\omega))\Omega D + (a_4+\varepsilon_4\alpha_4(\omega)) A\,\Omega\, D^{3/2}\, h^{1/2} + (a_5+\varepsilon_5\alpha_5(\omega))A\, D^{-1/2}\, h^{1/2}\, v \tag{2}$$

$$W_b = (a_6+\varepsilon_6\alpha_6(\omega))Y + (a_7+\varepsilon_7\alpha_7(\omega))\, v^{1/2}\, h^{1/4}\, D^{1/4} \tag{3}$$

The above equations relate the technological parameters:

$$(h,\ v,\ \Omega,\ D,\ A) \tag{4}$$

to the quantities Y, W, W_b through the deterministic parameters a_j and ε_j and the random constant parameters $\alpha_j(\omega)$.

Where a_j can be regarded as identification parameters in the deterministic case, $\alpha_j(\omega)$ can be regarded as random parameters for the identification of the fluctuations of the production process and ε_j parameters which identify the scale of the fluctuations.

The technological parameters have the following meaning:

h = wheel depth of cut;
v = workpiece velocity;
Ω = wheel angular velocity;
D = grinding wheel diameter;
A = wheel wear flat area (between wheel and workpiece).

I will not discuss here the model already proposed and discussed in the already cited paper [1]. I only want to point out that the optimization technique proposed in the next section is very general and modifications and improvements to the model do not modify the optimization technique.

One can reasonably argue on the utility of working on a stochastic model. On the other hand, I have to acknowledge that fluctuations and uncertainties are consistent in almost all production processes and consequently it appears worthwhile to the author constructing a pertinent optimization technique in the stochastic case.

3. Optimization Process and Reliability Analysis

In general an optimization process should provide to an increase of the product and to some minimization of the costs. In addition, as indicated in reference [1], if the system is characterised by random parameters, the optimization process has to reduce also the fluctuations in the production due to the inner stochasticities of the system itself.

I refer to the model proposed in the preceding section with the purpose of optimizing the choice of the vector:

$$\underline{x} = (x_1 = h, \quad x_2 = v, \quad x_3 = \Omega) \tag{5}$$

namely of the variable technological parameters at fixed characteristics of the machining process, i.e. D and A. Nevertheless, the method which is here proposed is flexible enough to include also D and A in the optimization.

The line which will be followed is classical and essentially consists in the following two steps:

(a) Defining a function $J = J(\underline{x})$ to be maximized, with the aim of increasing Y and reducing W as well as the fluctuations of Y and W, by a suitable choice of $\underline{x}$.

(b) Defining a suitable optimization procedure to realize the objective defined in step (a).

Keeping this in mind, consider the following averaged quantities:

$$E(Y)(\underline{x}) = \int Y(\underline{x},\alpha_1)\, P(\alpha_1)\, d\alpha_1 \tag{6}$$

$$V(\underline{Y})(\underline{x}) = \int [Y(\underline{x},\alpha_1) - E(\underline{Y})(\underline{x})]^2\, P(\alpha_1)\, d\alpha_1 \tag{7}$$

$$E(W)(\underline{x}) = \int W(\underline{x},\alpha_2,\ldots,\alpha_5)\, P(\alpha_2,\ldots,\alpha_5)\, d\alpha_2\ldots d\alpha_5 \tag{8}$$

$$V(W)(\underline{x}) = \int [W(\underline{x},\ldots) - E(W)(\underline{x})]^2\, P(\alpha_2,\ldots,\alpha_5) d\alpha_2,\ldots d\alpha_5 \tag{9}$$

as well as the following additional quantities:

$$W_M = \max_{\alpha_2,\ldots,\alpha_5} W(\underline{x}, \alpha_2,\ldots,\alpha_5) \tag{10}$$

$$W_{bm} = \min_{\alpha_6,\alpha_7} W_b(\underline{x}, \alpha_6, \alpha_7) \tag{11}$$

A functional $J(\underline{x})$ can now be defined according to point (b), i.e.

$$E(Y)\uparrow,\ V(Y)\downarrow,\ E(W)\downarrow,\ V(W)\downarrow \ \rightarrow \ J\uparrow \tag{12}$$

with the additional constraint:

$$W_M \leq W_{bm} \tag{13}$$

Of course the choice is not unique and can be, at least in part, the result of conjecture of the operator. I suggest the following expression:

$$J(\underline{x}) = E(Y)\exp(-K_1 E(W) - K_2(V(Y) + V(W))) \tag{14}$$

where K_1 and K_2 are positive constants with the meaning of weight functions.

It is plain that the expression (13) satisfies condition (12), condition (13) has to be verified at each step of the optimisation process. Independently on the choice of the weights K_1, the pertinent method, i.e. the "univariate search method" can be summarized in order to indicate how the practical optimisation of J can be realized.

This method [6] avoids the execution of analytic derivatives which are not useful in this case. The steps of the univariate search are the following:

1. Fix x_2 and x_3, $x_2 = x_{20}$, $x_3 = x_{30}$ (arbitrary values) and variate x_1 until the maximum of J is attained:

$$x_1 = x_{11} \rightarrow \max_{x_1} J(x_1, x_{20}, x_{30}) = J_{11}$$

2. Fix: $x_1 = x_{11}$, $x_3 = x_{30}$ and variate x_2 until the maximum of J is attained:

$$x_2 = x_{21} \rightarrow \max_{x_2} J(x_2, x_{11}, x_{30}) = J_{12}$$

3. Fix: $x_1 = x_{11}$, $x_2 = x_{21}$ and variate x_3 until the maximum of J is attained:

$$x_3 = x_{31} \rightarrow \max_{x_3} J(x_3, x_{21}, x_{31}) = J_{13}$$

4. Repeat steps 1,2,3 until the following inequality is satisfied:

$$|J_{n_3} - J_{(n-1)_3}| \leq \varepsilon$$

with ε error bound fixed *a priori*.

Generally the above search gives a satisfactory result after two or three iterations. A simple software can be organized so that the operator can realize an optimum choice of the working conditions.

In addition, it has to be mentioned that the optimization process which has been indicated above needs the control of condition (13) which indicates the damage. Even if W_b has been defined as a random function, it is more convenient considering its lowest value:

$$W_{bm} = \min_{\alpha_i} W_b$$

Then it is possible introducing the concept of *"Damage probability"* through the probability:

$$P_d \; (W \leq W_{bm}) \; (\underline{x}) \tag{15}$$

where, of course, the limit conditions hold

$$W_{bm} \geq W_M \rightarrow P_d = 0 \tag{16}$$

$$W_{bm} \leq W_m \rightarrow P_d = 1 \tag{17}$$

The determination of P_d requires the calculation of the probability density P(W) joined to the random variable W starting from its expression (2) and from the probability distribution of the random constant variables $\alpha_2,\ldots,\alpha_7$. This is a standard calculation of probability theory [7] when the probability density of α_2 and α_7 is known. On the other hand, as also discussed in [5] the identification of the random variables α_i has practically been realized only in mean value and variance so that only an analogous calculation can be realized for P(W). Keeping this in mind, introduce the random variable $\beta = \beta(\omega)$ defined by

$$\beta = \frac{W - W_m}{W_M - W_m}, \qquad 0 \leq \beta \leq 1 \tag{18}$$

such that the damage is attained if $W_m \leq W_{bm} \leq W_M$ for

$$\beta \geq \beta_d = \frac{W_{bm}-W_m}{W_M-W_m} \tag{19}$$

Then $P(\beta)$ can be simulated by a Beta-type distribution of probability density

$$P(\beta) = \frac{\Gamma(a_1+a_2)}{\Gamma(a_1)\ (a_2)} \beta^{(a_1-1)} (1-\beta)^{(a_2-1)} \tag{20}$$

where Γ classically denotes the Gamma function and a_1 and a_2 are identified by the mean value and variance of β

$$E(\beta) = \frac{a_1}{a_1+a_2} \tag{21a}$$

$$V(\beta) = \frac{a_1\ a_2}{(a_1+a_2)^2(1+a_1+a_2)} \tag{21b}$$

where $E(\beta)$ and $V(\beta)$ are related to $E(W)$ and $V(W)$ by

$$E(\beta) = \frac{E(W)-W_m}{W_M - W_m}, \qquad V(\beta) = \frac{V(W)}{(W_M - W_m)^2} \tag{22}$$

So that the identification is complete and P_d is given as a function of $E(W)$ and $V(W)$, formally

$$P_d = \int_{\beta_d}^{1} P(\beta;E(W),V(W))d\beta = 1 - \int_{0}^{\beta_d} P(\beta;E(W),V(W))d\beta \tag{23}$$

where $P(\beta)$ is given by equation (20), where a_1 and a_2 are computed through equations (21-22) by $E(W)$ and $V(W)$.

In practice the reliability analysis consists in joining at each step of the optimization process 1,...,4 the estimate (and analysis) of P_d through the following calculations:

(i) Compute, for each $\underline{x}$, $E(W)$ and $V(W)$ by equations (8-9) as well as $E(\beta)$ and $V(\beta)$ by equations (22).
(ii) Identify a_1 and a_2 by equations (21) and $P(\beta)$ by equation (20).
(iii) Compute P_d $(E(W), V(W))$ by equation (23).

The analysis of the strategy to be joined to the estimate of P_d needs to be related to production strategies which are not here discussed.

4. Application and Discussion

Some practical analysis is here realized in this final section continuing the experiments already proposed in references [1] and [5]. Some details will be given here on the relevant differences with respect to papers [1,5]. I refer to the experiments by the grinding machine 3G7IM (Soviet made) operating with ceramic β46/K1 wheel. The quantitave identification of the terms $a_i, \varepsilon_i, E(\alpha_i)$ and $V(\alpha_i)$, in fact the random variables as already mentioned in the preceding section are identified in mean value and variance, is the one of ref. [1] and is reported in the following table:

i	a_i	ε_i	$E(\alpha_i)$	$V(\alpha_i)$
1	0.95	0.010	0.30	0.010
2	0.0125	0.010	0.50	0.005
3	9.6×10^{-7}	0.010	0.30	0.001
4	8.5×10^{-6}	0.005	0.35	0.001
5	2	0.020	0.50	0.020
6	0.0055	0.010	0.50	0.010
7	0.0068	0.020	0.50	0.010

Table 1

In addition, I have to recall that lengths and time for the above model are measured in millimeters and seconds respectively.

Table 1 provides the identification of the random variables in terms of mean values and variance, consequently the expressions of the quantities defined in equations (6)-(9) become:

$$E(Y) = (a_1+\varepsilon_1 E_1)x_1x_2 \tag{24}$$

$$E(W) = (a_2+\varepsilon_2E_2)E(Y) + (a_3+\varepsilon_3E_3)x_3D + (a_4+\varepsilon_4E_4) Ax_3D^{3/2}\sqrt{x_1} + (a_5+\varepsilon_5E_5)AD^{-1/2}x_2\sqrt{x_1} \tag{25}$$

$$V(Y) = (x_1x_2)^2 \varepsilon_1 V_1 \tag{26}$$

$$V(W) = (a_2+2\varepsilon_2E_2)V(Y) + \varepsilon_2^2(V_2(V(Y)+E(Y)) + E_2^2 V(Y) + D^2\varepsilon_3^2V_3^2x_3^3 + A^2D^3\varepsilon_4V_4x_3^2 + \frac{A^2}{x_3} x_1x_2^2\varepsilon_5V_2 \tag{27}$$

The values of W_M are obtained for the largest values of α_i and W_m is obtained for the smallest values of α_6 and α_7.

The calculations and table which have been provided in this section can be used, even if the table applies only to the grinding machine, which has been mentioned above, for simulation and experiments. The experiments which have been realized by the author have shown that the indications above are sufficient to obtain satisfactory improvements of the production and a detailed estimate of the damage probability. In addition, it has to be mentioned that the objective function defined in equation (14) appears more efficient than the one proposed in [5]. In fact, the identification of the terms K_i is more realistic.

Experiments have been realized within the following ranges of working conditions:

$0 < x_1 = h \leq 0.15$ mm

$0 < x_2 = v \leq 1000$ mm/sec

$0 < x_3 = \Omega \leq 500$ rad/sec.

Some practical indications, even if detailed and particular analysis (which have also to face particular practical problems in real working conditions) are left to operators, can be given to conclude this paper:

- Increasing/decreasing h (at fixed Ω) moves optimum working conditions towards large/small values of v .
- Increasing/decreasing Ω (at fixed h) moves optimum working conditions towards small/large values of v .
- Damage probability increases, increasing the values of (h,Ω,v).

Experience suggests to identify the constants k_1 and k_2 through practically selected optimum working conditions, i.e. selecting k_i in a fashing that J is maximum for optimal working conditions.

References

1. G. Giorleo and R. Pasquino. On the Steel Grinding Process. Proceedings Int.Symp. on Design and Synthesis, J. Soc. Precision Engineering, Tokio, (1984) 80-85.
2. R. Meyne and S. Malkin, Parameter Optimization on the Steel Grinding Process. J. Eng. Ind. ASME, 98 (1976) 1048-1952.
3. S. Malkin. Grinding Cycle Optimization. Ann. of CIRP, 30 (1981) 223-230.
4. G. Amitey, S. Malkin and Y. Koren. Adaptive Control Optimization of Grinding. J. Eng. Ind. ASME, 103 (1981) 103-108.
5. G. Giorleo and R. Pasquino. On the Stochastic Simulation and Optimization in Production Engineering, in Complex and Distributed Systems: Analysis, Simulation and Control. S. Tzafestas and P. Borno Eds., North Holland, Amsterdam (1986), 9-14.
6. L. Cooper and D. Steinberg. Introduction to Methods of Optimization. J. Wiley, New York, (1975).
7. A. Papoulis, Probability Theory, Random Variables and Stochastic Processes. McGraw Hill, New York, (1965).

A Hierarchical Multiple Model Adaptive Control for Stochastic Systems with Instrument Failures

Keigo Watanabe* and Spyros G. Tzafestas**

*College of Engineering, Shizuoka University, Johoku 3-5-1, Hamamatsu 432, Japan
**Department of Electrical Engineering, National Technical University, Zografou, Athens 15773, Greece

ABSTRACT

A new hierarchical multiple model adaptive control (MMAC) approach is described for the control problem of stochastic distributed systems with unknown instrument failures. The global and local hypotheses on the unknown actuator and/or sensor failures are assumed to be introduced to the decentralized multiple model structure. This leads to a flexible design algorithm for passively adaptive control strategies in steady-state. Furthermore, the coordinator algorithm in evaluating the global a posteriori probability is relatively simple to implement.

KEYWORDS

Adaptive control; Bayes methods; failure detection; hierarchical decision making; Kalman filter.

1. INTRODUCTION

The design of reliable, fault-tolerant, digital control systems requires that any failure of an actuator or sensor must be detected and identified immediately after failure and corrective action must be taken within a very short time to reconfigurate the control system to accomodate the failure. An integral part of such a system is usually called as a failure detection and identification (FDI) system. A recent survey or trend of different approaches to the FDI problem can be found in Isermann (1984) or Basseville and Benveniste (1986).

Among existing methods, we concentrate in this work on a design method of FDI systems based on the multiple model adaptive filtering (MMAF) technique (Lainiotis, 1976a). Several successful investigations have been already reported by Gustafson and co-workers (1978) and Willsky and co-workers (1980), etc. The MMAF method involves the use of a bank of Kalman filters -- one for each of the hypothesized failure modes and one for the normal mode. The a posteriori probabilities for all hypothesized modes can be directly used as "decision statistics". The associated decision making (or decision logic) can also be easily constructed such that the hypothesis whose probability exceeds a threshold is true. Furthermore, we can simply estimate or control the state of the system under the failure mode.

In this paper we develop a hierarchical multiple model adaptive control (MMAC) algorithm for controlling stochastic systems with actuator and/or sensor failures. A major assumption in our results is that the global and any local processors have different knowledge of the hypotheses on the unknown instrument failures. Namely, the local hypothesis in the ith local processor can only treat with the hypothesized modes which are involved in the corresponding local model, and therefore any local hypothesis is assumed to be included as an element of the global hypothesis. Furthermore, the coordinator algorithm for reconstructing the global a posteriori probability is much simpler than that due to other results (e.g., Teneketzis, Castanon and Sandell (1985)), instead of using the local a posteriori probabilities, by applying the probability densities for the innovations of the local Kalman filters.

2. PROBLEM STATEMENT AND ASSUMPTIONS

Consider the following discrete-time stochastic system

$$x(k+1) = Ax(k) + \sum_{i=1}^{m} B_i(\gamma_i)u^{(i)}(k) + Gw(k) \tag{2.1}$$

which is controlled by m actuators with scalar controls $u^{(i)}(k) \in R^1$ and is being observed by m sensors such as

$$z_i(k) = C_i(\theta_i)x(k) + v_i(k), \qquad i = 1,\dots,m \tag{2.2}$$

where $x(k) \in R^n$, $w(k) \in R^p$, $z_i(k)$, $v_i(k) \in R^1$, and A, $B_i(.)$, G and $C_i(.)$ are matrices with appropriate dimensions. The noises $\{w(k), v_1(k),\dots,v_m(k)\}$ are assumed to be zero-mean white Gaussian processes with covariances $Q \geq 0$ and $R_i > 0$, $i = 1,\dots,m$, and to be independent of the initial state x(0). γ_i and θ_i denote the unknown failure mode for the ith actuator and sensor, respectively, where they are restricted to take a value from the finite set,

$$\Omega_c^{(i)} \triangleq \{\gamma_i : \gamma_i = 1,\dots,M_c^{(i)}\}, \qquad \Omega_s^{(i)} \triangleq \{\theta_i : \theta_i = 1,\dots,M_s^{(i)}\} \tag{2.3}$$

in which it is assumed that their parameters are simply modeled as random variables.

Let us now introduce two m-tuple index set

$$\Theta_c \triangleq \{i_c : i_c = \gamma_1 \times \dots \times \gamma_i \times \dots \times \gamma_m ;\ \gamma_i \in \Omega_c^{(i)}\} \tag{2.4}$$

$$\Theta_s \triangleq \{i_s : i_s = \theta_1 \times \dots \times \theta_i \times \dots \times \theta_m ;\ \theta_i \in \Omega_s^{(i)}\} \tag{2.5}$$

where Θ_c is the index for the permutation of m unknown failures for the actuators and similarly Θ_s is the index for the permutation of m unknown failures for the sensors. In addition, we define 2-tuple two set such as

$$\Theta_{sc} \triangleq \{i_{sc} : i_{sc} = \theta_i \times i_c ;\ \theta_i \in \Omega_s^{(i)},\ i_c \in \Theta_c\} \tag{2.6}$$

$$\Theta \triangleq \{j : j = i_s \times i_c ;\ i_s \in \Theta_s,\ i_c \in \Theta_c\} \tag{2.7}$$

Then, we can also define two hypotheses representing unknown event as

1) H_j, $j \in \Theta$: It is a global hypothesis since it concerns the all unknown failures of actuators and/or sensors.
2) $\bar{H}^i_{j(i)}$, $j(i) = \ell \in \Theta_{sc}$: It is a local hypothesis since it only concerns the unknown failures of all actuators and the ith sensor.

It should be noted that the global hypothesis can be represented, using all local hypotheses, such as

$$H_j = (\bar{H}^1_{j(1)}, \bar{H}^2_{j(2)}, \dots, \bar{H}^m_{j(m)}) \tag{2.8}$$

where the local a priori probabilities $P_r^{(i)}(\bar{H}^i_{j(i)})$ and the global a priori probabilities $P_r(H_j)$ are assumed to be available.

The control problem formulated here is to find control sequences $\{u^{(i)}(k)\}$, i = 1,...,m which minimize a quadratic cost functional

$$J = \lim_{N \to \infty} \frac{1}{N} E\{ \sum_{k=0}^{N-1} [x^T(k)Q^C x(k) + \sum_{i=1}^{m} u^{(i)^T}(k)R_i^C u^{(i)}(k)]\} \tag{2.9}$$

subject to the stochastic system (2.1) and (2.2), which can be considered as a distributed system consisting of m local stations, where $Q^C \geq 0$ and $R_i^C > 0$, i = 1,...,m. When defining the local measurement history as $Z_k^i \triangleq \{z_i(j),\ 0 \leq j \leq k\}$, i = 1,...,m and the global measurement data as $Z_k \triangleq \{Z_k^1, \dots, Z_k^m\}$, and viewing the above problem as a team decision problem (Ho, 1980) with classical information pattern, we expect that the controllers are of the form

$$u^{(i)}(k) = \Phi\{k,\ Z_k, \{u^{(i)}(j),\ 0 \leq j \leq k-1\},\ i = 1,\dots,m\} \tag{2.10}$$

where Φ denotes a measurable function of the global measurement data and all past control inputs. Since the problem under consideration does not give an optimal solution, we must resort to suboptimal control strategies. In the following, we shall develop an MMAC technique to the "distributed system" (2.1) and (2.2) under the hypotheses 1) and 2). Before proceeding the further discussions, we state the following assumptions.

A1) The local measurement $z_i(k)$, i = 1,...,m are mutually independent. This means that $Z_k^1, \dots, Z_k^m$ are mutually independent and also $z_i(k)$ and Z_k^j are mutually independent, where $i \neq j$.

A2) The local measurement $z_i(k)$, i = 1,...,m are also conditionally independent given H_j and Z_{k-1}.

A3) The pairs $(A, \tilde{G})$ and $(C_i(\theta_i), A)$ are respectively stabilizable and detectable for $\theta_i \in \Omega_s^{(i)}$, where $\tilde{G}\tilde{G}^T = GQG^T$ and $\tilde{G}$ is a matrix of full rank such that rank $\tilde{G}$ = rank GQG^T.

A4) The pairs $(A, \tilde{G})$ and $(C(j(i_s)), A)$ are respectively stabilizable and detectable, where $C^T(j(i_s)) \triangleq [C_1^T(j(\ell,1)), \dots, C_m^T(j(\ell,m))]$ for $i_s \in \Theta_s$, $\ell \in \Theta_{sc}$, $j \in \Theta$, where $j(i_s)$ denotes the index $i_s \in \Theta_s$ for the permutation of m unknown sensor failure modes under the global hypothesis H_j and $j(\ell,i)$ included in $C_i(.)$ denotes the value of $\theta_i \in \Omega_s^{(i)}$ at the local hypothesis $\bar{H}^i_{j(i)=\ell}$ under the global hypothesis H_j.

A5) The pairs $(A, B(j(i_c))$ and (C_c, A) are respectively stabilizable and detectable, where $B(j(i_c)) \triangleq [B_1(j(\ell,1)),\ldots,B_m(j(\ell,m))]$ for $i_c \varepsilon \Theta_c$, $\ell \varepsilon \Theta_{sc}$, $j \varepsilon \Theta$, where $j(i_c)$ denotes the index $i_c \varepsilon \Theta_c$ for the permutation of m unknown actuator failure modes under the global hypothesis H_j, $j(\ell,i)$ included in $B_i(.)$ indicates the value of $\gamma_i \varepsilon \Omega_c^{(i)}$ at the local hypothesis $\bar{H}^i_{j(i)=\ell}$ under the global hypothesis H_j, and $Q^c = C_c^T C_c$, where C_c is a matrix of full rank such that rank C_c = rank Q^c.

3. A MULTIPLE MODEL ADAPTIVE FILTERING

3.1 A Parallel Processing Kalman Filter

The optimal (minimum variance) state estimate of x(k) given $\{Z_k^1,\ldots,Z_k^m\}$ and the global hypothesis H_j, i.e., $\hat{x}(k/k,j) \triangleq E[x(k)/Z_k^1,\ldots,Z_k^m, H_j]$, $\forall j \varepsilon \Theta$ can be obtained by

$$\hat{x}(k/k,j) = \hat{x}_*(k/k,j) + y(k,j(i_c)) \tag{3.1}$$

where $\hat{x}_*(k/k,j)$ is given by a steady-state Kalman filter

$$\hat{x}_*(k/k,j) = \hat{x}_*(k/k-1,j) + P_f(j(i_s)) \sum_{i=1}^{m} C_i^T(j(\ell,i))R_i^{-1} \times[\tilde{z}_i(k,j_{sc}) - C_i(j(\ell,i))\hat{x}_*(k/k-1,j)] \tag{3.2}$$

$$\tilde{z}_i(k,j_{sc}) \triangleq z_i(k) - C_i(j(\ell,i))y(k,j(i_c)) \tag{3.3}$$

$$\hat{x}_*(k+1/k,j) = A\hat{x}_*(k/k,j) \tag{3.4}$$

provided that the assumption A4) is satisfied, in which $P_f(j(i_s)) \triangleq \lim_{k\to\infty} P(k/k,j(i_s)) \triangleq E\{[x(k) - \hat{x}(k/k,j)][x(k) - \hat{x}(k/k,j)]^T/Z_k^1,\ldots,Z_k^m, H_j\}$ is subject to the equation

$$P_f^{-1}(j(i_s)) = P^{-1}(j(i_s)) + \sum_{i=1}^{m} C_i^T(j(\ell,i))R_i^{-1}C_i(j(\ell,i)) \tag{3.5}$$

Here, $P(j(i_s)) \triangleq \lim_{k\to\infty} P(k/k-1,j(i_s)) \triangleq E\{[x(k) - \hat{x}(k/k-1,j)][x(k) - \hat{x}(k/k-1,j)]^T/Z_{k-1}^1,\ldots,Z_{k-1}^m,H_j\}$ is a unique stabilizing solution (Kučera, 1972) to the following algebraic Riccati equation:

$$P(j(i_s)) - AP(j(i_s))A^T + AP(j(i_s))\{\sum_{i=1}^{m} C_i^T(j(\ell,i))[R_i + C_i(j(\ell,i))P(j(i_s)) \times C_i^T(j(\ell,i))]^{-1}C_i(j(\ell,i))\}P(j(i_s))A^T - GQG^T = 0 \tag{3.6}$$

Furthermore, $y(k,j(i_c))$ is generated by

$$y(k+1,j(i_c)) = Ay(k,j(i_c)) + \sum_{i=1}^{m} B_i(j(\ell,i))u^{(i)}(k) \tag{3.7}$$

It is then understood that equations (3.2) and (3.4) represent a steady-state Kalman filter conditioned on $\tilde{Z}_k = \{\tilde{Z}_k^1,\ldots,\tilde{Z}_k^m\}$, $\tilde{Z}_k^i \triangleq \{\tilde{z}_i(t,j_{sc}),\ 0 \le t \le k\}$ and H_j, in which $\tilde{z}_i(k,j_{sc})$ are assumed to be generated from a system

$$x_*(k+1) = Ax_*(k) + Gw(k) \tag{3.8}$$

$$\tilde{z}_i(k,j_{sc}) = C_i(\theta_i)x_*(k) + v_i(k), \quad i = 1,\ldots,m \tag{3.9}$$

3.2 Local Steady-State MMAF

The derivation of the local steady-state MMAF for the model (3.8) and (3.9) is routine, and therefore see Lainiotis (1976a) and Watanabe, Yoshimura and Soeda (1981) for details.

If the assumption A3) is satisfied, then the conditional estimates, $\hat{x}_*^{(i)}(k/k,j(i)=\ell) \triangleq E[x_*(k)/\tilde{Z}_k^i,\bar{H}_{j(i)}^i]$, $i = 1,\ldots,m$ may be obtained from the steady-state Kalman filter. That is,

$$\hat{x}_*^{(i)}(k/k,\ell) = \hat{x}_*^{(i)}(k/k-1,\ell) + K^{(i)}(j(\ell,i))[\tilde{z}_i(k,j_{sc}) - C_i(j(\ell,i))\hat{x}_*^{(i)}(k/k-1,\ell)], \quad \ell \in \theta_{sc} \tag{3.10}$$

$$\hat{x}_*^{(i)}(k+1/k,\ell) = A\hat{x}_*^{(i)}(k/k,\ell) \tag{3.11}$$

$$K^{(i)}(j(\ell,i)) = P^{(i)}(j(\ell,i))C_i^T(j(\ell,i))\tilde{P}^{(i)-1}(j(\ell,i)) \tag{3.12}$$

$$\tilde{P}^{(i)}(j(\ell,i)) = C_i(j(\ell,i))P^{(i)}(j(\ell,i))C_i^T(j(\ell,i)) + R_i \tag{3.13}$$

where $K^{(i)}(j(\ell,i))$ is the local elemental filter gain, and $P^{(i)}(j(\ell,i))$ is a unique positive semidefinite stabilizing solution to the following algebraic Riccati equation

$$P^{(i)}(j(\ell,i)) - AP^{(i)}(j(\ell,i))A^T + AP^{(i)}(j(\ell,i))C_i^T(j(\ell,i)) \times\tilde{P}^{(i)-1}(j(\ell,i))C_i(j(\ell,i))P^{(i)}(j(\ell,i))A^T - GQG^T = 0 \tag{3.14}$$

Furthermore, $\tilde{P}^{(i)}(j(\ell,i))$ is the steady-state covariance for the innovation process:

$$\nu_i(k,\ell) \triangleq \tilde{z}_i(k,j_{sc}) - C_i(j(\ell,i))\hat{x}_*^{(i)}(k/k-1,\ell) \tag{3.15}$$

The local a posteriori probability on the hypothesis $\bar{H}_{j(i)}^i$, i.e. $P_r^{(i)}(\ell/k) \triangleq P_r^{(i)}(\bar{H}_{j(i)=\ell}^i/\tilde{Z}_k^i)$ can also be computed by using the information supplied by these Kalman filters, as

$$P_r^{(i)}(\ell/k) = \frac{L^{(i)}(\ell/k)P_r^{(i)}(\ell/k-1)}{\sum_{j \in \theta_{sc}} L^{(i)}(j/k)P_r^{(i)}(j/k-1)} \tag{3.16}$$

where $L^{(i)}(\ell/k) \triangleq (2\pi)^{1/2} P_r^{(i)}(\tilde{z}_i(k,j_{sc})/\tilde{Z}_{k-1}^i, \bar{H}_{j(i)=\ell}^i)$ and

$$L^{(i)}(\ell/k) = [\tilde{P}^{(i)}(j(\ell,i))]^{-1/2}\exp\{-\frac{1}{2}\ \nu_i^2(k,\ell)\tilde{P}^{(i)-1}(j(\ell,i))\} \tag{3.17}$$

4. A HIERARCHICAL MULTIPLE MODEL ADAPTIVE FILTER AND CONTROL

4.1 A Hierarchical Multiple Model Adaptive Filtering

The centralized estimate $\hat{x}(k/k,j)$ of $x(k)$ given $\{Z_k, H_j\}$ can be reconstructed by

$$\hat{x}(k/k,j) = \sum_{i=1}^{m} [r^{(i)}(k,j) + V^{(i)}(j(i_s))\hat{x}_*^{(i)}(k/k,j(i)) + \frac{1}{m}\ y(k,j(i_c))] \tag{4.1}$$

$$V^{(i)}(j(i_s)) \triangleq P_f(j(i_s))P_f^{(i)-1}(j(\ell,i)) \tag{4.2}$$

$$P_f^{(i)}(j(\ell,i)) = [P^{(i)-1}(j(\ell,i)) + C_i^T(j(\ell,i))R_i^{-1}C_i(j(\ell,i))]^{-1} \tag{4.3}$$

Furthermore, the compensator $r^{(i)}(k,j)$ satisfies the following linear equation,

$$r^{(i)}(k,j) = A^*(j(i_s))r^{(i)}(k-1,j) + K^{*(i)}(j(i_s))\hat{x}_*^{(i)}(k/k-1,j(i)) \tag{4.4}$$

$$A^*(j(i_s)) \triangleq P_f(j(i_s))P^{-1}(j(i_s))A \tag{4.5}$$

$$K^{*(i)}(j(i_s)) \triangleq A^*(j(i_s))V^{(i)}(j(i_s)A^{-1} - P_f(j(i_s))P^{(i)-1}(j(\ell,i)) \tag{4.6}$$

where A is assumed to have no eigenvalues equal to zero.

The derivation of (4.1)-(4.6) may be adapted from a steady-state version of usual decentralized Kalman filters without controllers (Speyer, 1979; Watanabe, 1986). See Watanabe (1987) for more details.

The global a posteriori probability $P_r(j/k) \triangleq P_r(H_j/Z_k)$ can be recursively computed by

$$P_r(j/k) = \frac{\prod_{i=1}^{m} L^{(i)}(j(i)/k)P_r(j/k-1)}{\sum_{s\,\varepsilon\,\Theta} \prod_{i=1}^{m} L^{(i)}(s(i)/k)P_r(s/k-1)} \tag{4.7}$$

This derivation is accomplished by applying Bayes rule and assumptions A1) and A2).

4.2 A Hierarchical Multiple Model Adaptive Control

Let $u^{(i)}(k,j)$ denote the optimal control at the ith local processor which would result if H_j were true model, $\forall j \in \Theta$. Extending the steady-state version of the passive type MMAC algorithm (Lainiotis, 1976b; Watanabe, 1984) to the present problem, the control law is defined as

$$u^{(i)}(k) = \sum_{j\,\varepsilon\,\Theta} u^{(i)}(k,j)P_r(j/k) \tag{4.8}$$

The "elemental" control law $u^{(i)}(k,j)$ at the ith local processor is of the form

$$u^{(i)}(k,j) = F^{(i)}(j(i_c))\hat{x}(k/k,j) \tag{4.9}$$

where $\hat{x}(k/k,j)$ is given by equation (4.1) and the elemental control gain is

$$F^{(i)}(j(i_c)) = -[R_i^c + B_i^T(j(\ell,i))S(j(i_c))B_i(j(\ell,i))]^{-1} \times B_i^T(j(\ell,i))S(j(i_c))A \tag{4.10}$$

The positive semidefinite matrix $S(j(i_c))$, $\forall i_c \in \theta_c$ is a unique stabilizing solution to the algebraic Riccati equation

$$S(j(i_c)) - A^T S(j(i_c))A + A^T S(j(i_c))\{\sum_{i=1}^{m} B_i(j(\ell,i))[R_i^c + B_i^T(j(\ell,i)) \times S(j(i_c))B_i(j(\ell,i))]^{-1}B_i^T(j(\ell,i))\}S(j(i_c))A - Q^c = 0 \tag{4.11}$$

if assumption A5) is satisfied.

5. FAILURE DETECTION AND IDENTIFICATION MECHANISMS

In this section, we shall give a way to design an instrument FDI system by using the hierarchical MMAC technique developed in the previous sections. It is here assumed that the FDI system is designed for either sensor failures or actuator failures.

5.1 Local Decision Mechanism

The local a posteriori probability $P_r^{(i)}(\ell/k)$ given by (3.16) is adopted as a local detector. It reduces to a local detector for the ith sensor for the sensor FDI system, while it reduces to a local detector at the ith processor for all actuators for the actuator FDI system. The associated decision making can be constructed by simply choosing the local hypothesis whose probability exceeds a threshold. Note that in this paper we have used $P_r^{(i)}(\ell/k) = 0.9$ as a threshold.

In order to avoid that one of the probability becomes very small (or very large), we must set a bound on the probability as suggested in Willsky and co-workers (1980), because the unknown parameters γ_i and θ_i, $i = 1,\ldots,m$ are not modeled by Markov or semi-Markov jump parameters (Tugnait, 1983).

5.2 Global Decision Mechanism

The global a posteriori probability provided by (4.7) can be adopted as a global detector, which detects a failure and identifies the failed sensor among m sensors or the failed actuator among m actuators. The global decision making consists of two stages (as described in the case of a decentralized MMAF (Watanabe, 1987)): preparatory decision and final decision. The preparatory decision rule is simply to choose the global hypothesis whose probability exceeds a threshold (e.g., $P_r(j/k) = 0.9$), as expressed in the local decision making. This rule can be written as

1) The hypothesis H_j is true if $P_r(j/k) \geq 0.9$.

The final decision rule is assumed to be defined as

2) The hypothesis H_j is true only if $P_r(j/k) \geq 0.9$ and all $P_r^{(i)}(j(i)/k) \geq 0.9$, $i = 1,...,m$.

6. CONCLUSIONS

A hierarchical methodology has been presented to derive a passively adaptive control algorithm for linear discrete-time stochastic systems with unknown instrument failures. It has been then shown that the distributed multiple model structure introduced here is much more flexible and the resultant coordinator algorithm for reconstructing the global a posteriori probability is also extremely simpler than those studied by Teneketzis, Castanon and Sandell (1985). Therefore, the present control algorithm is thought to be a very practical one to approximate the LQG control problem with unknown actuator and/or sensor failures.

REFERENCES

Basseville, M. and A. Benveniste (1986). Detection of Abrupt Changes in Signals and Dynamical Systems, Springer-Verlag, New York.

Gustafson, D.E., A.S. Willsky, J.-Y. Wang, M.C. Lancaster and J.H. Triebwasser (1978). ECG/VCG rhythm diagnosis using statistical signal analysis -- I. Identification of persistent rhythms. IEEE Trans. bio-med. Engng., **BME-25**, 344-353.

Ho, Y.-C. (1980). Team decision theory and information structures. Proc. IEEE, **68**, 644-654.

Isermann, R.(1984). Process fault dtection based on modeling and estimation methods -- A survey. Automatica, **20**, 387-404.

Kučera, V. (1972). The discrete Riccati equation of optimal control, Kybernetika, **8**, 430-447.

Lainiotis, D.G. (1976a). Partitioning: A unifying framework for adaptive systems I: Estimation. Proc. IEEE, **64**, 1126-1143.

Lainiotis, D.G. (1976b). Partitioning: A unifying framework for adaptive systems II: Control. Proc. IEEE, **64**, 1182-1198.

Speyer, J.L. (1979). Computation and transmission requirements for a decentralized Linear-Quadratic-Gaussian control problem. IEEE Trans. Aut. Control, **AC-24**, 266-269.

Teneketzis, D., D.A. Castanon and N.R. Sandell, Jr. (1985). Distributed estimation for large-scale event-driven systems. In C.T. Leondes (Ed.) Control and Dynamic Systems, Vol.22, pp.1-45.

Tugnait, J.K. (1983). Control of stochastic systems with Markov interrupted observations. IEEE Trans. Aero. and Electron. Syst., **AES-19**, 232-239.

Watanabe, K., T. Yoshimura and T. Soeda (1981). A discrete-time adaptive filter for stochastic distributed parameter systems. Trans. ASME J. Dynamic Syst., Measure., and Control, **103**, 266-278.

Watanabe, K. (1984). A passive-type multiple model adaptive control (MMAC) of linear discrete-time stochastic systems with unknown observation subsystems. Int. J. Syst. Sci., **15**, 647-659.

Watanabe, K. (1986). Decentralized two-filter smoothing algorithms in discrete-time systems. Int. J. Control, **44**, 49-63.

Watanabe, K. (1987). A multiple model adaptive filtering approach to fault diagnosis in stochastic systems. In R.J. Patton et al. (Ed.) Fault Diagnosis in Dynamic Systems: Theory and Applications. Ch.10, Prentice-Hall International (UK) Ltd.

Willsky, A.S., E.Y. Chow, S.B. Gershwin, C.S. Greene, P.K. Houpt and A.L. Kurkjian (1980). Dynamic model-based techniques for the detection of incident on freeways. IEEE Trans. Aut. Control, **AC-25**, 347-360.

A Comparison of Two Techniques for Eigenstructure Assignment

S.K. Mudge and R.J. Patton

Department of Electronics, University of York
Heslington, York, YO1 5DD, England

Abstract

Given the linear state-space representation of a dynamic system, a method for computing allowable eigenvector sub-spaces using Singular Value Decomposition for both real and complex eigenvalues is illustrated. Once these spaces have been determined, it is possible to assign the eigenvectors in two ways, one which performs desirable weightings of the system states for each mode permitting system decoupling and the second which assigns eigenvectors iteratively to make the corresponding eigenvalues as insensitive to perturbations in the system matrices as possible. A computational procedure for each of these techniques is described. The work is illustrated using the stability augmentation system control design problem for the lateral motion model of a non-linear aircraft system. The modal requirements for this problem are well known and this enables a comparison of the two methods to be made. Conclusions can then be drawn about the robustness of general control system design using eigenstructure assignment.

1. Introduction

The degrees of freedom available in eigenstructure assignment using state feedback control are well known. Consider the linear time-invariant system described by the state equation:

$$\dot{\underline{x}} = A\underline{x} + B\underline{u} \tag{1.1}$$

where $A \in R^{nxn}$ $B \in R^{nxm}$. Applying linear state feedback $\underline{u} = F\underline{x}$ with $F \in R^{mxn}$, it was shown by Wonham in 1967 [1] that an F can be found to an arbitrary self-conjugate set of eigenvalues if the system is controllable. In 1976, Moore [2] described the freedom available to assign eigenvectors for an arbitrary self-conjugate set of eigenvalues using state feedback. He gives both necessary and sufficient conditions for a feedback matrix F to exist which satisfies:

$$(A + BF)\ \underline{v}_i = \lambda_i \underline{v}_i . \quad i = 1, \ldots, n \tag{1.2}$$

These are:

(i) The $\underline{v}_i$ are linearly independent vectors
(ii) $\underline{v}_i = \underline{v}_j^*$ whenever $\lambda_i = \lambda_j^*$
(iii) $\lambda_i \in$ span (N(I - A)).

Furthermore, if such a feedback matrix F exists and rank (B) = m then F is unique.

Once a set of eigenvalues has been chosen to obtain the desired damping and rise time of the system response, a decision then has to be made as to how the corresponding eigenvectors can be chosen. If information is known about a desirable weighting of the system states for each mode, it is possible to choose a desired eigenvector specification and project this in a least squares sense into the allowable eigenvector subspace [3]. It is of interest to note that modal assignment (based on a different approach) was of considerable interest a decade ago [4]. This is very desirable in such applications as aircraft control where the dynamics are well known and result in a well defined modal structure [4,5,6,7] which can be successfully shaped using eigenstructure assignment. If, however, such desired weightings are unknown or are unimportant, an alternative method of eigenvector assignment is available. This permits assignment of eigenvectors such that the eigenvalues are insensitive to perturbations in the system matrices [8].

This paper outlines a computational method for constructing allowable eigenvector subspaces using singular value decomposition. Once the allowable subspaces have been determined, the procedures for implementing both modal assignment and robust eigenstructure assignment are presented. The techniques are then illustrated and compared using as an example the design problem of a stability augmentation system for the lateral motion model of an aircraft. The desired modal structure of the closed-loop system response is well known, and so a modally shaped design is possible. A robust eigenstructure design is also performed assuming no prior knowledge of the coupling requirements. These two designs are then compared using a full force and moment nonlinear simulation in the presence of wind gusts. The modally shaped design will be insensitive to parameter variations because a known structure has been exploited. This may be used as a standard to assess the robust assignment technique. From these results some useful conclusions about the more general problem of eigenstructure assignment to control system designs are drawn.

2. The Computation of Allowable Eigenvector Subspaces

Let it first be assumed that $\lambda_i \in R$ so that $\underline{v}_i \in R^n$. Equation 1.2 may be rewritten as:

$$(\lambda_i I-A)\,\underline{v}_i - BF\underline{v}_i = 0 \tag{2.1}$$

Associate with each λ_i a matrix $S_{\lambda i} \in R^{n\times(n+m)}$ such that:

$$S_{\lambda i} = [\lambda_i I-A : B] \tag{2.2}$$

and a compatibly partitioned matrix

$$K_{\lambda i} = [N_{\lambda i}, M_{\lambda i}]^T \tag{2.3}$$

where the columns of $K_{\lambda i}$ form a basis for the null-space of $S_{\lambda i}$. Any vector $\underline{k}_i$ which post-multiplies $K_{\lambda i}$ gives a resulting vector which lies in the null-space of $S_{\lambda i}$. Thus:

$$[\lambda_i I-A:B]\begin{bmatrix} N_{\lambda i} \\ M_{\lambda i} \end{bmatrix}\underline{k}_i = \mathbf{0} \tag{2.4}$$

A comparison of (2.4) with (2.1) yields:

$$\underline{v}_i = N_{\lambda i}\,\underline{k}_i \tag{2.5}$$

with $\underline{v}_i \in$ Span $[N_{\lambda i}]$ and $\underline{k}_i$ determining where in the allowable subspace $\underline{v}_i$ exists.

If $\lambda_i \in \mathbf{C}$ the analysis is a little more complicated. Write:

$$\lambda_i = \lambda_{re} + i\lambda_{im} \tag{2.6}$$

Let the corresponding eigenvector $\underline{v}_i$ be written:

$$\underline{v}_i = \underline{v}_{re} + i\underline{v}_{im} \tag{2.7}$$

This eigenvector must satisfy (1.2) and hence:

$$(A + BF)\underline{v}_{RE} = \lambda_{re}\,\underline{v}_{re} - \lambda_{im}\,\underline{v}_{im} \tag{2.8}$$

$$(A + BF)\underline{v}_{im} = \lambda_{re}\,\underline{v}_{im} - \lambda_{im}\,\underline{v}_{re} \tag{2.9}$$

Equation (2.8) may be rewritten in matrix form as:

$$[\lambda_{re}I-A : \lambda_{im}\, I : B]\begin{bmatrix} \underline{v}_{re} \\ -\underline{v}_{im} \\ -F\underline{v}_{re} \end{bmatrix} = 0 \tag{2.10}$$

Similarly for equation (2.9)

$$[\lambda_{re}I-A : \lambda_{im}I : B]\begin{bmatrix} \underline{v}_{im} \\ \underline{v}_{re} \\ -F\underline{v}_{im} \end{bmatrix} = 0 \tag{2.11}$$

Thus for complex eigenvalues $S_\lambda \in \mathbf{R}^{nx\,(2n+m)}$ is defined by:

$$S_\lambda = [\lambda_{re}I-A : \lambda_{im}I : B] \tag{2.12}$$

with a compatibly partitioned K_λ whose columns constitute a set of basis vectors for the null space of S_λ.

$$K_\lambda = [\; N \;\; P \;\; M\;]^T \tag{2.13}$$

Combining the information from (2.10), (2.11) and (2.13):

$$\begin{pmatrix} \underline{v}_{re} \\ \underline{v}_{im} \end{pmatrix} \in \text{span}\begin{pmatrix} N \\ -P \end{pmatrix} \cap \text{span}\begin{pmatrix} P \\ N \end{pmatrix} \tag{2.14}$$

or

$$\in \text{span}\,(\alpha^T) \qquad \text{span}\,(\beta^T) \tag{2.15}$$

Let K_α and K_β be the matrices whose columns constitute bases for the null spaces of α and β respectively. Also let:

$$\gamma = [K_\alpha : K_\beta]^T \tag{2.16}$$

and K_γ be the matrix whose columns consitute a basis for the null-space of γ. K_γ is thus a basis for span $(\alpha^T) \cap$ span (β^T) and so:

$$\begin{pmatrix} \underline{v}_{re} \\ \underline{v}_{im} \end{pmatrix} = K_\gamma\, \underline{w}_T \tag{2.17}$$

The required null-spaces are computed using singular value decomposition [9]. Let $C \in R^{mxn}$ with $n > m$. The singular value decomposition of C is given by:

$$C = U \ \mathrm{diag}\,(\sigma_1, \ldots, \sigma_m)\ V^T \tag{2.18}$$

where $U \in R^{m\times m}$ and $V \in R^{n\times n}$ are orthogonal matrices and $\sigma_1 \geqslant \sigma_2 \geqslant \ldots \geqslant \sigma_m \geqslant 0$. Equation (2.18) may be re-expressed as :

$$C = U \begin{array}{c} \begin{array}{ccc} \quad m \quad & & \quad n-m \end{array} \\ m\left[\begin{array}{cccc c c} \sigma_1 & & & & : & \\ & \sigma_2 & & & & \\ & & & & : & 0 \\ & & & \sigma_m & : & \end{array}\right] \end{array} V^T$$

Post-multiplying both sides of equation (2.19) by V, it is noted that the last (n-m) columns of the CV product are null and so an appropriate null-space of $S_{\lambda i}$ is found using these columns. This method has the advantage that existing poles of the open-loop system or values near to them may still be chosen as eigenvalues of the closed-loop system. This would be important for example in the eigenvector assignment of a partially controllable system. Methods which utilise $(\lambda I-A)^{-1}$ in the subspace calculation may lead to numerical problems.

3. Closed-Loop Modal Assignment (Method 1)

If enough is known about the desired weightings of the various system modes, it is now possible to assign desired eigenvectors to accomplish the required decoupling. For a given problem only a few of the components of a desired eigenvector $\underline{v}_i^d$ are specified. To account for this a row re-ordering operation is computed [3] such that:

$$\mathrm{RTRANS}\begin{pmatrix} \underline{v}_{re}^{d} \\ \underline{v}_{im}^{d} \end{pmatrix} = \begin{pmatrix} \underline{v}_{re}^{*} \\ v \\ \underline{v}_{im}^{*} \\ v \end{pmatrix} \tag{3.1}$$

where $\underline{v}_{re}^{*}$ and $\underline{v}_{im}^{*}$ are the specified subvectors of real and imaginary components, respectively, and the v's denote unspecified elements.

The eigenvector $\underline{v}_i$ must be in the subspace spanned by K_γ. Thus

$$\begin{pmatrix} \underline{v}_{re} \\ \underline{v}_{im} \end{pmatrix} = \mathrm{RTRANS}\ K_\gamma \begin{pmatrix} \underline{\delta}_{re} \\ \underline{\delta}_{im} \end{pmatrix} \tag{3.2}$$

$$= \begin{bmatrix} P \\ Q \\ R \\ S \end{bmatrix} \begin{pmatrix} \underline{\delta}_{re} \\ \underline{\delta}_{im} \end{pmatrix} \tag{3.3}$$

and $\underline{\delta}_{re}$ and $\underline{\delta}_{im}$ are the projections into the allowable subspaces of $\underline{v}_{re}^{*}$ and $\underline{v}_{im}^{*}$, respectively. This is done in a least-squares sense and so:

$$\underline{\delta}_{re} = (P^T P)^{-1} P^T \underline{v}_{re}^{*} \tag{3.4}$$

and

$$\underline{\delta}_{im} = (R^T R)^{-1} R^T \underline{v}_{im}^{*} \tag{3.5}$$

It should be noted that if $\lambda_1, \lambda_2 \in \mathbf{C}$ then $\lambda_1 = \lambda_2^{*}$ because the eigenvalues form a self-conjugate set. Furthermore, from the results of Moore [2] $\underline{v}_1 = \underline{v}_2^*$ and so the specification of one complex eigenvalue and eigenvector contains all the essential information of the complex-conjugate pair. Thus for an n-state system with one complex eigenvalue pair the eigenvector matrix V is given by:

$$V = [\underline{v}_{re}, \underline{v}_{im}, \underline{v}_3, \underline{v}_4, \ldots, \underline{v}_n] \tag{3.6}$$

Once V has been determined, the computation of a feedback matrix F is well known and described elsewhere [8, 10, 11].

4. The Technique of Robust Eigenstructure Assignment (Method 2)

It is required that the closed-loop eigenstructure should be minimally sensitive to perturbations in the system matrices. Hence in this second method a control law is calculated by assigning linearly independent eigenvectors corresponding to the desired eigenvalues such that the matrix of eigenvectors is well-conditioned. The work described is an extension of the work of Kautsky, Nichols and Van Dooren. Their "Method 0" [8] has been augmented here to enable complex eigenvalue assignment using the allowable subspaces described in section 2 of this paper. This provides an algorithm which is simple to implement and is, although not guaranteed to converge, adequate to provide considerable improvement in conditioning, usually with little effort.

If $\underline{v}_i$ and $\underline{y}_i$ are the right and left eigenvectors of the closed loop system matrix A + BF corresponding to the eigenvalue λ_i the sensitivity of λ_i to perturbations in A, B and F is dependent upon the condition number c_i [12]:

$$c_i = \frac{\|\underline{y}_i\|_2 \, \|\underline{v}_i\|_2}{|y_i^T v_i|} \geqslant 1. \tag{4.1}$$

A bound on these sensitivities is given by:

$$\max \; c_i \leqslant \kappa_2(V) = \|V\|_2 \, \|V^{-1}\|_2 \tag{4.2}$$

where $\kappa_2(V)$ is the condition number of the matrix of right eigenvectors. The objective of "Method 0" is to choose a set of right vectors such that each is maximally orthogonal to the space spanned by the remaining vectors. This is equivalent to choosing $\underline{v}_i$ and the normalised vector $\underline{y}_i$ orthogonal to the space $l_i = \langle \underline{v}_j : j \neq i \rangle$ for all i. This is accomplished using an iterative process in which each vector $\underline{v}_i$ is replaced by a vector with maximum angle to the current space l_i for each i = 1, ...,n in turn. The new vector is obtained using the QR method. The decomposition of V_i is found as follows:

$$V_i = [\underline{v}_1, \ldots \underline{v}_{i-1}, \underline{v}_{i+1}, \ldots \underline{v}_n] \tag{4.3}$$

$$= [Q_i \; \underline{y}_i] \begin{bmatrix} R_i \\ O^T \end{bmatrix} \tag{4.4}$$

Thus the vector $\underline{y}_i$ orthogonal to l_i is found. This vector is then projected into the allowable subspace $N_{\lambda i}$ to give

$$= N_{\lambda i} \frac{N_{\lambda i}^T \, \underline{y}_i}{\|N_{\lambda i}^T \, \underline{y}_i\|_2} \tag{4.5}$$

This vector has minimum angle to $\underline{y}_i$.

Some modification of the above is required for the case of complex-conjugate eigenvalues. This is because for a complex eigenvector the real and imaginary parts are related via:

$$\begin{pmatrix} \underline{v}_{re} \\ \underline{v}_{im} \end{pmatrix} = K_{\gamma} \; \underline{w}_T \tag{4.6}$$

$$= \begin{pmatrix} S_{re} \\ S_{im} \end{pmatrix} \underline{w}_T \tag{4.7}$$

Thus if we take:

$$V = [\underline{v}_{re}, \underline{v}_{im}, \underline{v}_3, \ldots, v_n] \tag{4.8}$$

where $\underline{v}_3, \ldots, \underline{v}_n$ are real, then remove $\underline{v}_{re}$ and proceed as before to find the new real component which has minimum angle to the vector which is orthogonal to the set $< \underline{v}_{im}, \underline{v}_3, \ldots, \underline{v}_n >$ ie.

$$\underline{v}_{re}^{new} = S_{re} \frac{S_{re}^T \underline{y}_1}{\| S_{re}^T \underline{y}_1 \|_2} \tag{4.9}$$

It is necessary from (4.7), to update $\underline{v}_{im}$ such that:

$$\underline{v}_{im}^{new} = S_{im} \frac{S_{re}^T \underline{y}_1}{\| S_{re}^T \underline{y}_1 \|_2} \tag{4.10}$$

Now proceed with the sweep and update $\underline{v}_3, \ldots, \underline{v}_n$ in turn.

Any set of linearly independent eigenvectors belonging to the required allowable subspaces can be taken to give an initial eigenvector matrix. It has been found in practice to be better to update V after the (n-1)th iteration as:

$$V_n = V_{n-1} + (1-\alpha) V_{n-2} \qquad 0.5 < \alpha < 1. \tag{4.11}$$

This is because the imaginary parts of the eigenvector are not updated using the maximum orthogonality criterion but are predetermined from the corresponding real part. Some random oscillation of successive V_k results which is smoothed using (4.11).

5. A Practical comparison of the two techniques of Eigenstructure Assignment

The design of a stability augmentation system for the lateral motion of a light aircraft is now considered. The equations of motion of a fixed wing aircraft are usually decoupled into two sub-systems; the longitudinal motion is concerned with forward velocity and pitching excursions whilst the lateral motion relates to roll, yaw and sideslip velocity. For the vehicle under consideration the equations of motion and the aerodynamic parameters are well defined [13]. The lateral and longitudinal motions are assumed decoupled for control system design and a "stick-fixed" linearisation corresponding to a $33ms^{-1}$ airspeed of the non-linear system gives the following lateral system linear model:

$$\dot{\underline{x}} = A\underline{x} + B\underline{u} \tag{5.1}$$

where

$$\underline{x}^T = (v, p, r, \varphi, \psi, \zeta, \xi)$$

= (sideslip velocity,roll rate,yaw rate,roll angle,yaw angle,rudder angle,aileron angle) (5.2)

$$\underline{u} = \begin{pmatrix} \zeta_c \\ \xi_c \end{pmatrix} \begin{matrix} \text{-rudder angle demand} \\ \text{-aileron angle demand} \end{matrix}$$

$$A = \begin{bmatrix} -0.277 & 0.000 & -32.900 & 9.810 & 0.000 & -5.432 & 0.000 \\ -0.103 & -8.325 & 3.750 & 0.000 & 0.000 & 0.000 & -28.640 \\ 0.365 & 0.000 & -0.639 & 0.000 & 0.000 & -9.490 & 0.000 \\ 0.000 & 1.000 & 0.000 & 0.000 & 0.000 & 0.000 & 0.000 \\ 0.000 & 0.000 & 1.000 & 0.000 & 0.000 & 0.000 & 0.000 \\ 0.000 & 0.000 & 0.000 & 0.000 & 0.000 & -10.000 & 0.000 \\ 0.000 & 0.000 & 0.000 & 0.000 & 0.000 & 0.000 & -5.000 \end{bmatrix} \quad (5.3)$$

$$B^T = \begin{bmatrix} 0.000 & 0.000 & 0.000 & 0.000 & 0.000 & 20.000 & 0.000 \\ 0.000 & 0.000 & 0.000 & 0.000 & 0.000 & 0.000 & 10.000 \end{bmatrix}$$

The lateral motion is characterised by three modes; roll, Dutch roll and the spiral mode. These modes are required to be such that the spiral mode is very slow , the roll made is fast and stable and the Dutch roll mode has 0.7 damping ratio. The rudder and aileron actuator modes are determined by the actuator limitations. A set of suitable eigenvalues for these closed-loop modes have been chosen as:

$$\begin{aligned} \lambda_1, \lambda_2 &= -2.0 \pm j\,1.0 && \text{(roll mode)} \\ \lambda_3, \lambda_4 &= -1.5 \pm j\,1.5 && \text{(Dutch roll mode)} \\ \lambda_5 &= -0.05 && \text{(Spiral mode)} \\ \lambda_6 &= -15.0 && \text{(rudder actuator mode)} \\ \lambda_7 &= -10.0 && \text{(aileron actuator mode)} \end{aligned} \quad (5.4)$$

For the modally assigned controller design (Method 1) it is necessary to choose a set of desired eigenvectors. The criteria for doing so are well known. For example, it is necessary to decouple the rolling motions from the yawing motions; if the aircraft is banked a side force is created due to the tilting of the lift vector which will create sideslip or yaw. The converse is also true. These desired eigenvectors are chosen to produce the required handling qualities [14] through the following structure:

$$\underline{v}_1, \underline{v}_2 = \begin{matrix} 0 & 0 \\ 1 & * \\ 0 & 0 \\ * & 1 \\ * & * \\ * & * \\ * & * \end{matrix}, \quad \underline{v}_3, \underline{v}_4 = \begin{matrix} 1 & * \\ 0 & 0 \\ * & 1 \\ 0 & 0 \\ * & * \\ * & * \\ * & * \end{matrix} \quad \underline{v}_5 = \begin{matrix} 0 \\ * \\ * \\ 1 \\ * \\ * \\ * \end{matrix} \quad \underline{v}_6 = \begin{matrix} * \\ * \\ * \\ * \\ * \\ 1 \\ * \end{matrix} \quad \underline{v}_7 = \begin{matrix} * \\ * \\ * \\ * \\ * \\ * \\ 1 \end{matrix} \quad (5.5)$$

where * denotes that the magnitude of the element is unimportant. The desired eigenvectors are projected in a least-squares sense into the computed allowable subspace and yield the following eigenvector matrix [$\underline{v}_1, .., \underline{v}_7$]:

$$\begin{bmatrix} 1.7E-02 & -3.60E-03 & 1.0 & 33.14 & 0 & 1.74 & 6.88 \\ 0.96 & -2.70 & 0 & 0 & -5.00E-02 & -9.64E-02 & 14.53 \\ -0.27 & 0.25 & 1.49 & 1.0 & 0.30 & 0.62 & 1.34 \\ -0.92 & 0.89 & 0 & 0 & 1.0 & 6.42E-03 & -1.45 \\ 0.16 & -4.66E-02 & -0.16 & -0.83 & -6.03 & -4.11E-02 & -0.13 \\ -1.17E-02 & 6.48E-02 & 0.33 & 1.13 & -1.87E-02 & 1.0 & 1.59 \\ 0.34 & 0.60 & 0.19 & 1.14E-02 & 5.39E-02 & 5.20E-02 & 1.0 \end{bmatrix} \quad (5.6)$$

For the matrix V_1:

$$\kappa_2(V_1) = 123.7 \qquad (5.7)$$

and the individual eigenvalue sensitivities are given in the second column of table 1 below:

λ_i	c_i(Method 1)	c_i(Method 2)
$-2.0 \pm j1.0$	5.93	6.59
$-1.5 \pm j1.5$	15.93	15.38
-0.05	1.12	1.13
-10	5.06	4.57
-15	2.79	2.63

Using the robust assignment technique a second full-state feedback matrix is generated. For the corresponding eigenvector matrix V_2:

$$\kappa_2(V_2) = 43.2 \qquad (5.10)$$

and the individual eigenvalue sensitivities are shown in the final column of table 1.

There is little difference between the individual eigenvalue sensitivities for Method 1 and Method 2. This is to be expected, as by modally shaping this system one must be rendering it insensitive to some degree to perturbations. Figure 1 shows a comparison of the sideslip velocity and roll angle responses for both modally decoupled and robustly assigned controllers when applied to the linear system with an initial condition on the heading angle of $\psi = 0.02$ radians. As it to be expected some coupling into sideslip velocity is found with the Method 2 control system, but as for the bank angle the other state trajectories follow very closely for both controllers.

The two control systems are now tested more realistically with a non-linear simulation of the unmanned aircraft. This simulation incorporates both the full force and moment lateral and longitudinal dynamics together with cross-coupling effects. To realistically challenge the robustness of the controllers wind turbulence is applied to the sideslip velocity in the form of a Dryden Spectrum [15] . Figure 2 shows the v and φ states for this non-linear simulation. The state trajectories for both controllers are seen to track. Using equation (5.3) the non-linear model was linearised at each simulation sample point and the eigenstructure of the closed-loop matrix so formed investigated. Figure 3 shows the eigenvalue variation for the real part of the Dutch Roll mode which was seen to be the worst conditioned in both cases. In the presence of unknown disturbancs applied to a non-linear model the eigenvalues were seen to hold very well. It can thus be seen that the Robust eigenstructure assignment design performs as well as the modally shaped design.

6. Conclusion

This paper has presented two alternative methods of eigenstructure assignment both of which are computationally straightforward to implement in practice. The methods have been illustrated using as an example the design of a stability augmentation system for the lateral motion of an aircraft. The required damping and modal assignment for this problem is well known. This enables a comparison as to how satisfactorily the robust eigenstructure assignment method performs. From non-linear simulation results in the presence of external disturbances it can be concluded that the two methods produce very similar, stable system responses holding the assigned closed-loop eigenvalues well. In general one can say that if the desired weightings are well-known a closed-loop design can be determined easily. However, if such knowledge is unavailable the technique of robust eigenstructure assignment will provide a very satisfactory controller for the price of an iterative, albeit straightforward, design procedure

7. Acknowledgements

This work has been supported by an SERC CASE Studentship to Miss S.K.Mudge in collaboration with British Aerospace, Aircraft Group, Brough, North Humberside,U.K.

8. References

1. Wonham, W.M., On Pole Assignment in Multi-input, Controllable Linear Systems, IEEE Trans, 1967, AC-12, pp. 660-665.
2. Moore, B.C., On the Flexibility Offered by State Feedback in Multivariable systems Beyond Closed Loop Eigenvalue Assignment, IEEE Trans, 1976, AC-21, pp. 659-692.
3. Harvey, C.A, and Stein, G., Quadratic Weights for Asymptotic Regulator Properties, IEEE Trans, 1978, AC-23, pp.378-387.
4. Porter,B. and Crossley, T.R., Modal Control Theory and Applications, Taylor and Francis Ltd.,1972.
5. Sobel, K.M., and Shapiro, E.Y., Eigenstructure Assignment for Design of Multi-Mode Flight Control Systems, IEEE Control Systems Magazine, 1985, 5, pp. 9-15.
6. Sobel, K.M., and Shapiro, E.Y. Application of Eigenstructure Assignment to Flight Control Design: Some Extensions, Journal of Guidance, Control and Dynamics, 1987, 10, pp. 73-81.
7. Alag, G.S., and Duke, E.L., Development of Control Laws for a Flight Test Maneuver Autopilot, Journal of Guidance, Control and Dynamics, 1986, 9, pp. 441-445.
8. Kautsky, J., Nichols, N.K., and Van Dooren, P., Robust Pole Assignment in Linear State Feedback, Int. J. Control, 1985, 41, pp. 1129-1155.
9. Golub, G., and Van Loan, C., Matrix Computations, North Oxford Academic,1983.
10. D'Azzo, J.J. and Houpis, C.H., Linear Control System Analysis and Design, McGraw-Hill Book Company, 1975.
11. Mielke, R.R., Tung, L.J., and Carraway, P.I., Design of Multivariable Feedback Control Systems Via Spectral Assignment using Reduced-order Models and Reduced-order Observers, Nasa Contractor Report 3889, 1985.
12. Wilkinson, J.H., The Algebraic Eigenvalue Problem, Oxford University Press, 1965.
13. Mcruer, D., Ashkenas, I., and Graham D., Aircraft Dynamics and Automatic Control, Princeton University Press, 1973.
14. MIL - F 8785C, Flying Qualities of Piloted Vehicles, 1980.
15. Gault,J.D. and Gunter,D.E.,Atmospheric Turbulence Considerations for Future Aircraft Designed to Operate at Low Altitudes, J.Aircraft,5,(2),1968.

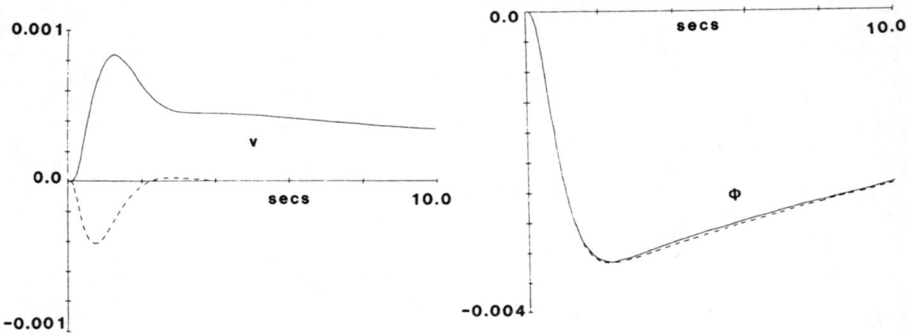

Figure 1 Linear Simulation

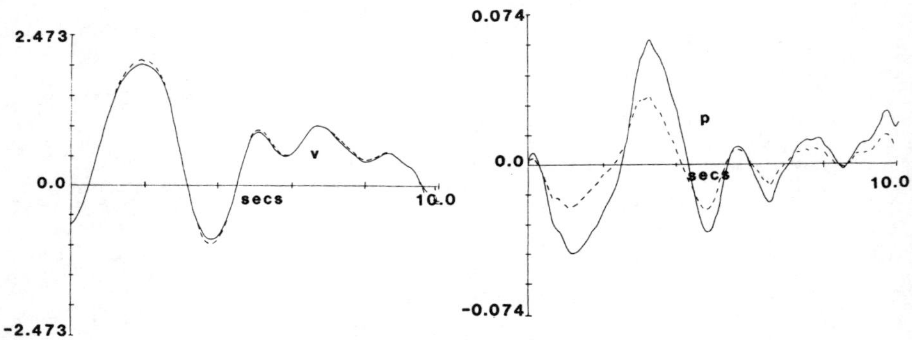

Figure 2 Non-linear Simulation

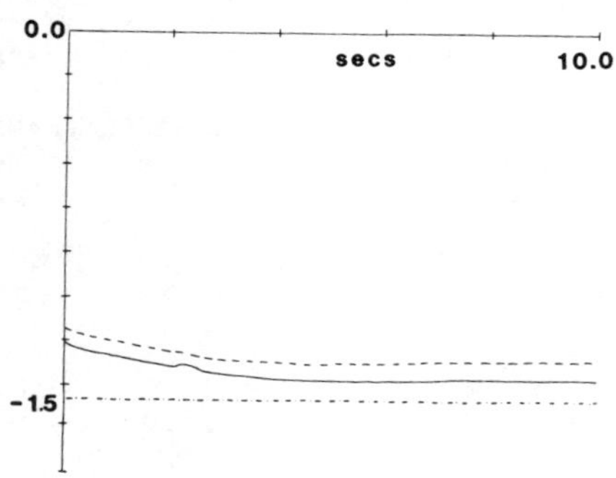

Method 1 - - - - -

Method 2 ———

Figure 3 Variation of Real Component

of Dutch Roll Eigenvalue

IV. Fault Tolerant Digital Techniques

Petri Nets Approach to Modelling and Logical Fault Detection of MOS LSI Circuits

Ewa Idzikowska

Computer Science Centre, Technical University of Poznań, pl. M. Skłodowskiej-Curie 5, 60-965 Poznań, Poland

ABSTRACT

In this paper the problems of modelling nMOS LSI/VLSI circuits and of their physical faults, as well as fault detection in these circuits are considered from a new perspective. Petri nets have been used as the modelling tool. This paper describes a method of modelling MOS circuits with the help of Petri nets, as well as of modelling classical (stuck-at and bridging faults) and nonclassical faults by means of Petri nets. The nonclassical faults change the structure of MOS cells and are characteristic of MOS LSI technology. The problem of detecting all types of physical failures has also been considered. The correct and faulty Petri nets models of the circuit are the starting point of the process of test generation procedures. The test sets generated in this way detects all detectable single and multiple faults.

KEYWORDS

Petri nets; circuit modelling; fault models; fault detection; test generation; LSI/VLSI testing.

INTRODUCTION

The advent of LSI/VLSI technology has made fault modelling and the generation of test sets for detecting physical failures a complex problem. Logical modelling of these failures depends on the technology being used. The single stuck-at model is the simplest and most widely accepted fault model. However, many physical failures, like opens and shorts in nMOS circuits, cannot be modelled in this way. In this paper the method of modelling shorts and opens between the interconnection lines presented by Bhattacharya and Gupta (1984) has been adopted, which exactly simulates the behaviour of the faulty circuits. As was also the case in the above mentioned work, our considerations have been limited to MOS complex cells which consist of series-parallel connections of FET's, and to physical failures which do not disturb this structure.
Recently, research has been carried out on the application of Petri

nets in test generation. Different approaches to digital circuits modelling by means of Petri nets have also been observed. These nets are a simple and convenient formal device used in studying information flow, being intrinsically able to represent both concurrent and conflicting events, both static and dynamic situations. Petri nets can represent digital circuits at various levels. In many works (Kha and others, 1981; Costagnolo, Corsi, 1983; Hura, 1984) the circuits have been modelled at gate level, and in the work of Alukaidey and Musgrave (1984) - at system level.
In this paper a new approach to fault modelling and detection in MOS complex cells using Petri nets has been presented. MOS cells are modelled on the basis of their electrical diagrams. First, models of correct cells are created, and then - models with given faults. Thes models are used in the generation of tests detecting all these fault

DEFINITIONS AND NOTATION

The basic terminology and notations to be used in this paper are defined as follows:

Def. 1. (Best, Fernandez, 1986)
A triple $N=(S,T,F)$ is called a Petri net (PN) iff:

a/ $S \cap T=0$,
b/ $F \subseteq (S \times T) \cup (T \times S)$,
c/ $S \cup T \neq 0$,
d/ $dom(F) \cup cod(F)=S \cup T$,

where:

X - a set of elements of the net, $X=S \cup T$,
S - a set of places,
T - a set of transitions,
F - a flow relation between the elements of X,
$dom(F)=\{x \mid y: (x,y) \in F\}$,
$cod(F)=\{x \mid y: (y,x) \in F\}$,
$x,y \in X$.

We represent a PN by a digraph. The vertices of this graph are place and transitions represented by circles and dashes, respectively. The flow relation F is represented by arrows between the elements of X.

Def. 2. (Best, Fernandez, 1986)
Let $N=(S,T,F)$ be a Petri net, and $x \in X$. Then:

$^{\bullet}X=\{y \mid (y,x) \in F\}$ (Pre-set of $x \in X$),
$X^{\bullet}=\{y \mid (x,y) \in F\}$ (Post-set of $x \in X$).

Def. 3. (Best, Fernandez, 1986)
A function $M:S \rightarrow \mathbb{N}$ is called a marking of a Petri net $N=(S,T,F)$. One can represent the marking M by a vector $\vec{M}$, such that its i-th component is $M(s)$, $s \in S$.

Def. 4. (Best, Fernandez, 1986)
A function $M_0:S \rightarrow \mathbb{N}$ is an initial marking function.

Starting from the initial marking M_0, we build the marking class $\bar{M}_0$ of the net by iterative use of transition firing. A transition may be fired when it is enabled for the marking M.

Def. 5.
A transition $t \in T$ is enabled for the marking M if $\forall s \in {}^{\bullet}t$, $M(s)>0$. The firing of t for marking M changes this marking into the new marking M'. We may denote this fact by $M \xrightarrow{t} M'$.

Def. 6. (Peterson, 1981)
A marked PN is a doubled (N, M_0), where N is a PN and M_0 is an initial marking of N. A marked PN is said to be bounded if for every one of its places S there exists $k \in \mathbb{N}$, such that $\forall M \in \bar{M}_0$, $M(s)<k$.

From now on we will only be concerned with bounded Petri nets. Using thus defined Petri nets we will model MOS complex cells.

LOGICAL MODELLING OF PHYSICAL FAILURES IN MOS CELLS

A majority of physical failures that can occur in MOS complex cells is constituted by breaks and unwandet connections (opens and shorts, respectively) in metallization and diffusion layers (Galidy, Crouzet, Vergianault, 1980). They may involve a FET as well as the interconnection lines. Some of these failures can be easily modelled by stuck -at faults, but for others this type of modelling is not possible. For instance, consider the series-parallel MOS cell shown in Fig. 1.

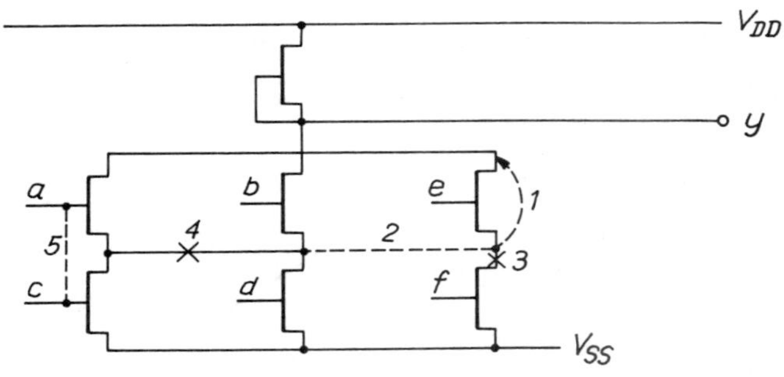

Fig. 1. Series-parallel complex cell.

Figure 1 shows the most frequent failures. Short 1 between the source and the drain of the FET can be modelled by stuck-at 1 fault at e-primary input. Similarly open 3 can be modelled by stuck-at 0 fault at f-primary input. Short 5 is functionally equivalent to the bridging fault between the primary inputs a and c, and can be modelled by AND bridging (for positive logic). Faults 2 and 4 - short and open faults between interconnection lines - cannot be modelled in this way. The correct function is given by $\bar{y}=(a+b)\cdot(c+d)+e\cdot f$, and the faulty functions y_2 and y_4 are given by $\bar{y}_2=(a+b+e)\cdot(c+d+f)$ and $\bar{y}_4=a\cdot c+b\cdot d+e\cdot f$. These failures cannot be modelled by any stuck-at faults. Bhattacharya and Gupta (1984) have shown that such failures in a series- parallel MOS complex cell can always be modelled by some bridging faults, provided the fault does not disturb the series -parallel structure of the cell. Open 4 can be modelled by AND bridging between lines a and c, or b and d, and denoted by: $*(a/c)$, or $*(b/d)$. Short 2 can be modelled by OR bridging and denoted by: $+(a+b/e)$, or $+(c+d/f)$.

PETRI NETS MODEL OF MOS COMPLEX CELLS

The Petri net of a series-parallel MOS complex cell can be obtained by representing series connections of FET's as Boolean AND, and parallel connections as Boolean OR functions of the FET's inputs, and adding an inverter at the primary output. The MOS cell can therefore be modelled using Petri-nets models of elementary logical functions - logical product, sum, and negation. The problem of modelling of logical functions is discussed in details by Idzikowska (1988). Let every line or logical variable x be represented by place x for the affirmated form and place $\bar{x}$ for the negated form. The logical value 1, /0/ of a path or a variable is modelled as $M(x)=1$, $M(\bar{x})=0$ /$M(\bar{x})=1$, $M(x)=0$/.

Logical Product

The Petri net shown in Fig. 2 models logical product function $f=\prod_{i=1}^{N} x_i$, $N \in \mathbf{N}$.

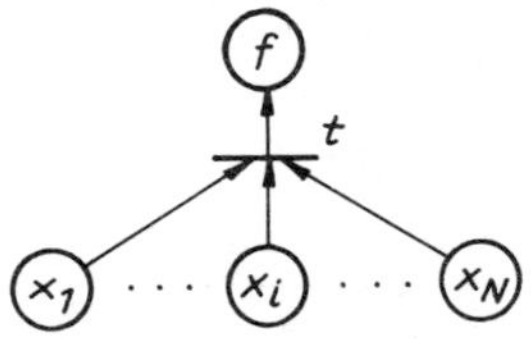

Fig. 2. Petri net model of the logical product function.

Def. 7.
The Petri net models logical product function $f=\prod_{i=1}^{N} x_i$, if the following relation is valid:

$$\exists f \in t^{\bullet},\ M(f)=1 \Leftrightarrow \forall x_i \in {}^{\bullet}t,\ M(x_i)=1.$$

This may be illustrated as follows:

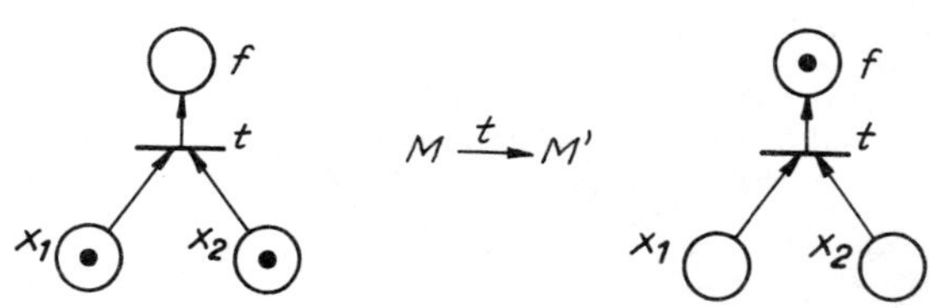

Fig. 3. Petri net model of the logical product function $f=x_1 \cdot x_2$.

Logical Sum

The net in Fig. 4 models logical sum function $f=\sum_{i=1}^{N} x_i$, $N \in \mathbf{N}$.

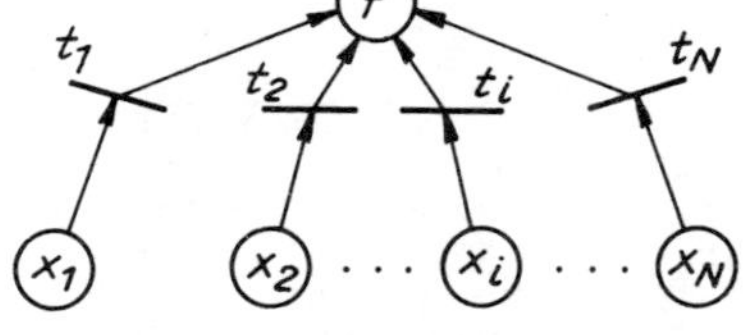

Fig. 4. Petri net model of the logical sum function.

Def. 8.
The Petri net models logical sum function $f=\sum_{i=1}^{N} x_i$, if there exist N transitions t_i T such that $\forall t_i$, $f \in t_i^{\bullet}$; and $M(f)=1$ if at least one of these transitions is enabled for marking M.

This definition is illustrated by the following example:

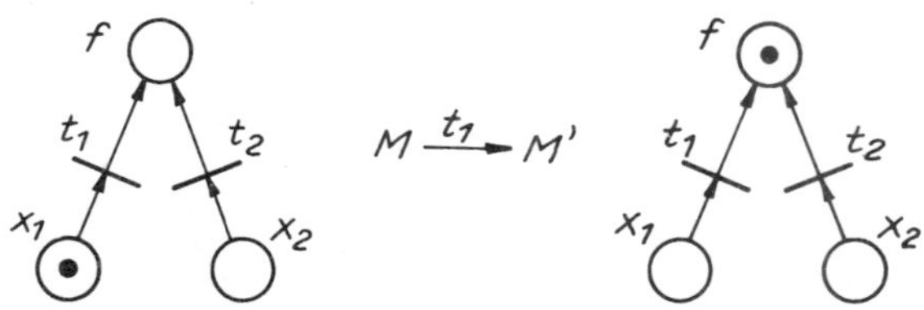

Fig. 5. Petri net model of the logical sum function $f=x_1+\bar{x}_2$.

According to Def. 5 transition t_1 is enabled for firing. As a result marking M changes to M'.

Logical Negation

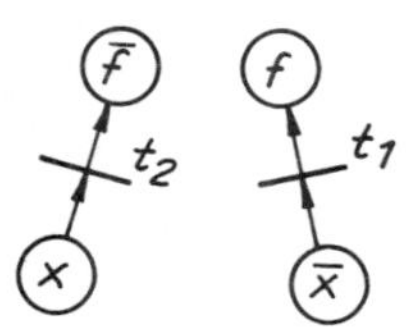

Fig. 6. Petri net model of the logical negation function $f=\bar{x}$.

Def. 9.
Let $\bar{f} \in t_2^{\bullet}$ and $x \in {}^{\bullet}t_2$, or $f \in t_1^{\bullet}$ and $\bar{x} \in {}^{\bullet}t_1$. The Petri net represents a logical negation if the relation $M(f)=1 \Leftrightarrow M(\bar{x})=1$ or $M(\bar{f})=1 \Leftrightarrow M(x)=1$ is valid.

Petri net model of nMOS complex cell from Fig. 1 is shown in Fig. 7.

PETRI-NETS FAULT MODELS

The already mentioned logical models of physical failures in a MOS cell, should now be represented by means of Petri net.

Stuck-at Faults

To make modelling of these failures possible, we replace the description of places x and $\bar{x}$ used so far, with the expression:

$$X=x \cdot x_n \oplus x_1 \quad \text{and} \quad \bar{X}=\bar{x} \cdot x_n \oplus x_0,$$

where:

x_n, x_1, $x_0 \in \{0,1\}$, and $x_n=1$ if x is non-faulty, $x_1=1$ if x is stuck-at 1, $x_0=1$ if x is stuck-at 0 (Friedman, Menon, 1974).

For short 1 $e_1=1$, hence E=1. This means that place E contains a token, regardless of the values assumed by e ($M(E)=1$, $M(\bar{E})=0$). Similarly, for open 3 $f_0=1$, hence $\bar{F}=1$ which means that regardless of the values assumed by f, place $\bar{F}$ contains a token but F cannot contain one ($M(F)=0$, $M(\bar{F})=1$). Figure 8.a shows a net which models the faulty circuit with fault 1.

Short Faults

A short between two or more primary input paths can always be modelled by AND bridging for positive logic. The Petri net model of the circuit must then be supplemented with an extra functor model between bridging paths.

Faults between Interconnection Lines

These failures are modelled by OR or AND bridging. The Petri nets model must also be supplemented with extra functor models. Thus, open 4 is modelled by AND bridging between paths a and c, or b and d. Figure 8.b shows the Petri net of this faulty circuit.

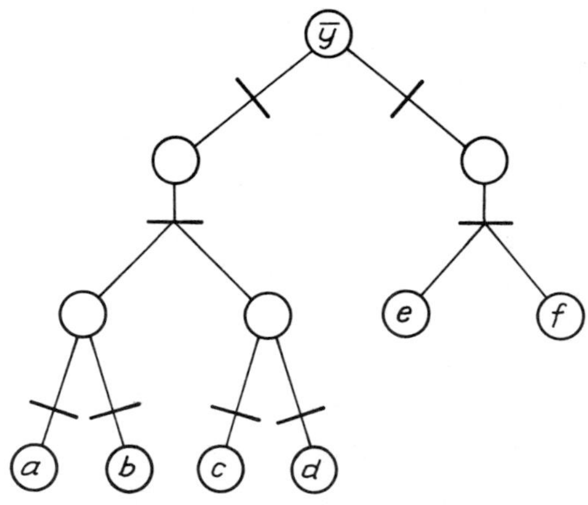

Fig. 7. The Petri net of the MOS cell from Fig. 1.

GENERATION OF COMPLETE TEST SETS

On the basis of the Petri net representing a MOS cell, its logical function may be determined. In doing so however, we should follow the principles of token transfer. For logical product and sum the following principles may be formulated:

Logical product: $\forall s_k \in \{t_i^{\bullet}\}$, $M(s_k)=1 \Leftrightarrow \forall s_j \in \{{}^{\bullet}t_i\}$, $M(s_j)=1$.

Logical sum: $M(s_k)=1 \Leftrightarrow \exists t_i$, $s_k \in \{t_i^{\bullet}\}$, $\forall s_j \in \{{}^{\bullet}t_i\}$, $M(s_j)=1$.

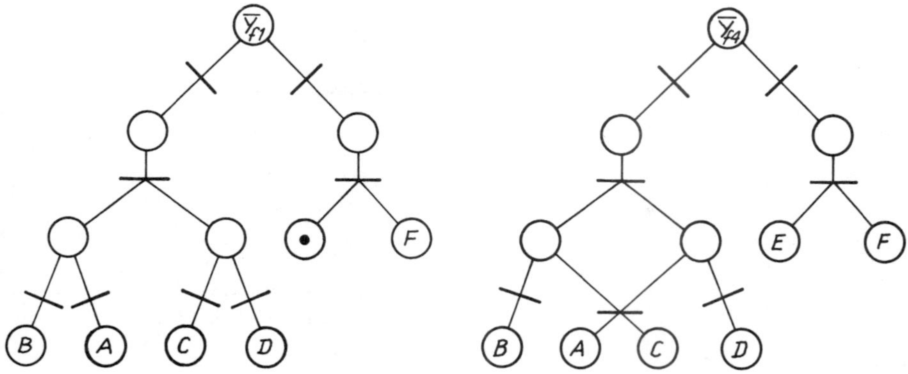

a/ b/

Fig. 8. Petri net models of a faulty MOS cell: short 1 /a/, and open 4 /b/.

To make the fault detectable, the output of the faulty circuit must have a value opposite to that of the output of the correct circuit. If Y denotes the functions of the correct circuit and Y_f of the faulty circuit, then test T which detects a given fault must satisfy the equation:

$$T = Y \oplus Y_f = Y \cdot \overline{Y}_f + \overline{Y} \cdot Y_f.$$

Y and Y_f are determined on the basis of the Petri nets models of the correct circuit and the circuit with a given failure. The function of the correct MOS cell determined on the basis of the model in Fig. 7 satisfies the equation: $\overline{y} = (a+b)(c+d)+ef$. For the short between the drain and source of FET fault 1, stuck-at 1 at e-input, $M(E)=1$ and $M(\overline{E})=0$. The net thus modified realizes the function $\overline{y}_{f1} = (a+b)(c+d)+f$. The test to detect this failure must satisfy the equation:

$$T_{f1} = y \oplus y_{f1} = \overline{a}\overline{b}\overline{e}f + \overline{c}\overline{d}\overline{e}f.$$

On the other hand, the faulty circuit with open 4 satisfies the equation $\overline{y}_{f4} = ac+bd+ef$, hence:

$$T_{f4} = y \oplus y_{f4} = (a\overline{b}\overline{c}d + \overline{a}bc\overline{d})(e+f).$$

The tests thus generated for all possible failures form a complete test set.

SUMMARY

The paper presents a method of modelling and testing series-parallel nMOS complex cells. For the purposes of testing, the cells are modelled by means of Petri nets. The Petri-nets model is formed on the basis of its electrical diagram. The complexity of Petri-nets models defined as the number of places, transitions, and arcs is lower than that of the models of analogous circuits (Costagnolo, Corsi, 1983). Physical failures of MOS cells are also easily modelled by means of Petri nets. The set of tests detecting a given failure is generated with the help of the nets models of the faulty and correct circuit. This method is particularly useful in automatic test generation. The paper presents a Petri nets approach to the problem of structural testing of MOS cells. The approach may also be applied in testing other digital circuits, and not only at structural but also at functional level.

REFERENCES

Alukaidey, T. (1984). Petri-net test generation on systems. Proceedings European Conference on Electronic Design EDA 84, 72-78.

Best, E., and C. Fernandez (1986). Notations and terminology on Petri net theory. Petri Net Newsletter, 23, 21-46.

Bhattacharya, B. B., and B. Gupta (1984). Logical modelling of physical failures and their inherent syndrome testability in MOS LSI/VLSI networks. Proceedings of 1984 International Test Conference, 847-855.

Costagnolo, B., and F. Corsi (1983). Modelling digital circuits with delays by stochastic Petri nets. Microelectron. Reliab., Vol. 23, 6, 1075-1086.

Friedman, A. D., P. R. Menon (1974). Wykrywanie Uszkodzeń w Układach Cyfrowych. WNT, Warszawa.

Galiay, J., Y. Crouzet, and M. Vergianault (1980). Physical versus logical fault models in MOS LSI circuits: impact on their testability. IEEE Trans. Comput., Vol. C-29, 6, 527-531.

Hura, G. S. (1984). A Petri net approach to the analysis of combinational logic circuits. Policy and Information, Vol. 8, 2, 95-101.

Idzikowska, E. (1988). Petri net modelling of logical circuits. Modelling, Simulation and Control, A, Vol. 15, 2, 1-15. Accepted for publication.

Khan, A. A., G. S. Hura, H. Singh, and N. K. Nanda (1981). State equation representation of logic operations through a Petri net. Proceedings of the IEEE, Vol. 69, 4.

Peterson, J. L. (1981). Petri Net Theory and Modeling of Systems. Prentice-Hall, Inc., Englewood Cliffs, N. J.

VLSI Wafers and Boards Diagnostics Using Multisignature Analysis

A. Hlawiczka

Technical University of Gliwice, Gliwice, Poland

ABSTRACT

The paper presents the idea of locating faulty chips on VLSI wafers and fault diagnostics on PCB's by means of multisignature analysis and the use of a built-in evaluator with MISR.

KEYWORDS

Multisignature analysis; VLSI wafer diagnostics; PCB diagnostics; built-in self-test; built-in evaluator; boundary scan path.

INTRODUCTION

Fault diagnostics is a very important problem at each level of hierarchy from chip prototypes, through VLSI wafers, Multichip Logic Modules, printed circuit boards (PCB), to a full digital system. To make these diagnostics simple, cheap, and effective, it is necessary to introduce into each of the above mentioned hierarchy levels a common framework of testing and design of testability as well as a standard that clarifies the link between the testability technology and testing technology. The LSSD and boundary scan path (Beenker and colleagues, 1986) techniques for structured implementation of testability, as well as the built-in self-test (BIST) with the random pattern generator (RPG) and compression of the circuit under test (CUT) response using the signature analysis (SA) make a framework which, from the author's point of view, satisfies almost all the above mentioned requirements.
A simplified scheme of VLSI circuit testing using such a framework is given in Fig. 1.

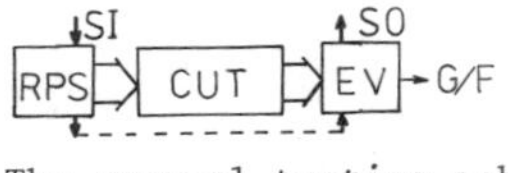

Fig. 1. The general testing scheme

where: SI - boundary scan path input, SO - boundary scan path output, EV - evaluator using signature analysis, G/F - evaluator output informing that CUT is good (G) or faulty (F).
The fault coverage of such a BIST technique depends on the effectiveness of the input test sequences generated by the test pattern generator in sensitizing the CUT faults, and on the fault escape caused by the errors being masked by the signature analyzers. Let us assume in this paper that the effectiveness of the input sequence

is guaranteed. The probability of fault escape in a single sequential or parallel signature analysis (1SA) depends on the length n of the applied Linear Feedback Shift Register (LFSR) or Multi Input Shift Register (MISR), and is equal to 2^{-n} (Bhavsar, Krishnamurthy, 1984). Due to this fault escape, some faults are not detected, and in connection with this, they are not located in, e.g. the traceback analysis or the shift-register-latch fault analysis (Curtin, Waicukauski, 1983), using the 1SA technique. Even if the 1SA technique gives high fault coverage, it can be found however that some faults, though distinguishable in conventional testing and detectable in the 1SA technique, are undistinguishable in the full fault dictionary analysis (Curtin, Waicukauski, 1983) using 1SA. This means that the distinction potential defined by David (1980) is not high, which gives in effect too low a diagnostic resolution.

The aim of this paper is to present various aspects of using the multisignature analysis for the recovery of faults (Hlawiczka, Kubica, 1981; Bhavsar, Krishnamurthy, 1984; Hlawiczka, 1986) and for increasing the diagnostic resolution, in VLSI wafer's and on printed circuit boards constructed from IC's.

MULTISIGNATURE ANALYSIS AND ITS INFLUENCE ON THE RECOVERY OF FAULTS AND ON DIAGNOSTIC RESOLUTION

The LFSR (MISR) extension, made in order to reduce the probability of fault escape, leads to the costly redundancy of the hardware of VLSI chips and PCB's (Hlawiczka, Kubica, 1981). The idea of heightening the detection effectiveness without the necessity of LFSR (MISR) extension consists in forming k different signatures of the same diagnostic data stream, where each signature uses the LFSR (MISR) with different feedbacks (Bhavsar, Krishnamurthy, 1984 ; Hlawiczka, 1986). This multisignature analysis is known as multiple compression or as the k-Signature Parallel Analysis (kSPA) (Hlawiczka, 1986). Each of k divisions in the kSPA analysis, using MISR with k internal Exclusive-OR type feedbacks, is represented as the following expression:

$$e_z(x)/p_j(x) = q_j(x) + r_j(x)/p_j(x) \tag{1}$$

where:

$$e_z(x) = \sum_{i=0}^{m-1} e_i(x)x^i; \quad j \in \{1,2,\dots,k\}$$

$e_i(x)$ is a polynomial of at most m-1 degree describing an error stream at input i of the MISR ($e_i(x) = u(x) + u_f(x)$ where $u(x)$ and $u_f(x)$ are polynomials describing fault-free and faulty responses on ith output of the CUT, respectively).

$p_1(x), p_2(x), \dots, p_k(x)$ are characteristic polynomials of the MISR k feedbacks, each being of n degree.

$q_j(x)$ is the quotient of the division described by expression 1.

$r_j(x)$ is the remainder of this division of at most n-1 degree which can be treated as a signature containing n bits.

The concatenation of such k signatures $[s_{e1}, s_{e2}, s_{e3}, \dots, s_{ek}]$ can be treated as one signature s_e containing kn bits.

The effectiveness of error detection provided by the above mentioned kSPA analysis results from the following polynomial division (Hlawiczka, 1986).

$$e_z(x)/\mathrm{LCM}[p_1(x), p_2(x), \dots, p_k(x)] = q(x) + r(x)/\mathrm{LCM}[p_1(x), p_2(x), \dots, p_k(x)] \tag{2}$$

where $r(x)$ is the remainder of at most kn-1 degree and $p(x) = \mathrm{LCM}[p_1(x), \dots, p_k(x)]$ is the characteristic polynomial of at most kn degree. Let us notice that if for example the polynomials $p_j(x)$ are primitive polynomials then

$p(x) = \prod_{j=1}^{k} p_j(x)$ is of kn degre. A single signature analysis using MISR with the characteristic polynomial p(x) will be treated as a substitute scheme of the kSPA analysis. Definition: The term kMISR(kLFSR) will refer to the MISR(LFSR) with k internal Exclusive-OR type feedback described by such characteristic polynomials $[p_1(x), p_2(x), \ldots, p_k(x)]$, each being of n degree that the polynomial

$p(x) = \mathrm{LCM}[p_1(x), p_2(x), \ldots, p_k(x)]$ is of kn degree.

It is evident from 1 and 2 that for the kMISR(kLFSR) register the following conclusion is true:

$$r(x) = 0 \Longleftrightarrow \{r_1(x) = 0 \wedge r_2(x) = 0 \wedge, \ldots, \wedge r_k(x) = 0\}$$

So, $s_e = 0 \Longleftrightarrow \{s_{e1} = 0 \wedge s_{e2} = 0 \wedge, \ldots, \wedge s_{ek} = 0\}$

For further simplification of the discussion carried on in this chapter let us assume that in the CUT all fault derived errors are uniformly mapped into all signatures produced by kMISR. Then the probability E of fault escape, using kMISR and m>>n, is equal to 2^{-kn} which, versus the probability $E = 2^{-n}$ in a single signature analysis using MISR, gives very good chances of fault recovery, and due to this, fault location. The discussion in (Hławiczka, 1986) presents a good example of fault recovery using kMISR. It is clearly indicated there that the kSPA technique enables to detect the fault derived kn,d burst errors at each input of the applied n stage MISR. Meanwhile, the same MISR in the one-signature analysis detects only the fault derived n,d burst errors. So, using the kSPA technique in the traceback analysis and/or in the shift-register-latch fault analysis we can increase the diagnostic resolution. But the influence of the kSPA technique on the increase in the diagnostic resolution in the full fault dictionary analysis needs a somewhat different approach to the discussion.

Let F be the set of faults in the CUT detected by an input test sequence and U_f be the set of faulty responses of the CUT corresponding to the faults in F. Let U be the set of all possible error responses. Let S_k be the set of all signature concatenations $s = [s_1, s_2, \ldots, s_k]$ likely to obtain in the kSPA analysis. So, $|S_k| = 2^{kn}$. Let us assume that all fault derived errors in U_f are distinguishable, i.e. $|F| = |U_f|$. Let us notice that the diagnostic resolution depends on the number of different signatures which can be observed. Hence, in order to distinguish faults from the set F using signatures in 1SA, it is assumed that $|F| \leqslant |S_1|$, which means that $|F| \leqslant 2^n$. David defined (1980) the distinction potential for LFSR as the probability that $s^f \neq s^g$ for any fault f and g in the set F. Let $D(|F|, |S_1|)$ denote the probability for the |F| signatures obtained using MISR and a being distinct. This probability can be expressed, as in (1980), using the following formula:

$$D(|F|, |S_1|) = d^{|F|} |S_1|!/(|S_1| - |F|)!$$

where: d - is a priori probability of the signature $s^f \in S_1$. Due to the uniformity of the mapping provided by MISR, $d = 2^{-n}$. As at the same time $|S_1| = 2^n$, therefore

$$D(|F|, 2^n) = 2^{-n|F|} 2^n!/(2^n - |F|)!$$

If we assume, for the time being, that $|F| = |S_1|$, then even for $n \geqslant 5$, $D(2^n, 2^n) \cong 0$. Increeased distinction potential, at least to the level of probability of fault detection obtaind in the 1SA technique, can be achieved using the kSPA analysis. Then, having assumed the number of faults |F|, it can be determined for which number of k the inequality below is satisfied:

$$D(|F|, |S_k|) \geqslant 1 - E$$

For $n \geqslant 8$, $|F| = 2^n$ and $k \in \{2,3,4\}$ the following values of the distinction potential have been obtained:

$$D(2^n, 2^{2n}) \cong 0.6, \quad D(2^n, 2^{3n}) \cong 1-2^{-n}, \quad D(2^n, 2^{4n}) \cong 1$$

THE EVALUATOR FOR kSPA ANALYSIS

The evaluator in the BIST structure performs the two following basic functions:

- compresses into a signature s responses of CUT to a test,
- compares the measured signature s with the reference signature s_o and generates the signal G if $s = s_o$ or F if $s \neq s_o$.

The evaluator carrying out the kSPA analysis must perform k times the above mentioned functions using, among other things, kMISR and/or kLFSR. Such an evaluator will be abbreviated to kEV. The problem is how to design the kEV in order to make the circuit overhead cost as low as possible and to make the number of the integrated circuit additional pins reduced to a minimum. The solution to the former problem can be found by way of the following approaches:

a) proper selection of the k characteristic polynomials $p_1(x), p_2(x), \dots, p_k(x)$,
b) using of only one built-in checking of the correct self-test signature (Hlawiczka, 1987),
c) application of the evaluator operation mode as a shift register in the boundary scan path.

The latter problem can be solved using the boundary scan path to perform the following functions:

d) scanning in, through the SI input, the information about the evaluator operation mode,
e) scanning out, through the SO output, the information: good/faulty (G/F).

Let us begin from a. The question is how to properly select k characteristic polynomials $p_j(x) = 1 + a_{j1}x + a_{j2}x^2 + \dots + a_{ji}x^i + \dots + a_{jn-1}x^{n-1} + x^n$ so that the set of their coefficients a_{ji} may imply a minimum overhead cost of realization of internal Exclusive-OR type feedback of kMISR. It is very difficult to give any precise criteria of selection of the above mentioned polynomials. This problem ought to be solved. Let us however try to analyze this problem in a very simplified manner.

Figure 2 illustrates the ith line of the kMISR feedback. The cost of scanning in the kMISR of such an ith feedback line depends on the cost of the Boolean expression performing the following function $G_i = q_i \oplus g_i(t_1, t_2, \dots, t_p)q_n$ where:

$$g_i(t_1, t_2, \dots, t_p) = a_{ji} = \begin{cases} 0\text{-denotes the lack of the ith line in the jth feedback} \\ 1\text{-denotes the presence of the ith line in the jth feedback} \end{cases}$$

The function $g_1(t_1, t_2, \dots, t_p)$, performed by the combinational circuit CC_i, according to the state of the variables $\langle t_1, t_2, \dots, t_p \rangle = T_j$ by means of which the jth feedbacks are formed described by the polynomials $p_j(x)$, takes the values of the coefficients $\{a_{1i}, a_{2i}, \dots, a_{ji}, \dots, a_{ki}\}$.

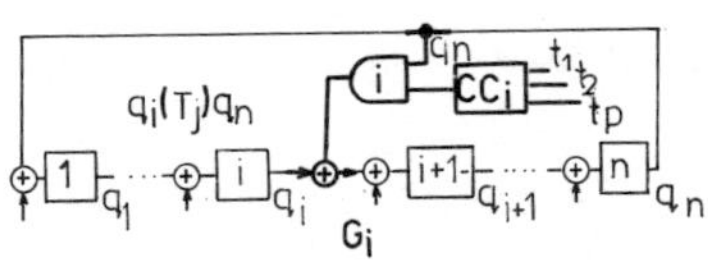

Fig. 2. The simplify kMISR scheme

Let us denote by $\{T_j\}$ the set of k possible states at the input of CC_i. Let us notice that if $g_i(T_j) = 0$ for each T_j, then it means the lack of the ith line ($a_{ji} = 0$ for each j) in each of k feedbacks. In such a case, $G_i = q_i$ does not involve any overhead costs either. If in turn $g_i(T_j) = 1$ for each T_j, then it means that the ith line ($a_{ji} = 1$ for each j) is present in each of k feedbacks. In this case, $G_i = q_i \oplus q_n$ involves the cost of one XOR gate. In other cases, in which $G_i = q_i \oplus g_i(T_j)q_n$, the cost of the Boolean expression representing the function $g_i(T_j)$ depends on the values of the coefficients $a_{1i}, a_{2i}, \dots, a_{ki}$ of k characteristic polynomials $p_j(x)$ chosen to perform kMISR. On this basis, two following rules can be proposed to select feedbacks for 2MISR, 3MISR, and 4MISR:

- the polynomials $p_1(x), p_2(x), \dots, p_k(x)$ should guarantee kn degree for the polynomial $p(x) = \mathrm{LCM}[p_1(x), p_2(x), \dots, p_k(x)]$,
- the polynomials $p_1(x), p_2(x), \dots, p_k(x)$ should force the highest possible number of the functions $g_i(T_j) = 0$ for each T_j and the lowest possible number of the

functions $g_i(T_j) = 1$ for each T_j, as well as at most k functions $G_i = q_i \oplus g_i(T_j)q_n$. By way of these rules, the polynomials of 7th and 8th degree were chosen to carry out the 2SPA, 3SPA, and 4SPA analysis. The examples of these polynomials are presented below.

$$\left.\begin{array}{ll} 1 + x^7 ; & 1 + x^2 + x^3 + x^6 + x^8 \\ 1 + x^3 + x^7 ; & 1 + x^2 + x^5 + x^6 + x^8 \\ 1 + x^4 + x^7 ; & 1 + x^3 + x^5 + x^6 + x^8 \\ 1 + x^3 + x^4 + x^5 + x^7 ; & 1 + x^2 + x^3 + x^5 + x^8 \end{array}\right\} \begin{array}{l} \text{2SPA (first two rows)} \\ \text{3SPA (first three rows)} \\ \text{4SPA (all four rows)} \end{array}$$

Verification of k signatures by kEV with conventional circuit solutions would resolve itself to the necessity of memorizing, by kEV, k pattern signatures and generating the G/F signal k times. Such an approach results in unnecessarily high overhead costs. Another approach is presented in (Hlawiczka, 1987). In this papers simple procedure were described for finding the initial value of the MISR (LFSR) register used for compression responses in BIST, so the final good-machine signature is always unique, say all ones or zeros. The idea of such a compression technique results from the following expressions:

$$u_z(x)/p_j(x) = q_{zj}(x) + r_{zj}(x)/p_j(x); \quad x^m h_j(x)/p_j(x) = q_{hj}(x) + r_{hj}(x)/p_j(x) \qquad (3)$$

where: $u_z(x)$ is the polynomial representing the MISR input pattern, $h_j(x)$ is the polynomial representing initial value of MISR in jth measurement. First equation 3 presents an algebraical description of the compression of the MISR input pattern different from zero with the initial value equal to zero. Second equation 3 presents an algebraical list of the MISR operation with the initial value different from zero and the MISR input pattern equal to zero. The addition of the sides of these equations results in the equation describing the MISR compressor operation with the initial value different from zero. Let us notice that the initial state $h_j(x)$ can be selected in such a way that $r_{zj}(x) + r_{hj}(x) = 0$ or $\sum_{i=0}^{n-1} x^i$. Introducing in each of k measurements in the kSPA technique, through the boundary scan path input SI in kMISR (kLFSR), a different initial value which is however such that in each measurement it forces always the same unique signature it is possible to use, in the BIST structure, the cheap kEV (Hlawiczka, 1987).

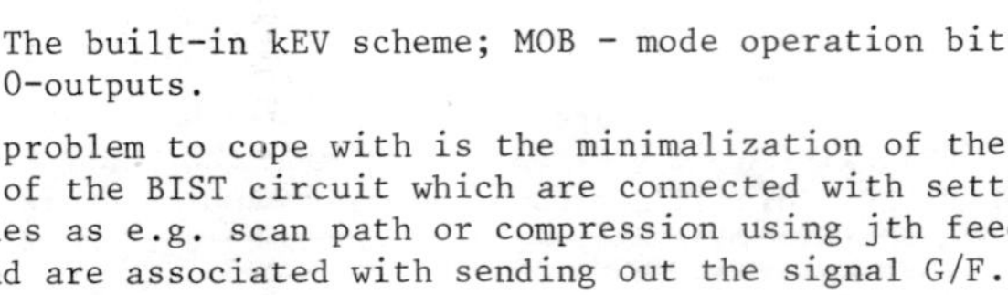

Fig. 3. The built-in kEV scheme; MOB - mode operation bits, 0-outputs.

Another essential problem to cope with is the minimalization of the number of input/output pins of the BIST circuit which are connected with setting in kEV of such operation modes as e.g. scan path or compression using jth feedback of kMISR $(t_1, t_2, \ldots, t_p)$, and are associated with sending out the signal G/F. This can be executed using operation mode type cells (Beenker and colleagues, 1986) which, together with data bit cells, form the boundary scan path architecture in the VLSI circuit. Due to this technique, e.g. $t_1, t_2, \ldots, t_p$ bit stream can be shifted, through input SI, to the shift-register stages which hold operation mode bits. Besides this, the G/F signal can be sent, through multiplexer and output SO, out of the VLSI circuit, as is illustrated in Fig. 3. Let us notice that the additional mode bit w, applied by SI, controls the above mentioned multiplexer, enabling it to send out, to output SO, either the G/F signal or a signature from kMISR together with other data bits included in the scan path. The above described kEV can be built in the VLSI circuit involving low overhead costs. Further in the paper, such a kEV will be abbreviated to built-in kEV (BIKEV). Basic cells of the kEV illustrated in Fig. 3 must also have the architecture enabling pin testing in a package (Kommonytsky, 1983). So, five package dedicated pins with the functions TEST, CLOCK, LOAD/SHIFT, SO, and SI which usually serve the purpose of testing the BIST circuit by the boundary scan standard (Beenker and colleagues, 1986) are sufficient to perform all the kEV operation modes.

VLSI WAFER AND BOARD DIAGNOSTICS

During the manufacture of the standard VLSI circuits only the parameters of the circuits on a wafer are tested. With the appearance of application-specific integrated circuits (ASIC) their manufacturers have attempted to test the functionality of the devices on a wafer before its breaking up for packaging. Some companies perform a self-test of the entire wafer before it is cut. Other companies apply testers which test device reliability using needles that access nodes on the silicon wafer. If the BIST circuits, which are placed on a wafer, make use of the boundary scan path architecture, then the five package dedicated pins mentioned in the previous chapter are sufficient to initiate a test in each device on the wafer and to state whether each circuit has passed the test. So, each device on a wafer can be tested functionally by an external tester, using five needles. If each BIST circuit on a wafer contains BIKEV, then the external tester becomes a very simple diagnostic device. Its functions are limited only to the scanning in, through node SI, of the initial value for kMISR, the generating of the signals TEST, CLOCK, SHIFT/LOAD, and the reading out of the signal G/F on node SO. In the case of the kSPA analysis this process should be repeated k times for each device on a wafer. Taking into account additional costs of each repeated test in the kSPA analysis, the manufacturer can limit himself only to the carrying out of 1SPA. On special demand of the customer, in order to obtain ASICs of high reliability, the manufacturer can additionally, before cutting the wafer, carry out the 2SPA, 3SPA or 4SPA analysis. Then however, he can include in the calculation of the ASICs price, the cost of each of the additional tests carried out in the kSPA technique. If the self-test control unit (STCU), performing functions which are similar to those of the above mentioned wafer, is fabricated on a wafer containing ICs with BIKEV, then STCU carries out at the same time the kSPA analysis of all these digital ICs. Typical wafer probers are not required for the tests. The signals G/F from each chip after finishing the test are parallelly written in the cells of the special shift register stages of STCU, from where they are shifted outside STCU through one output. Test start as well as readout of the test results can, in this case, be, for example initiated from outside of the wafer by means of special electron beams directed to two CMOS inverter pads.

The test task, in the case of a circuit constructed from a collection of modules, e.g. PCB, can be decomposed into several phases. Among these phases there are the following two:

a) confirmation that each module performs its required function; interior test in (Beenker and colleagues, 1986), and
b) confirmation that the modules are interconnected in a correct manner; an exterior test in (Beenker and colleagues, 1986) and that they can be realized by means of the BIKEVs which are present in each module and in a special Universal Test Controller Chip (UTCC) dedicated to board self-test (Hlawiczka, Badura, 1987).

In each of these phases, the boundary scan paths of each integrated element on a board are connected into one long ring whose SI input and SO output are connected with UTCC as it is illustrated in Fig. 4. The primary inputs and outputs of a board are also connected with UTCC. The interior test as well as the exterior test are controlled by UTCC. In phase a, UTCC loads initial values into each BIKEV in each integrated element through the boundary scan ring, and then, after finishing the interior self-test, it sends out in the same way the signals G/F from each module.

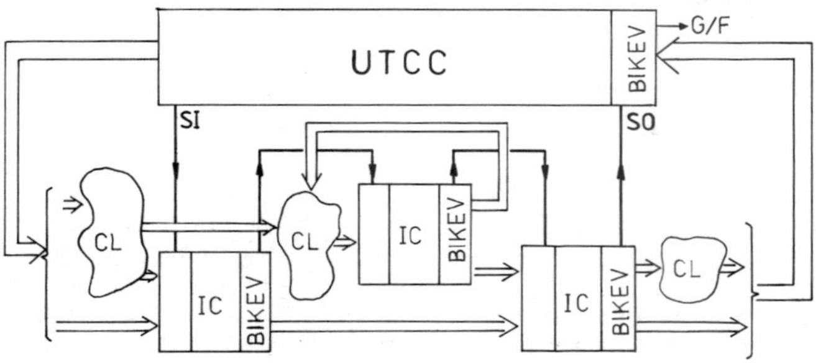

Fig. 4. The PCB self-test scheme; CL - combinational logic

Owing to the BIKEVs included in the integrated elements under test, UTCC can control the performance of the kSPA technique during the interior self-test. It is particularly useful in the case of service diagnostics where the time of the interior test k times longer does not involve high costs. In phase b, the boundary scan ring is used for testing the circuits and interconnections assembled outside the integrated elements packages. The cells of this ring, connected with output pins, are used for sending out test stimuli, whereas the cells connected with input pins are used for capturing test results. In this exterior test, the BIKEV becomes useful, contained in UTCC, which carries out alternately compression, in the kLFSR mode, of test response appearing at the boundary scan ring output, and compression, in the kMISR mode, of test response appearing at the board prime outputs. The final effect of so carried out a compression is the G/F signal at UTCC output and a signature containing information about circuits and interconnections outside the modules forming the boundary scan ring. The G/F signal can be the interrupt signal for the external diagnostic processor containing the full fault dictionary of tested PCB. Then the G/F signal, signalling faults, simultaneously initiates, in the diagnostic processor, the interrupt service program whose basic task is readout of a signature from BIKEV and stating which fault occured in the tested PCB, on the basis of the full fault dictionary containing the set of all characteristic signatures.
In the case of detection, by the diagnostic process, of non-distinguishable faults in the 1SPA technique, the kSPA technique can be applied. An example of such an analysis repeated k times is the table presented below in which, by s_{ij}, jth signature is denoted obtained in the ith step of the 3SPA technique.

TABLE 1 Fault Dictionary for 3SPA Technique

Fault	f_1	f_2	f_3	f_4	f_5	f_6	f_7
step 1	$s_{1,1}$	$s_{1,2}$	$s_{1,3}$	$s_{1,3}$	$s_{1,3}$	$s_{1,8}$	$s_{1,8}$
step 2	-	-	$s_{2,4}$	$s_{2,5}$	$s_{2,5}$	$s_{2,9}$	$s_{2,10}$
step 3	-	-	-	$s_{3,6}$	$s_{3,7}$	-	-

Let us notice that the faults f_3, f_4, and f_5 are non-distinguishable in the first step. In the second step on this group of faults also the faults f_4 and f_5 are non-distinguishable. It is in the third step that these faults are distinguishable. The application of the 3SPA technique enables in this example to distinguish all faults. Attention should be paid to the fact that the full fault dictionary for this example does not contain 3 x 7 signatures and only 10 signatures. So, the location of each fault, though three times longer, requires however to memorize three signatures more than in the case of the 1SPA technique.

CONCLUSION

The basic drawback of the kSPA analysis technique is testing time k times longer. But the obtaining of results in the one-signature analysis technique similar to the results of the analysis in the kSPA analysis technique has required so far to build the MISR with a length kn, which increased the cost of the measurement hardware realization by approximately k times. Therefore, the kSPA analysis, particularly when using the built-in kEV, described in the chapter 3, is a good optional solution to the problem of increasing the diagnostic resolution for the case of fault diagnostics on VLSI wafers, multichip logic modules, and PCBs. It is especially important where time is not a bottle-neck in the process of diagnostic testing.

REFERENCES

Beenker, F. P. M., C. M. Maunder, and C. Vivier (1986). A standard boundary scan architecture. Joint Test Action Group, Draft 3, September.

Bhavsar, D. K., and B. Krishnamurthy (1984). Can we eliminate fault escape in-self testing by polynomial division (signature analysis)?. Proc. of Int'l Test Conf., 135-139.

Curtin, J. J., and J. A. Waicukauski (1983). Multi-chip test and diagnostic methodology. IBM Jl Res. Dev. vol 27, No 1, January, 27-34.

David, R. (1980). Testing by feedback shift register. IEEE Trans. Comput. (USA) C-29, No 7, July, 668-673.

Hlawiczka, A., and H. Kubica (1981). The method of increasing the probability of detecting errors for signature analysis. Proc. of 4th Int'l FTSD Conf., Brno, 273-277.

Hlawiczka, A. (1986). Parallel multisignature analysis of faults in the multi-output digital system. Proc. of 16th FTCS Conf., Vienna, 398-403.

Hlawiczka, A. (1987). BIST structure using multiinput shift register with initial value. Proc. of 10th Int'l FTSD Conf., Sofia. Accepted for publication.

Hlawiczka, A., and D. Badura (1987). Universal test controler chip for board self test. Proc. of 10th Int'l FTSD Conf., Sofia. Accepted for publication.

Kommonytsky, D. (1983). Synthesis of techniques creates complete system self-test. Electronics, March No 10, 110-115.

Error Detection and Correction in Real-Time Processing of Measurement Signals on Highly Concurrent Computing Structures

G.D. Hassapis

Dpt. of Electronic and Computer Engineering
Faculty of Electrical Engineering, School of Engineering,
Aristotelian University of Thessaloniki, 540 06 Thessaloniki, Greece

ABSTRACT

Highly concurrent computing structures, such as linear processor arrays, may be used to process at high speed signals received from intelligent sensors. Any functional error, however, may jeopardize the operation of the physical process which is measured by these sensors. Consequently, some form of error detection and correction mechanism must be included in the system to ensure that the results of long computations are valid. This paper presents the application of a new coding scheme, called "weighted checksum code" to detect and correct errors appearing during the execution of the signal processing algorithms, when a fault in the computing structure occurs. This coding scheme is based on the observation that the required signal processing tasks can be reduced to a set of basic matrix operations for which a set of checksum vectors can be easily derived.

KEYWORDS

Fault diagnosis; intelligent instrumentation; process control; concurrent computing; data coding.

INTRODUCTION

The manufacturing and process industries are making significant efforts in our days to improve the availability and efficiency of their plants, reduce the energy consumption and improve the quality of the finished products. For this reason, they install very expensive and sophisticated computerized control systems, the safe and efficient operation of which depends however, on the integrity, reliability and repeatability of the instruments measuring various physical variables of the plant.

One of the possible methods for improving the plant measurement system is to adopt some of the techniques used in the avionics and nuclear industries, where triplicate systems or two out of three voting is applied. However, this approach is innapropriate for the manufacturing and process industry where there is pressure to reduce the purchase cost, the cost of installation and the cost of maintenance. One approach that is attracting attention is the development and use of active sensors in place of traditional passive devices. They provide output signals in the form of electrical pulses, the period of which follows the changes of the measured quantity. These signals can be easily digitized by timing the interval between successive zero crossings. Then, signal analysis techniques can be implemented by microprocessors or other computing structures enabling information regarding the operating status of the sensors as well as the operation of the plant to be determined. It is, however, necessary the corresponding signal processing algorithms to be executed fast. In this way it will be known by the time the inferred measurement value is going to be used in subsequent control computations whether it corresponds to the actual plant situation or it is an erroneous value because of a sensor malfunction.

With the advent of VLSI technology, large numbers of processing elements which cooperate with each other at high speed have become economically feasible. It is therefore, possible to utilize such a computing structure for executing the signal processing algorithms. This is usually less expensive than the use of triplicate or two out of three voting configurations of passive sensors. Since any functional error in the computer system may jeopardize the operation of the system and its data integrity, some level of fault detection and correction must be incorporated in the system to ensure valid results of the long computations on the basis of which the correct operation of the sensors is determined.

The most intensive computations for verifying the correct operation of a sensor are those corresponding to the digital filtering of the incoming signals and the evaluation and comparison of their power spectrums. These calculations, however, can be formulated as matrix-vector multiplications which can become fault tolerant by using the Weighted Checksum Encoding Technique . According to this technique the input data to an algorithm are encoded in a form of error detecting and correcting code and the algorithm is redesigned to operate on the encoded data. Next, this matrix encoding technique is presented and the error correction procedure is outlined. Then, it is described how the computations for checking the status of a sensor can be carried out fast by a set of linear processor arrays and how the computer control system, which is linked with these arrays, can detect a fault in an array and correct the erroneous result. At the end of the paper an application example is given.

THE WEIGHTED CHECKSUM ENCODING SCHEME

Let us consider the matrix equation:

$$C = B.A \qquad (1)$$

where C is a n by 1 vector, B is a n by n matrix and A is a n by 1 vector. Further, if we define the 1 by n vectors:

$$e_1^T = [\ 1,\ 1,\ \dots\ 1\ ]$$

$$e_2^T = [\ 1,\ 2,\ \dots\ 2^{n-1}\ ]$$

$$\vdots$$

$$e_t^T = [\ 1,\ 2^{t-1},\ \dots\ (2^{n-1})^{t-1}\ ]$$

and the matrices:

$$C_c = \begin{bmatrix} C \\ e_1^T.C \\ \cdot \\ \cdot \\ \cdot \\ e_t^T.C \end{bmatrix}$$

$$B_c = \begin{bmatrix} B \\ e_1^T.B \\ \cdot \\ \cdot \\ \cdot \\ e_t^T.B \end{bmatrix}$$

then one can easily prove that:

$$C_c = B_c.A \qquad (2)$$

It can be also proved (Jing, 1986) that the vectors e_1^T ... e_t^T are linearly independent. Then, one can define the quantities:

$$\begin{aligned} S_1 &= e_1^T.C - WCS_1 \\ S_2 &= e_2^T.C - WCS_2 \\ &\cdot \\ &\cdot \\ &\cdot \\ S_t &= e_t^T.C - WCS_t \end{aligned} \qquad (3)$$

where $WCS_1=e_1^T.B.A$ ... $WCS_t=e_t^T.B.A$. Obviously there will be no errors in the computation of the elements of the C matrix if at least one of these quantities is zero. The S_i quantities, i=1, 2, ... t, which are not zero indicate that the corresponding WCS_i quantity is erroneous. If, however, there is not even one S_i quantity equal to zero, then one can consider that certain elements of C have been wrongly computed. If we consider that the correct values of the wrongly computed elements of C are c_{i_1} ... c_{i_m} and the actually computed values are c'_{i_1} ... c'_{i_m} where $m < t$, then the relationships (3) take the form.

$$e_1^T.C-WCS_1=(c'_{i_1}-c_{i_1})+(c'_{i_2}-c_{i_2})+ \ldots +(c'_{i_m}-c_{i_m})$$

$$e_2^T\text{-}WCS_2 = 2^{i_1-1}.(c'_{i_1} - c_{i_1}) + 2^{i_2-1}.(c'_{i_2} - c_{i_2}) + \ldots + 2^{i_m-1}.(c'_{i_m} - c_{i_m}) \quad (4)$$

$$\vdots$$

$$e_t^T.C\text{-}WCS_t = (2^{i_1-1})^{t-1}.(c'_{i_1} - c_{i_1}) + \ldots + (2^{i_m-1})^{t-1}.(c'_{i_m} - c_{i_m})$$

This is a system of equations with a number of 2m unknown variables, that is the $i_1 \ldots i_m$ and $c'_{i_1} \ldots c_{i_m}$ variables. Therefore, if m=t/2 the correct values of the C matrix can be found by solving this system of equations.

The findings of this analysis, as it will be demonstrated in the following, provide a powerful tool to check the correct operation of a computing structure by using another computing device and identify and correct erroneous results.

FAULT-TOLERANT CONCURRENT COMPUTING STRUCTURE

The bandwidth of the active sensors, which produce output signals in the form of a frequency is by far greater than the bandwidth of the signals produced by passive devices. This means that useful information which arrives at the interface between the sensor and the process and concerns not only the operation of the associated process plant but also the operating status of the sensors is not lost. After filtering out the possible noise this information may be analyzed by discrete Fourier transform techniques. For the reasons explained earlier, these techniques must be implemented on a computing structure which can detect and correct errors attributed to faults of the computing structure, within the response time of the associated control loop.

A given digital filter can be represented in state space form as:

$$X(k+1) = A.X(k) + B.u(k)$$

$$y(k) = C.X(k) + d.u(k) \quad (5)$$

where A is a mxm matrix, C is a 1xm matrix, B is a mx1 matrix, u(k) is the kth sample of the input signal, y(k) is the kth sample of the filter output and d is a scalar. The order of the filter is m. By using the previously explained encoding scheme we can convert relationship (5) to the coded matrix equations:

$$\begin{bmatrix} x_1(k+1) \\ x_2(k+1) \\ \vdots \\ x_m(k+1) \\ y(k) \\ WCSA_1 \\ WCSA_2 \\ \vdots \\ WCSA_r \end{bmatrix} = \begin{bmatrix} a_{11} & a_{12} & \ldots & a_{1m} & b_1 \\ a_{21} & a_{12} & \ldots & a_{2m} & b_2 \\ & & & & \\ a_{m1} & a_{m2} & \ldots & a_{mm} & b_m \\ c_1 & c_2 & \ldots & c_m & d \\ a_{i1} & \ldots & & & b_i+d \\ a_{i1} & \ldots & 2^m & & b_i+d \\ a_{i1} & \ldots & 2^{m(r-1)} & & b_i+d \end{bmatrix} \begin{bmatrix} x_1(k) \\ x_2(k) \\ \\ x_m(k) \\ u(k) \end{bmatrix} \quad (6)$$

The discrete form of the Fourier transform of the filter output will be:

$$
\begin{aligned}
Y(k,n) &= \sum_{i=0}^{k} y(i)\exp(-jnwiT) = \sum_{i=0}^{k} y(i)\cos(nwiT) + \\
&+ j\sum_{i=0}^{k} y(i)\sin(nwiT) = Y_1(k,n) + jY_2(k,n)
\end{aligned}
\qquad (7)
$$

where $w=2\pi/NT$, k and n denote discrete values in time and frequency domain respectively, and T is the signal sampling period. Taking the squares of the magnitude of the Fourier transforms for a wide range of n one can obtain the power spectrum of the filtered input signal. Any variation of this spectrum from the powwer spectrum of the sensor taken at the manufacturing stage can be used to provide a warning of a sensor malfunction. The computations involved in finding and comparing the spectrums can be also formulated as a set of coded matrix-vector multiplications, that is:

$$
\begin{bmatrix} Y_1(k,1) \\ Y_2(k,1) \\ Y_1(k,2) \\ Y_2(k,2) \\ \vdots \\ Y_1(k,n) \\ Y_2(k,n) \\ WCSB_1 \\ WCSB_2 \\ \vdots \\ WCSB_s \end{bmatrix}
=
\begin{bmatrix}
\cos wT & \cos(w.2T) & \dots & \cos(w.kT) \\
\sin wT & \sin(w.2T) & \dots & \sin(w.kT) \\
\cos 2wT & \cos(2w.2T) & \dots & \cos(2w.kT) \\
\sin 2wT & \sin(2w.2T) & \dots & \sin(2w.kT) \\
 & & & \\
\cos nwT & \cos(nw.2T) & \dots & \cos(nw.kT) \\
\sin nwT & \sin(nw.2T) & \dots & \sin(nw.kT) \\
g_{11} & g_{12} & & g_{1k} \\
g_{21} & g_{22} & & g_{2k} \\
 & & & \\
g_{s1} & g_{s2} & & g_{sk}
\end{bmatrix}
\begin{bmatrix} y(1) \\ y(2) \\ \cdot \\ \cdot \\ \cdot \\ y(k) \end{bmatrix}
\qquad (8)
$$

where $$g_{pq} = \sum_{1<i<n} 2^{(q-1)(p-1)} (\cos iwqT + \sin iwqT) \qquad (9)$$

$$
\begin{bmatrix} D_1 \\ D_2 \\ \vdots \\ D_n \\ WCSC_1 \\ \vdots \\ WCSC_t \end{bmatrix}
=
\begin{bmatrix}
Y_1(k,1) & y_2(k,1) & Y_d(k,1) & 0 & 0 & 0 \dots & 0 \\
0 & 0 & 0 & Y_1(k,2) & Y_2(k,2) & Y_d(k,2) \dots & 0 \\
 & & & & & & \\
0 & 0 & 0 & 0 \dots Y_1(k,n) & Y_2(k,n) & Y_d(k,n) & \\
Y_1(k,1) & Y_2(k,1) & & \dots & & & Y_d(k,n) \\
 & & & & & & \\
Y_1(k,1) & & & \dots & & 2^{(k-1)(t-1)} Y_d(k,n) &
\end{bmatrix}
\begin{bmatrix} Y_1(k,1) \\ Y_2(k,1) \\ Y_d(k,1) \\ \vdots \\ Y_1(k,n) \\ Y_2(k,n) \\ Y_d(k,n) \end{bmatrix}
\qquad (10)
$$

where $Y_d^2(k,i)$, i=1, 2, ... n denotes the magnitude of the ith component of the power spectrum taken during the manufacturing stage of the sensor.

D_i denotes the difference $Y_1^2(k,i)+Y_2^2(k,i)-Y_d^2(k,i)$

To carry out these calculations as fast as possible the use of a linear processor array for each matrix equation is considered. The operation of the array is as follows. Assume that we want to calculate the equation $A_{n+2,n} \cdot X_{n,1} = B_{n+2,1}$ The matrix element $a_{i,j}$ is stored in the local memory of the ith processor at the jth time step, and x_j is broadcast to each processor at the jth time step. Each processor multiplies n pairs of $a_{i,j}$ and x_j and accumulates the products of a register. Each processor thus calculates one element of the result vector and a faulty processor affects only one element. All the arrays are linked to the computer control system which sequentially reads the outputs of them, computes the values of the $S_1 \ldots S_t$ quantities, identifies possible erroneous outputs of each array and corrects them, if their number is half or less than t/2.

AN APPLICATION EXAMPLE

The turbine flowmeter is a typical instrument generating an output signal in the frequency form. Specifically, during a full rotation of the turbine meter blade a fixed number of short duration electrical pulses is generated, each pulse having unit magnitude. The frequency of occurence of these pulses depends on the rate of the flow passing through the turbine meter. For this type of output signal the Fourier transform in (7) reduces to:

$$Y(k,i) = Y_1(k,i) + Y_2(k,i) = \sum_{0<k<N-1} \cos iwt_k + j\sum_{0<k<N-1} \sin iwt_k \qquad (11)$$

where t_k has replaced kT because the time interval between successive pulses may vary. This may occur in the case of a meter malfunction or change in the process plant characteristic. Let us assume that N=5. Damage of the the flowmeter itself will always appear as a change in the relative power of specific subarmonics of the rotor frequency for a particular flowrate(Higham,1986) If we further assume that the frequencies of these subharmonics are $w/8\pi$ and $w/16\pi$ then the structure of the linear arrays which can implement the corresponding matrix computations will have the form shown in Fig. 1. The connected with these arrays computer verifies the correct sensor operation by executing the following procedure:

1. Compute $S_1 = \sum_{0<k<4} (\cos wt_k/4+\sin wt_k/4) + \sum_{0<k<4} (\cos wt_k/8 + \sin wt_k/8) - WCSB_1$

   $S_2 = \sum_{0<k<4} 2^{k-1}(\cos wt_k/4+\sin wt_k/4) + \sum_{0<k<4} 2^{k-1}(\cos wt_k/8 + \sin wt_k/8) - WCSB_2$
2. Check whether $S_1=S_2=0$. If this is true then there is not any fault in the first array.
3. If $S_1 \neq 0$ and $S_2=0$ then the $WCSB_1$ is erroneous.
4. If $S_1=0$ and $S_2 \neq 0$ then the $WCSB_2$ is erroneous.

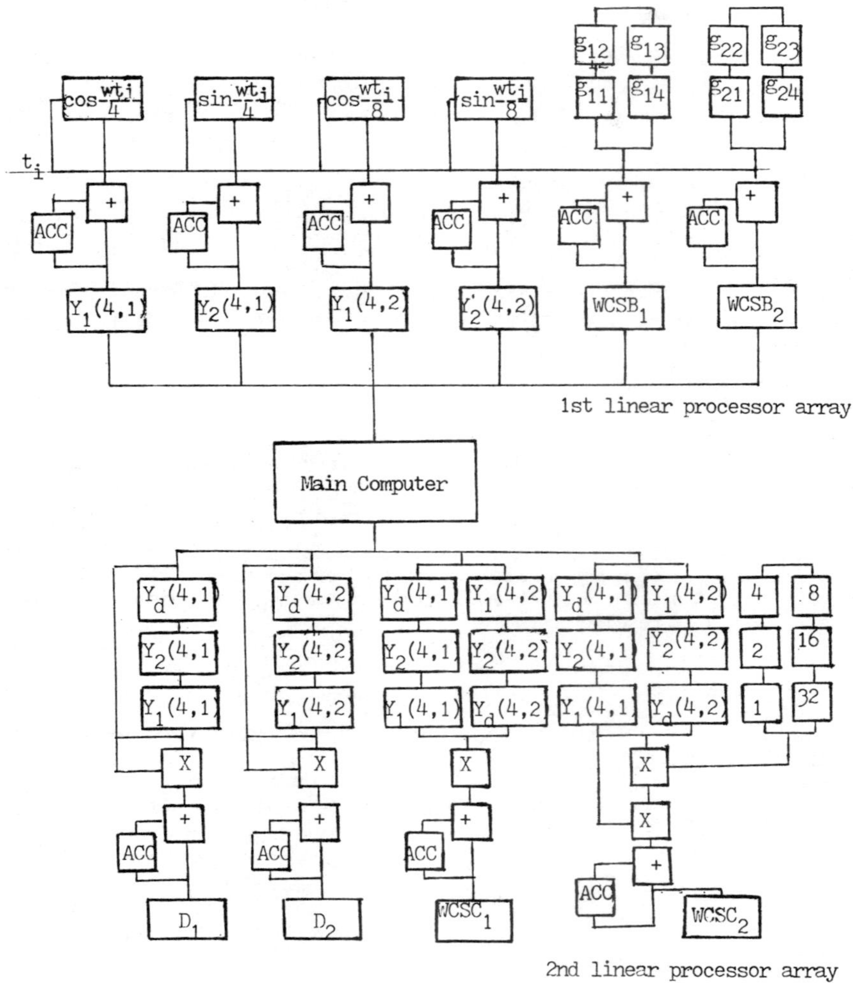

Fig. 1. The concurrent computing structure for the turbine meter signal processing.

5. If $S_1 \neq 0$ and $S_2 \neq 0$ then the ratio $S_2/S_1=2^{i-1}$ is computed and the index of the wrongly computed element of the output vector is estimated. Then the computer can find the correct value of the element by using the relationship $c_i=c_i'-S_1$ and load with it the proper register of the second array . Next, the output of the second array will be loaded in the computer and after checking the correctness of this output vector by applying a similar with the just described procedure, it can proceed with the verification of the correct sensor operation.

This procedure assumes that a single fault can occur in each array during the complete computation cycle. If a larger number of faults is assumed then a rather complicated system of S_i equations has to be solved.

CONCLUSIONS

It has been shown that the computational requirements for verifying the correct operation of instruments used to measure industrial process variables can be formulated as a set of basic matrix operations. These operations can become fault-tolerant if a weighted checksum coding technique, explained in this paper, is applied to the data. For fast execution of the operations on the coded data the use of a set of linear processor arrays is proposed. The correctness of the results is checked by a computer connected with the arrays and executing a proposed procedure.

REFERENCES

Higham, E. H. , R. Fell, and A. Ajaya (1986). Signal analysis and intelligent flowmeters. Measur. + Control J.,19, 47-50

Jing-Yang Jou, and J. A. Abraham. (1986). Fault-tolerant matrix arithmetic and signal processing on highly concurrent computing structures. IEEE Proceedings, 74 No 5, 732-741.

Transient Fault Tolerance in Microprocessor Controllers

J. Sosnowski

Institute of Computer Science, Warsaw Technical University
Nowowiejska 15/19, Warszawa 00-665, Poland

ABSTRACT

The paper deals with mechanisms of transient fault tolerance in microprocessor controllers. These mechanisms are discussed in context of various classes of applications. Error coverage, error latency and hardware/software overhead are considered. A new approach to self-checking programs with programmable watchdog timers and program tags is presented.

KEYWORDS

Transient faults, error detection mechanisms, fault tolerance, watchdog timers, self-checking programs.

INTRODUCTION

Many sources of transient faults in microprocessor controllers are possible e.g. supply line transients, ground noise, electrostatic discharges, lightning effects, cosmic radiations. Transient faults result in errors at the circuit level and failures at the system level. We distinguish three classes of transient fault effects (Sosnowski,1986a): permanent, steady-state and transient effects. Permanent errors and failures are relevant to circuit damages and will not be discussed. Steady-state errors are latched-up faults e.g. false setting of a flip-flop, entry of CPU into HLT state or false infinite loop, wrong operation mode of I/O circuit. These errors can be removed by system reset signal. Transient effects refer to temporal perturbations in the system. In practice transient and steady-state failures are more probable than permanent failures, moreover they are recoverable. Hence, we must treat them in a special way (Yak and others,1985).

Designing carefully controllers we can avoid some errors (e.g. false loops) or decrease the probability of their occurrence (Glaser and Masson,1982; Sosnowski,1986a). In fact, transient fault tolerance can be achieved using various error detection mechanisms. A systematic overview of these mechanisms is given taking into account various classes of critical errors. We present also a new approach to self-checking programs which is based on programmable watchdog timers and program tags.

ERROR DETECTION MECHANISMS

We distinguish two approaches to error detection: functional and structural check. Functional check verifies the correctness of realized functions (algorithms). Structural check deals with the validity of data, program control flow etc.

Functional check can be accomplished sequentially or concurrently. Sequential check is based on time redundancy. Concurrent check requires some hardware redundancy (triple or dual modular redundancy, watchdog processors). The verification that the realized function has properly been carried out by the controller is based on result checking and it can be performed as one process or a set of processes. Full or simplified checking of results is possible. In the sequential approach full checking is implemented by the repetition of task or subtask execution and comparison of results (software triple modular redundancy - Schuette and others, 1986). For example in a voting machine the state of voting keys can be checked several times in one ballot. In concurrent check the same procedure is realized by two or more hardware modules. In order to tolerate common errors it is recommended to use different algorithms for these procedures.

The hardware and software overhead in functional check can be decreased using assertions about the verified procedures. These assertions should be simple and strong enough to catch most errors. They are based either on the property of the problem itself or the property of algorithms, data structures etc. Inverse relationships are very useful as assertions if they are simpler than original algorithms. For example if the program finds the solution of N linear equations than it can be checked whether the solution satisfies one of the equations or not (Namjoo and others,1982; Mahmood,1983). Quite often the problem can be transformed into a simpler one e.g. checking high precision calculations with the use of low precision arithmetics. Sometimes it is possible to check the reasonableness of results such as their ranges, relations between variables, their dynamic behaviour. In the voting machine the ballot results can be verified by checking the sum of results. In a railway weighing system with four strain gauges mounted in rails the results delivered by these gauges ought not to differ significantly. In concurrent check assertions can be verified using simple watchdog processors. It is also possible to model the watchdog as a watchdog process communicating with the main process.

Functional check covers most data and program flow errors.However, sequential check may fail in the case of steady-state errors (e.g. false loop, HLT state). Hence, the use of a watchdog timer is required. Detection of steady-state errors in concurrent check is useful also. It allows to avoid an idle state of the watchdog processor or simultaneous inactivity of all modules. In the case of systems comprising several processors (tightly or loosely coupled) mutual functional check can be easily included. Sometimes concurrent check can be performed with the help of the controlled object. It is also reasonable to check the watchdog processor by the controller. Sophisticated functional check allows to achieve error coverage exceeding 99% at the cost of high overhead (up to 300%) and high error latency (Schuette,1986).

Data check is based on some redundancy in data structures, checksums, range verification etc. Examples of redundant data structures are: duplicated data, data with embeded error detection codes. In the first case it is reasonable to use the true and complementary values of data. This allows to cover common errors. Choosing error detection codes we have to take into account appropriate class of errors e.g. multiple unidirectional errors. We distinguish critical and not critical data. Errors within not critical data do not influence the operation of the controller or cause only temporal (and acceptable) system degradation. Critical data errors lead to wrong results, unsafe operation, steady-state failures etc. Quite often critical data are relevant to program control variables (e.g. buffer pointers, configuration tags, operation mode vectors, stack pointer, stack contents), output messages or signals. In the latter case it is reasonable to use fault tolerant communication protocols and complementary signals (e.g. start signal and device enable signal).

Program control flow checking is performed with the help of various error detection mechanisms: detectors of illegal opcodes, invalid fetch or read/write addresses, non-existent/unused memory references; capability checkers (Namjoo and McCluskey,1982); watchdog processors; test sequences in programs. These mechanisms detect program execution abnormalities such as false sequencing and false interpretation of instructions. False sequencing is relevant to following errors: <u>illegal branches</u> - branches which do not appear in the graph associated to the program, <u>wrong branches</u> - branches which are included in the graph and which do not represent the correct control transfer for the given program inputs, <u>infinite loops</u> and <u>deadlocks</u> (Sosnowski 1986b). Error coverage for simple detection mechanisms is 5÷50% and for sophisticated watchdog processors it exceeds 80%, error latency ranges from one instruction cycle to several thousands (Schmid and others,1982; Li and others 1984; Schuette and others,1986).

The basic idea of checking program flow with watchdog processors is to monitor the sequence of addresses relevant to fetched instructions, detect all branches and check them. In the scheme proposed by Namjoo (1982) valid branch exit and entry points are tagged in the program memory and appropriate checkwords associated with these points are embeded into the program. Illegal branches are detected by checking the tags and checkwords. For the valid branch the exit and entry point checkwords are the same. This scheme requires a sophisticated branch detector.

Watchdog processors based on path signature analysis are more effective (Namjoo,1982; Sridhar,1982; Shen,1983; Eifert,1984). They check the sequence and codes of instructions fetched in each executed program path. The checking procedure verifies signatures. At assembly time a signature is generated for each computation unit (e.g.branchless block, a sequence of such blocks). Signatures can be determined in many ways e.g. a cyclic code generated for the unit. The signatures must be embeded into the program. They are distinguished from instruction codes by tags in program memories or they are inserted just after pseudobranch instructions. In order to decrease the number of signatures they ought to be generated for sequences of branchless blocks. This leads to pseudosignature and branch address hashing concepts (Namjoo,1982; Shen,1983). During the run time the predetermind signature is catched by the watchdog processor and is compared with the actual signature computed by this processor. Signature checking covers illegal branches and opcode errors. Wrong branches are not detected.

The program flow can be checked by incorporating self-testing sequences into the program of the controller. Yau and Chen (1980) proposed a scheme with a data base containing a formal description of the program flow graph. The validity of program paths followed at run time is checked by comparing calculated checkwords with the information in the data base. The method proposed covers a class of illegal and wrong branches;however, it leads to significant software overhead. Much simpler scheme described by Tront and others (1985) has lower error coverage. Both algorithms do not deal with interrupts. More effective algorithm is given in this paper.

Various fault tolerance techniques are characterized by their error coverage, error latency and software/hardware overhead. The use of watchdog processors seems to be attractive. However, only functional check can be performed using off-the-shelf processors. The effectiveness of error detection mechanisms depends upon the function realized by the controller. Taking into account transient fault tolerance we distinguish three classes of controllers: A,B,C. In class A only steady-state failures are critical. It is typical for controllers operating in a cyclic way with time redundancy (e.g. data acquisition system). Steady-state failures and some specified transient failures are critical in class B controllers. Quite often these transient failures refer to illegal or wrong branches between program modules. Typically class B controllers operate in several modes (e.g. normal and test mode) or realize a specified sequence of tasks (e.g. satelite controllers - Tront and others,1985). False branches between modules covering various operation modes or tasks are critical. For example in a data acquisition system with programmed data processing parameters (stored in E^2PROM) an inadvertent branch into the software module changing these parameters leads to wrong results. Similarly, in an universal transmission controller false branch into a program module covering different protocol may be critical. Controllers of class C must cover steady-state failures and most transient failures (e.g. automatic weighing system). Realizing controllers tolerating transient faults we have to choose appropriate error detection mechanism. Often it is reasonable to use various mechanisms e.g. functional check of a critical program module and simple structural check of other modules.

WATCHDOG TIMERS

Time-out verification is very effective tool for detecting steady-state failures. It can be realized using a watchdog timer/counter (WDT) which is periodically initialized by reset sequences (WRI's) embeded in the program. If not initialized (due to an error) WDT generates count-out signal. The delay from the initialization until count-out is called WDT period (T). Watchdog timer may also cover some transient failures. Analysing the effectiveness of this mechanism we take into account program checked segments (CS). CS is defined as the sequence of program instructions which begins with WRI and ends immediately before the next WRI.

Consider two checked segments CS_i and CS_j with execution times t_i and t_j, respectively. The probability that an illegal branch from CS_i into CS_j will be detected by WDT with fixed time-out T can be estimated as:

$$p_{ij} = \frac{(t_i+t_j-T)^2}{2t_it_j} \text{ if } t_i+t_j>T; \quad p_{ij} = 0 \text{ if } t_i+t_j\leqslant T \qquad (1)$$

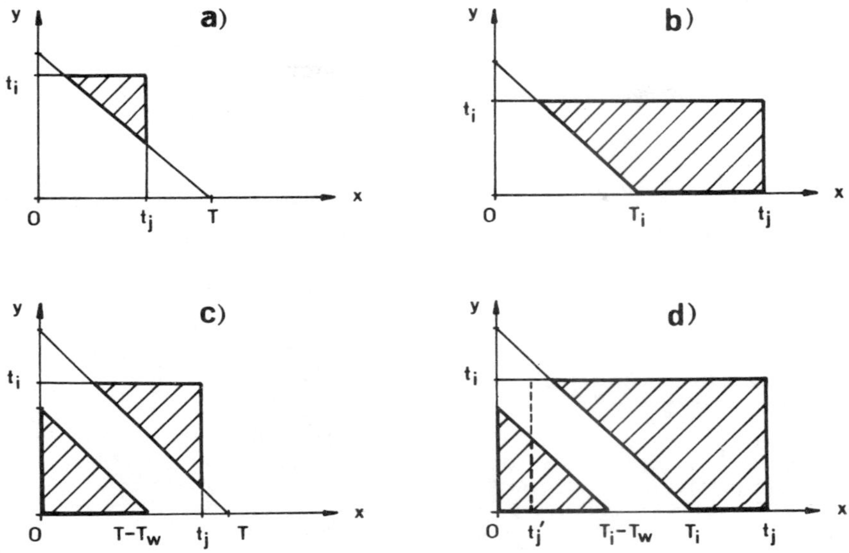

Fig. 1. Graphical illustration of error detection performance for watchdog timers.

This expression results from Fig. 1a and it can be interpreted as the ratio of dashed area to the area of the rectangle t_i,t_j. Variables y and x represent the exit and entry points of illegal branches, respectively. All possible exit and entry points are within the rectangle t_i,t_j. Variable y defines the execution time of a sequence of instructions within CS_i which begins with WRI and ends with the exit point of the illegal branch. Similarly, variable x defines the execution time of a sequence of instructions within the segment CS_j between the illegal branch entry point and the end of the segment. The dashed area refers to illegal branches which will cause time-out due to the fact that in this area x+y>T. For $t_i=t_j=T$ the probability p_{ij} achieves its maximal value 0.5.

In the case of WDT with dynamically programmed time-out for each checked segment the analysis is similar. Let T_i,T_j ($T_i<T_j$) denote time-out periods of WDT for segments CS_i and CS_j, respectively. The probability $p_{ij}(T_i)$ that an illegal branch from CS_i into CS_j will be detected by WDT is estimated by the ratio of the dashed area in Fig. 1b to the area of the rectangle t_i,t_j. Hence:

$$p_{ij}(T_i) = \frac{t_i(t_j-T_i) + t_i^2/2}{t_i t_j} \qquad (2)$$

The effectiveness of error detection may be higher than in the previous case. This results from the fact that low value of the preprogrammed time-out $T_i<T_j$ can be exceeded for many branch entry points within CS_j segment.

Checking execution times of program segments will lead to better error coverage than time-out checks. Assume that WDT checks the lower

$(T-T_w)$ and upper (T) bounds of segment execution time and that parameters T and T_w are fixed for all segments. The probability $p_{ij}(T_w)$ that an illegal branch from segment CS_i into segment CS_j will be detected by WDT can be estimated as the ratio of the dashed area in Fig. 1c to the area of the rectangle t_i, t_j :

$$p_{ij}(T_w) = \frac{(t_i+t_j-T)^2 + (T-T_w)^2}{2t_i t_j} \qquad (3)$$

The optimal situation arises for $t_i=t_j=T-T_w/2$. Similarly, we can estimate detection efficiency of WDT with dynamically programmed parameters (T,T_w) for all segments (see Fig. 1d). Narrowing the window T_w we may achieve quite good error coverage. Expressions defining p_{ij} and $p_{ij}(T_w)$ hold also for illegal branches within one segment $(t_i=t_j)$.

An illegal branch outside the program leads to a false sequence which will bump into the valid program or results in a false loop or other deadlock. The first case is equivalent to a branch into a CS_j^* segment with parameters t_j^*, T_j^*. The latter two cases ought to cause time-outs due to the fact that the probability of WRI within a random false sequence may be very low. To decrease this probability more complex WRI sequences can be used. Simple watchdog timers realized with non-programmed counters are free of steady-state errors and are very effective in covering CPU steady-state errors. However, in some situations these errors may be missed. For example in interrupt driven controllers infinite false loops will not be detected if interrupt handling routines comprise WRI sequences. Similarly, a false loop created within the program memory by RET instruction and an error in the stack may comprise WRI. Hence some extra checking (e.g. stack) is recommended (Sosnowski,1986a).

Inserting WDT reset instructions into the program we have to optimize this process taking into account error detection performance, error latency and software overhead. This can be done in an automatic way using a special program. This program finds the flow graph of the analysed program and inserts WRI's within the distance T'<T for all possible paths of that program. For simplicity WRI's are also inserted at the entry and exit points of called subroutines or interrupt handling routines and at the entry points of cycles. If these routines are relatively short this rule may be not used. Watchdog timers with programmable parameters T and T_w can be implemented with programmable timer/counters. Their use is discussed in the sequel.

CONTROL FLOW CHECKING PROGRAMS

In many real-time applications the controller accomplishes a sequence of tasks defined by software blocks. Illegal or wrong branches within these blocks may not be critical (e.g. in looping control systems). Hence, it will be sufficient to check false branches between blocks. The simplest way to do this is to insert some tags into the software blocks.

We introduce two types of tags: path tags and block tags. The path tag is used to check the validity of block sequence. The block tag allows to check if the block execution started from its entry point. Each block is assigned a unique number called its signature. Upon entering any block the block tag is set to the value of its signature. This tag is checked when the block is exited. The path tag is set at the end of the block to the value of the next block signature. This tag is checked upon entering any block. System reset signal sets the path tag to the value of the first block signature. In the case of blocks comprising loops it is recommended to check additionally the block tag for every loop execution. This will decrease error latency. Distribution of tag setting and checking instructions within a block is illustrated below:

```
SET BLOCK TAG
TEST PATH TAG
optional tests
      .
      .
The body of
the block
      .
      .
SET PATH TAG
TEST BLOCK TAG
```

Loop within the block

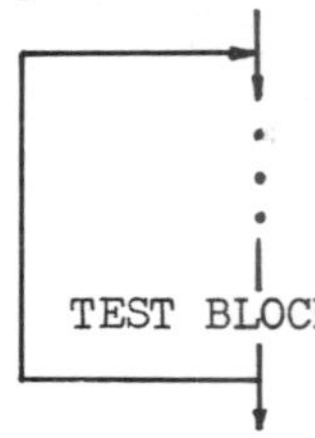

The given scheme detects illegal branches between blocks including branches into tag setting instructions. The latter error is not coverd in single tag approach (Tront,1985). Wrong branches can also be detected if we extend path tag check by adding path predicate check (e.g. testing of a branch control variable). We do not care about false branches within blocks. However, some of them may be critical e.g. a spurious bump into the path tag setting instruction gives no chance to accomplish the task assigned to the block (in fact, the probability of this situation is quite low). Critical branches within a block can be detected by further extension of path tag checking and adding a check of an assertion on the performed function by the previous block. Depending upon applications we can use very simple (e.g. check the number of loop executions) or sophisticated assertions. Quite good results we get with time verification performed by watchdog timer with programmed parameters T and T_w. These parameters are dynamically adjusted to blocks or their intervals. The function of the watchdog-timer can be performed by the program and system timer. Reading the state of the timer the program is able to check the execution time of the block or its interval.

Some comments must be given to called and interrupt activated routines. If they are simple and perform secondary functions only watchdog timer adjustment is required. Subroutines are called at well specified points so they can be easily included in calculated parameters T and T_w. Interrupt handling routines are initiated randomly. Hence, it is reasonable to adjust watchdog timer for every activation of these routines.

In the case of complex subroutines and interrupt handling routines (e.g. realizing a sequence of tasks) control flow check may be required within these routines. Taking into account that these routines are entered from various points of the program and are ended with RET instructions it is reasonable to use a stack for tags, state of the watchdog-timer etc. If during a block execution a CALL instruction is met the parameters checked in the block are pushed to the stack. When finishing the execution of the subroutine these parameters are popped back. In a similar way we deal with interrupt handling routines. It is also possible to check if the evoked routines were initiated from the correct blocks.

Critical output signals of the controller ought to be redundant (e.g. dual valves in a chemical reactor) and issued within different blocks. In this way if one signal were falsly issued the second one would not be (Tront and others,1985).

CONCLUSION

The proposed method of checking critical effects of transient faults seems to be attractive for many microprocessor controllers. It is due to low hardware/software overhead and high error coverage. Moreover, this method can be easily implemented in single chip-microcomputers comprising programmable timers (e.g. Intel 8048, 8051, 8096). Taking into account the possibility of simultaneous steady-state errors in CPU and the timer it is reasonable to add a watchdog timer realized with simple counter (free of steady-state errors). Such circuit is available within 8096 on-chip resources.

REFERENCES

Eifert,J.B., and J.P.Shen (1984). Processor monitoring using asynchronous signatured instruction streams. Proc. of IEEE FTCS-14.

Glaser,R.E., and G.M.Masson (1982). The containment set approach to crash-proof microprocessor controller design. Proc.of IEEE FTCS-12.

Li,K.W.,J.R.Armstrong, and J.G.Tront(1984). An HDL simulation of the effects of single event upsets on microprocessor program flow. IEEE Trans.on Nuclear Science, Vol.NS-31, No.6, 1139-1144.

Mahmood,A.,E.J.McCluskey,and D.J.Lu(1983). Concurrent fault detection using a watchdog processor and assertions.Proc.of Int.Test Conf.

Mahmood,A.,A.Ersoz,and E.J.McCluskey(1985).Concurrent system-level error detection using a watchdog processor. Proc.of Int.Test Conf.

Namjoo,M.,and E.J.McCluskey(1982). Watchdog processors and capability checking. Proc. of IEEE FTCS-12, 245-248.

Namjoo,M.(1982). Techniques for concurrent testing of VLSI processor operation. Proc. of Int. Test Conf., 461-468.

Schmid,M.E.,R.L.Trapp,A.E.Davidoff, and G.M.Masson(1982). Upset exposure by means of abstraction verification.Proc.of IEEE FTCS-12.

Schuette,M.A.,J.P.Shen,D.P.Siewiorek, and Y.X.Zhu(1986).Experimental evaluation of two concurrent error detection schemes. Proc. of IEEE FTCS-16, 138-143.

Shen,J.P., and M.A.Schuette(1983). On-line self-monitoring using signatured instruction streams. Proc.of Int.Test Conf.

Sosnowski,J.(1986a). Transient fault effects in microprocessor controllers. Proc.of REL-CON EUROPE 86, Danmark, 239-248.

Sosnowski,J.(1986b). Risky transient in microprocessor controllers. Proc.of 5th Int.Conf. on Rel.and Maint.,France, 641-646.

Sridhar,T.,and S.M.Thatte(1982). Concurrent checking of program flow in VLSI processors.Proc.of Int.Test Conf., 191-199.

Tront,J.G.,J.R.Armstrong,and J.V.Dak(1985). Software techniques for detecting single-event upsets in satelite computers. IEEE Trans.on Nuclear Science, Vol.NS-32, No.6, 4225-4228.

Yak,Y.W.,K.E.Forward, and T.S.Dillon(1985). Modelling the effect of transient faults in fault tolerant systems. IFAC SAFECOMP, Italy, 129-133.

Yau,S.S., and Fu-Chung Chen(1980). An approach to concurrent control flow checking. IEEE Trans. on Soft.Eng., Vol. SE-6, No.2, 126-137.

Systems Architectures and Reliability Statistics

Dr. D.E. Ventzas

System Engineer, Hellenic Aspropyrgos Refinery,
Kazantzaki 23, Ano Pefki - Athens 15 121, Greece

ABSTRACT

The requirements of the industry with regards to the safety from the instrumentation technology aspect, of electronics is discussed. Redundant codings protect against risks through defects in the hardware only, when the requirements of instrumentation technology functioning are fulfilled.

KEYWORDS

System architecture, reliability, redundancy, two out of three system, reliability model, availability.

INTRODUCTION

In a control system, the central computer subsystem is required to perform various real-time monitoring and control functions, support the man-machine interface and run off-line processing tasks needed for operation. The computer is further needed to support continuing system maintenance, testing, improvements and development of new functions. With all the above demands on the computer with emphasis on fast response and high availability of critical functions, especially during times of emergency, it has become necessary to improve system architecture.
The various generic types of CPU configurations which are used in control systems or which are offered by suppliers are given in fig.

1 to 14. There are various factors which may be considered in the technical evaluation of competitive configurations especially those using more than two processors; these are to compare them on the basis of "minimum system configuration" which we define as the smallest assemblage of computer resources which is capable of supporting the set of critical operating functions, or on the basis of performance response, availability and maintainability. Parameters that affect the above factors are the communication paths between CPUs and memories, the number of CPUs and integer memory blocks and the arrangement around the main communication bus.

CONTROL SYSTEM RELIABILITY

Improving performance of the controls impacts directly and heavily on profits. In a typical process industry instruments and control systems represent 8 to 10 % of the plant investment. Maintenance of this portion of the plant requires 15 to 20 % of the total mechanical manpower. We must expect an average of about five years between major maintenance work on a well designed device supplied by a reputable vendor. Maintenance people should be selected accordingly. Power sources must be duplicated, both with 100 % reliability. Reliable, high quality instrument air for pneumatic control and automatic valves is usually expensive, but the technology is more mature. The hardware selection should be based on all aspects simplicity and standardization. Vendors that use obsolete, commercial grade chips, bugs and transistors in their design should be rejected. Fig. 15 gives the cost for a hired person to be trained in automatic control systems maintenance until he retires. Fig. 16 shows the training productivity improvement by increased hands-on and concentrated training.

RELIABILITY CONCEPTS

Reliability (R) is the probability that a module will perform a required function under stated conditions for a specific period of time. Failures can be classified in: critical, major, minor ones. The bath-tub curve consists of the early life (infant mortality), useful life and wear out life. During the IC´s early life their failure rate is relatively high due to diffusion and oxidation, contamination and masking defects, assembly contact failures and electrostatic discharges, microcracks, etc. A typical value of the mean life of a

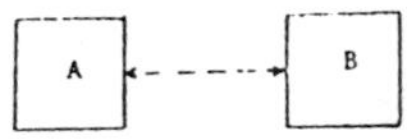

Fig. 1. The dual CPU configuration

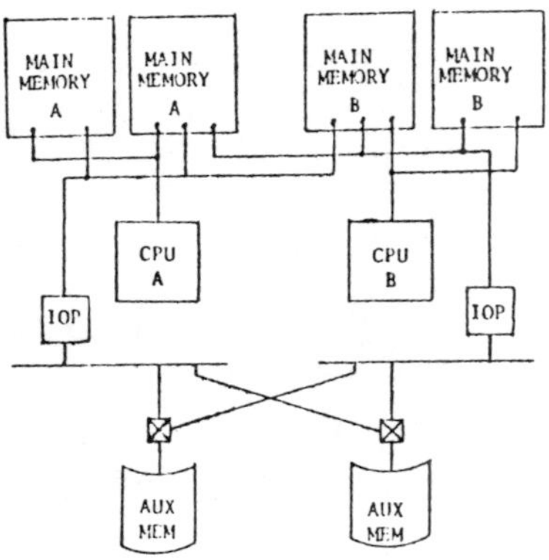

Fig. 2. A loosely-coupled configuration

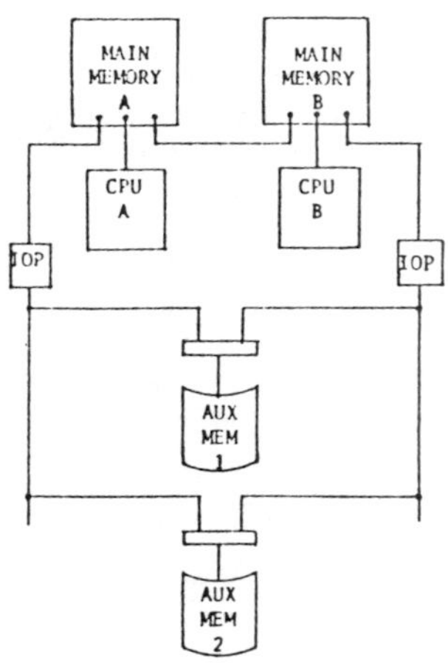

Fig. 3. A loosely-coupled configuration

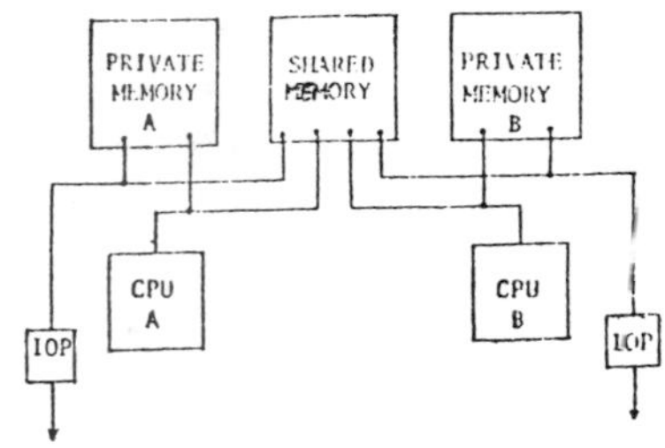

Fig. 4. A tightly-coupled configuration

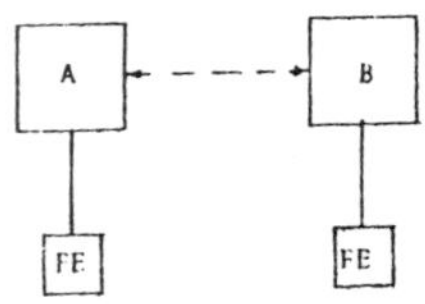

Fig. 5. A front-ended configuration

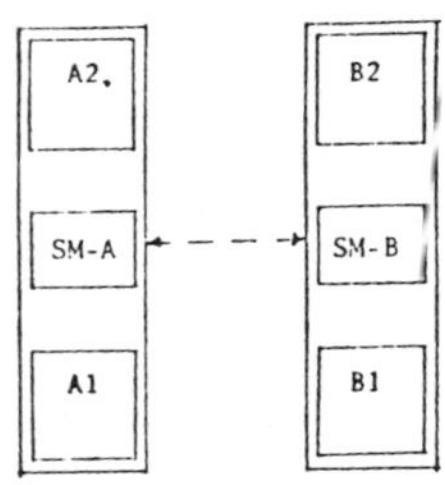

Fig. 6. The dual-duplex configuration (with shared memory)

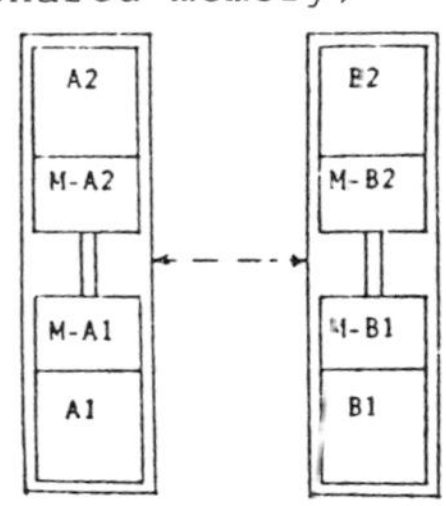

Fig. 7. The dual-duplex configuration (with shadow memory)

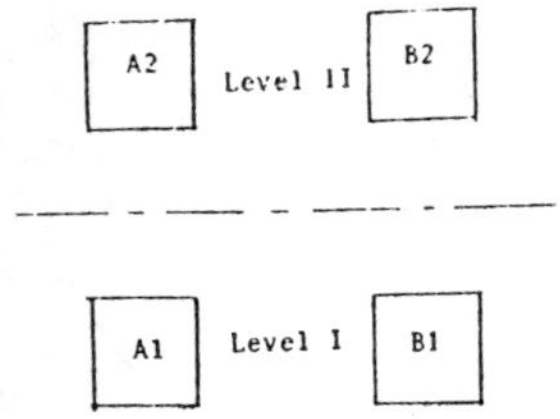

Fig. 8. The quad cofiguration

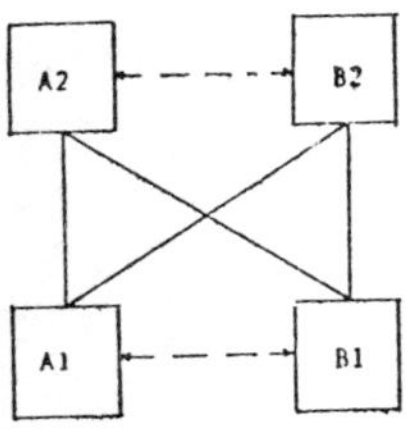

Fig. 9. The quad configuration (loosely-coupled)

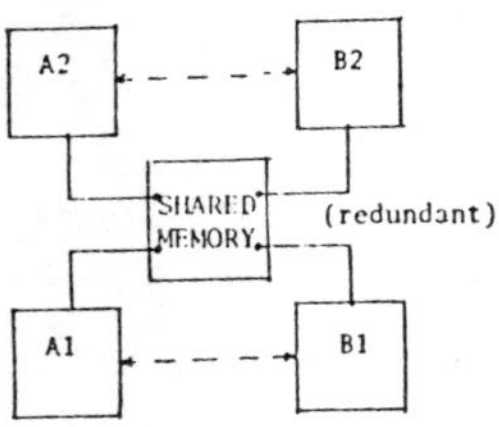

Fig. 10. The quad Configuration (tightly-coupled)

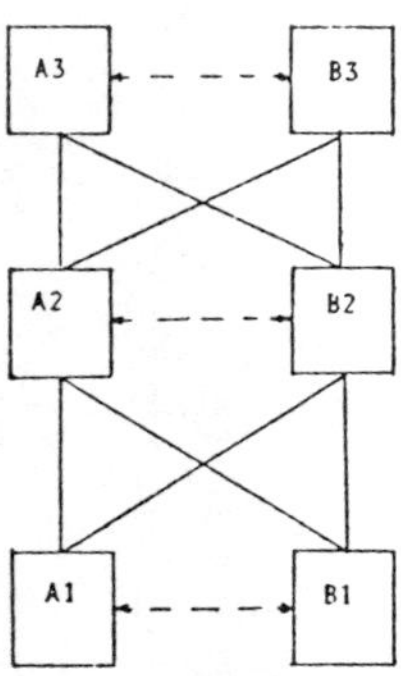

Fig. 11. A three-level system

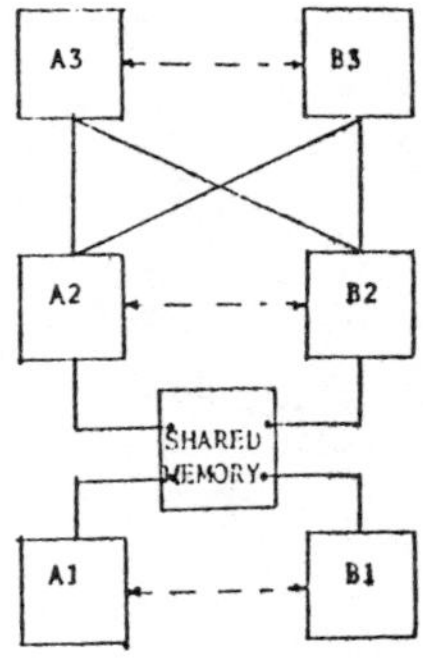

Fig. 12. A three-level system

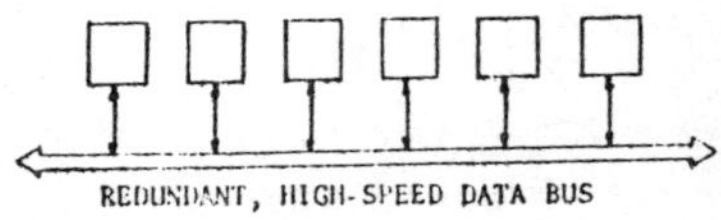

Fig. 13. A loosely-coupled distributed system

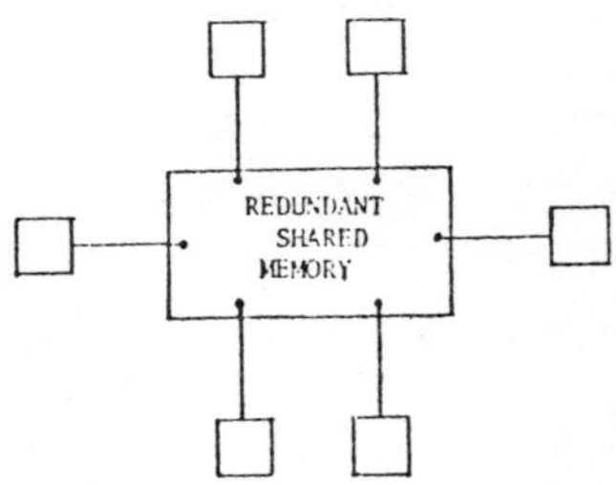

Fig. 14. A tightly-coupled distributed system

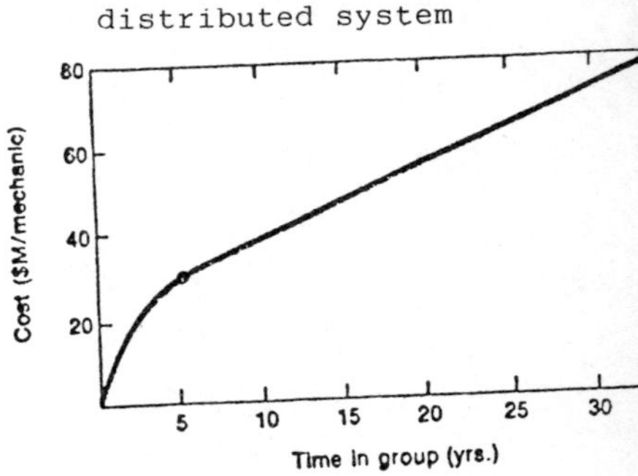

Fig. 15. Training costs

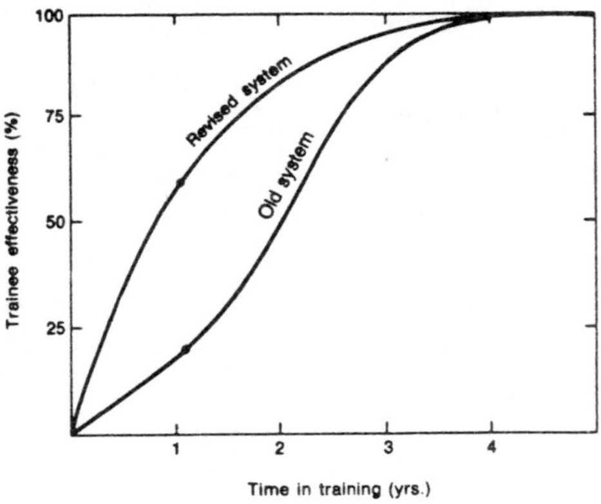

Fig. 16. Improving productivity

hermetically sealed MOS device is approximately 20 to 40 years. MTBF is the average time interval between failures for repairable modules:

$$MTBF = \int_{0}^{\infty} R\,dt = \int_{0}^{\infty} e^{-\lambda t}dt$$

where   is the failure rate. Finally

$$MTBF = 1/\lambda$$

A general equation for the failure rate of any part is: $\lambda_p = \lambda_b\,[\pi_E \cdot \pi_Q \cdot \pi_{other}]$

where: $\lambda_b = [S,T]$, S = component stress factor, T = temperature π_E = environmental factor, π_Q = quality factor, π_{others} = specific factors per device. A collection of components (module) has a failure rate based on the theory of Chain rule:

$$\lambda_{module} = N_R\lambda_R + N_C\lambda_C + N_{IC}\lambda_{IC} + \ldots$$

where N_R. N_C, N_{IC} are a quantity of resistors, capacitors, and integrated circuits involved in the module. MTBF values are valid only if we know the failure rate source, the underlying assumption, and the temperature. Another parameter is the availability (A) that considers both MTBF and Mean Time To Repair (MTTR):

$$A = MTBF / (MTBF + MTTR) = 1 / (1 + \lambda MTTR).$$

MTTR values depend on the location of repair personnel, skills, experience in diagnosis, spares availability, maintainability, serviceability. Redundancy is in principle the parallel connecting of identical functioning components. Binomial probability function is a good model for such systems.

PLCs are used extensively for complicated systems; this leads to difficulties in assessing their integrity. Functionally PLCs can achieve virtually all the operations traditionally performed by relay and solid state logic. The only exception is where there is a need for very high speed switching, because the scanning requirement of a PLC limits the total cycle time for detection of input change and setting the appropriate outputs. PLCs reliability features can be focused on spurious trips, system unavailability (dead time), components failures, interface problems. A fail danger rate of 0.05 per year may be expected as an acceptable fail unsafe condition; this is equal to the fail rate per year of trip valves, solenoid valves. The Fractional Dead Time (FDT) is a function of the failure rate λ and the testing interval T and is given by the equation $FDT = \frac{1}{2}\lambda T$ as long as $\lambda T << 1$and the frequency of testing is much greater than the frequency needed on the protective system.

The reliability evaluation of ICs consists of accelerating tests which include high temperature reverse bias (HTRB), pressure pot, thermal shock and temperature cycle tests. High temperature is known to accelerate many physical and chemical reactions in transistors leading to device failure. In order to express accelerated test results in terms of expected failure rate at actual use conditions (usually $+40^o$ and 70° C), semiconductors manufacturers commonly use the Arrhenius model. The Arrhenius model assumes that degradation of a performance parameter is linear with time, with the rate of degradation depending on the temperature stress. Using this model, at real operating conditions, failure rate can be estimated from:

$$\frac{FR1}{FR2} = \frac{MTBF2}{MTBF1} = e^{\frac{E_A}{k}\left(\frac{1}{T_2} - \frac{1}{T_1}\right)}$$

where: FR1 is the failure rate at transistor junction temperature T_1 (Test condition)

FR2	failure rate at transistor junction temperature T_2 (use condition)
MTBF1	mean time between failure at transistor junction
MTBF2	temperature T_1 and T_2 respectively
E_A	Activation energy (0.8 eV for FET transistors)
k	Boltzman´s constant (8.63×10^{-5} eV/^{o}K)
T_1, T_2	Transistor junction temperatures in ^{o}K
e	the natural antilog constant (2.7183)

The failure rates resulting from HTRB and HTGB tests present an average of the typical expected failure rate for a product, but has no statistical boundaries established. In order to apply the results of these tests to the whole population of the same devices, some statistical confidence factors should be accounted for. These factors depend on the number of rejects (RJ) and tested "chi square half" charts. Using this tables, failure rates (FR) can be predicted from:

$$FR = \left(\frac{x^2}{2}\right) \cdot \frac{1}{SS.H}$$

where

$x^2/2$	is the "chi square half" factor, for a specific number of rejects and for accepted confidence level
SS	Sample size
H	Test duration in hours.

Reliability growth models are used for extrapolating the results of earlytests or use to future times. They are used, for example, to predict the reliability of a system in use, when data are available from development tests, or to predict how much development testing might be required to ensure a particular value reliability. The best known of these models is the Duane one. This states:

$$M_C = M_O \left(\frac{T}{T_O} \right)^a$$

where: M_C = cumulative mean time between failures (MTBF) at time T

M_O = MTBF at time T = 0

a = constant

The value of a is related to the intensity of the effort put into diagnosing and correcting causes of failure as they arise. The Duane model is completely empirical. Other reliability growth models have been postulated, some initially from consideration of the processes that lead to faults in software operation.

If integrated circuits are to be used in fail-safe systems, then the only way to achieve safety against individual failures is by methods, in which the details of the inner structure of the circuits do not play a part. Therefore several integrated circuits must be allowed to co-operate in such a way, that they check one another and so that the failure of an IC cannot be dangerous. The most well-known way is the multi-channel system. An indispensable pre-condition for all these methods is: The circuits, whose collective failure could be dangerous, are uncoupled in such a way, that they must be considered as independent; this could prevent sequential and multiple effects of failures. When each individual failure remains harmless, multiple failures in an accidental or systematic manner can still be dangerous even when the equipment was originally free of defects. A risk arises in a systematic way, when a failure gives rise to further ones or groups. It is better to take care that systematic failures do not occur and that the effects of individual failures are limited to defined areas. Independence means distinct construction elements or function groups -if necessary- separated in such a way that a combined faulty functioning is excluded or that only controllable faulty functions are transmitted from one element to the other or from one function group to the other. Electronic safety circuits on the contrary in most cases operate with ordinary commercial construction elements, which are not manufactured specially for safety tasks. Fig. 17 shows a few typical places in a two-channel installation, which have an influence on the independence. It maybe, that a channel fails and that retroactive distortions of the input signals, in themselves free of faults, arise in this way, which finally affect the other

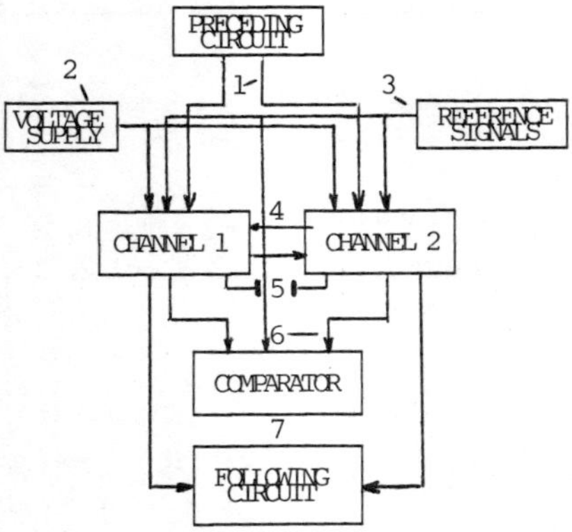

Fig. 17. Diagram of independence (numbers in figure show critical points)

channel. Suitable uncoupling connections, which consist of construction elements for which there is no need to reckon with critical short circuits, provide a remedy. A sensible remedy is to feed the two channels from separate voltage regulators. Besides galvanic couplings electromagnetic controls gain considerably in importance for the small capacities. All direct connections between two channels, which for reasons of synchronization, certain controls etc, cannot always be avoided, are to be considered as particularly critical. A failed and a trouble free channel produce unrecognizably false results. False items of information from one channel are not processed further in the other one, or they affect the following one only in a way, which is recognized as safe. "Correct" items of information at a wrong time are dangerous. The average time up to a double failure is estimated by the equation:

$$t_d = \frac{a1 + a2 + 1/to}{a1.a2}$$

In this a1 indicates the rate for the first failure and a2 that for the second failure. Quantity to is the revelation period of the first failure. The condition $a.t_o \leqq 10^{-3}$ where a is the sum of the failure rates of the units considered, whose common failure can be dangerous and t_o the permissible maximum revelation period. In case of a 2-versus-3 (voting) system the requirement is

$$a.t_o \leqq 5.10^{-4}$$

Five types of the revelation of an individual failure can be distinguished: spontaneous, conditioned by data flow, autonomous check, comparative check, inspection.

CONCLUSIONS

The requirement of the industry with regard to the safety from the instrumentation technology aspect of microelectronics was discussed. Individual components and devices reliability, which is combined in a way dependent on the system architecture to establish the overall system's reliability was quantitatively presented.

Subject Index